Interactive and Dynamic Dashboard

The text comprehensively discusses the representation of visual data and design principles of interactive and dynamic dashboards. It further covers the theoretical concept of inference and machine learning algorithms for making the concepts clear to the reader. The book illustrates important topics such as data testing a parametric hypothesis, data testing a non-parametric hypothesis, exploratory data analysis, outlier detection and interpretation.

This book:

- Covers various data analysis tools such as KNIME, RapidMiner, Rstudio, Grafana, and Redash
- Discusses the theoretical concept of inference and machine learning algorithms for designing dynamic dashboards
- Presents statistical modelling techniques with an emphasis on pattern mining, and pattern relationships
- Explains the problem of efficient retrieval of similar time series in large databases to enrich the knowledge of the readers to effectively handle various real-time datasets
- Illustrates dimensionality reduction techniques such as principal component analysis, linear discriminant analysis, singular value decomposition, and piecewise vector quantized approximation

It is primarily written for senior undergraduates, graduate students, and academic researchers in the fields of electrical engineering, electronics and communications engineering, computer science and engineering, and information technology.

Future Generation Information Systems

Series editor- Bharat Bhushan

With the evolution of future generation computing systems, it becomes necessary to occasionally take stock, analyze the development of its core theoretical ideas, and adapt to radical innovations. This series will provide a platform to reflect the theoretical progress, and forge emerging theoretical avenues for the future generation information systems. The theoretical progress in the Information Systems field (IS) and the development of associated next generation theories is the need of the hour. This is because Information Technology (IT) has become increasingly infused, interconnected and intelligent in almost all context.

Industry 6.0: Technology, Practices, Challenges, and Applications
Kishor Kumar Reddy C, Srinath Doss, Lavanya Pamulaparty, Kari J Lippert and Ruchi Doshi

Convergence of Deep Learning and Artificial Intelligence in Internet of Things
Ajay Rana, Arun Rana, Sachin Dhawan, Sharad Sharma and Ahmed A. Elngar

Next Generation Communication Networks for Industrial Internet of Things Systems
Sundresan Perumal, Mujahid Tabassum, Moolchand Sharma and Saju Mohanan

Interactive and Dynamic Dashboard
Design Principles

Edited by
A. Vadivel
K. Meena
P. Sumathy
Henry Selvaraj
P. Shanmugavadivu
Shaila S. G.

CRC Press is an imprint of the
Taylor & Francis Group, an **informa** business

Designed cover image: shutterstock

First edition published 2025
by CRC Press
2385 NW Executive Center Drive, Suite 320, Boca Raton FL 33431

and by CRC Press
4 Park Square, Milton Park, Abingdon, Oxon, OX14 4RN

CRC Press is an imprint of Taylor & Francis Group, LLC

ISBN: 978-1-032-74597-8 (hbk)
ISBN: 978-1-032-89421-8 (pbk)
ISBN: 978-1-003-54273-5 (ebk)

DOI: 10.1201/9781003542735

Typeset in Sabon
by SPi Technologies India Pvt Ltd (Straive)

Contents

About the Editors viii
List of Contributors xi

1 **Bibliometric analysis on visual data analysis and dynamic dashboard tools: A literature review** 1
SANDHYA RANI NALLOLA, VADIVEL A, SENTHIL KUMAR J. P., MEENA K, AND SUMATHY P

2 **Visual data analysis and inference through dimensionality reduction techniques** 21
JYOTHSNA V, SANDHYA E, BHASHA P, N. NAGA SWETHA, AND SAI DIVYA SREE T

3 **Visual data analysis of temperature, ground water level, precipitation for climate-driven socio-economic prediction** 69
S. SHOBA, MELVIN, SASITHRADEVI A, P. PRAKASH, AND DEEPA S

4 **AI-based online interview bot with an interactive dashboard** 86
RAKOTH KANDAN SAMBANDAM, DIVYA VETRIVEERAN, J. JENEFA, BALAMURUGAN M, AND THAIYALNAYAKI S

5 **Visualizing food quality and safety: A dynamic dashboard approach with near-infrared imaging and machine learning** 100
BRIGHTY EBENEZER L, SASITHRADEVI A, CHANTHINI BASKAR, AND KANIMOZHI S

6 Interactive dashboard and 3D visualization using t-SNE dimensionality reduction technique 119

PRAVEEN GUJJAR J, LUBNA AMBREEN, VANI HIREMANI, AND RAGHAVENDRA M DEVADAS

7 Dynamic dashboard creation for sales trends and optimize pricing strategies 135

SASITHRADEVI A, KANIMOZHI S, AND SRINATH SRINIVASAN

8 Scaling up the business with the aid of power query tool 146

NARAYANAN GANESH, MANICKAVASAGAM SURUTHI, AND GANESH HARIHARAN

9 Interactive visualization techniques for thermal imaging analysis in ophthalmology: Comparative insights and future directions 168

PERSIYA J, SASITHRADEVI A, SHOBA S, AND PRAKASH P

10 Mind scan: Dynamic brain cancer detection dashboard with MRI imaging 193

J. SUNEETHA, SMITA DARANDALE, NIRANJAN L, AND TANVIR H SARDAR

11 Interactive and dynamic stock market dashboard 206

PRASAN MITTAL, KANIMOZHI S, AND SAIRAMESH L

12 Performance analysis of hierarchical clustering and high-dimensional clustering algorithms on network IDS benchmark datasets using interactive dynamic dashboard 218

S. SRIVARSHINEE AND MENAKA PUSHPA A

13 Breaking boundaries: The next frontier in skin cancer diagnosis combining transfer learning and multi-scale deep learning 236

S. A. SAHAAYA ARUL MARY, SAMEER CHAUHAN, S. SANJITH SURYA, AND LUV SACHDEVA

14 Nourish net: Machine learning innovations in food recognition and calorie monitoring 254
RASHMI D AND POORANI M

15 Comprehensive study of coral reef assessment and colour correction using deep learning 267
PRAKASH P, KASTHURI P, SASITHRADEVI A, VIJAYALAKSHMI M, DIVYA P, JENNIE GRATIA FRANKLIN, AND SENGAMALI K N

Index 285

About the Editors

A. Vadivel received his master's degree in science from NITT. He pursued a Masters in Technology and Doctorate of Philosophy from the Indian Institute of Technology (IIT), Kharagpur, India. He has 13 years of Technical Experience in Network Engineering & Instrumentation Engineering in IIT-Kharagpur, and sixteen years of teaching experience. Currently he is working as Professor in Department of Artificial Intelligence and Data Science, GITAM School of Technology, GITAM University, Bengaluru, India. He has published papers in more than 120 international journals and conferences. His Research areas are Content-Based Image and Video Retrieval, Multimedia Information Retrieval from Distributed Environment, Medical Image Analysis, Object Tracking in Motion Video and Cognitive Science. He has received the Young Scientist Award by Department of Science and Technology, Govt. of India in 2007, Indo-US Research Fellow Award by Indo-US Science and Technology Forum in 2008. Obama-Singh Knowledge Initiative Award in 2013. Under his guidance, five PhDs were awarded, and five PhD scholars are pursuing their degree. He is recognized as research supervisor for GITAM University, National Institute of Technology (NIT), Trichy. He has four Indian patents to his credit.

K. Meena received her PhD from Manonmaniam Sundaranar University in 2014. She completed her M.E degree in CSE and B.E degree in ECE from Manonmaniam Sundaranar University in the year 2009 and 2002 respectively. She had 20 years of teaching experience in various reputed organizations. Currently, she is working as Professor in the Department of Computer Science and Engineering, GITAM School of Technology, GITAM University, Bengaluru, India. She has published papers in more than 110 international journals and conferences. Her research areas are Biometrics, Machine learning for information retrieval, Computer vision, and Medical Data Analysis. Under her guidance, four PhD were awarded, and five PhD scholars are pursuing their degree. She is recognized as research supervisor for GITAM University, Vel Tech University and Anna University. She has 6 Indian Patents and 1 Australian Patents to her credit.

P. Sumathy received her master's degree and bachelor's degree from Bharathidasan University, Trichy. She received her PhD from Gandhigram Rural Institute (Deemed to be University), Dindugal. She has 18 years of teaching experience in various reputed organizations. Currently she is working as Assistant Professor of Computer Science Engineering and Applications, Bharathidasan University. She has published papers in more than 70 international journals and conferences. Her Research areas are Content-Based Image and Video Retrieval, Multimedia Information Retrieval from Distributed Environment, Object Tracking in Motion Video and Cognitive Science. Under her guidance, two PhDs were awarded, and five PhD scholars are pursuing their degree. She is recognized as research supervisor for Bharathidasan University.

Henry Selvaraj is working as Professor in the Department of Electrical and Computer Engineering, University of Nevada, Las Vegas, USA. His research interests include Logic Synthesis, Digital Design, Programmable Devices, Artificial Intelligence, Multiple Valued Functions, Digital Signal Processing, Biomedical Image Processing, Networks, Path Planning and Object Tracker in Video Sequence. He has 36 years of teaching and research experience. He guided five research scholars in various domains such as Artificial Intelligence, Digital design and video analytics.

P. Shanmugavadivu is currently the Professor and Head, Department of Computer Science and Applications, at The Gandhigram Rural Institute (Deemed to be University), and is involved in Teaching, Research, and Extension. She has 31 years of academic experience and 22 years of research experience. Her research areas include Medical Image Analysis, Healthcare Analytics, Parallel Computing, Software Engineering and Content-Based Image Retrieval. Under her guidance, TEN PhDs were awarded; presently EIGHT PhD scholars are pursuing their degree. She has published 51 research articles in International Journals and 87 International/National Conferences and 19 book chapters. She has authored two research books and three edited volumes. She serves as an Editorial Board Member of two Journals. She is sanctioned with funded major research projects of UGC, DST, ICMR and PMMMNMTT for an outlay of Rs.2.39 Cr. She serves as the coordinator for her department's funding schemes, under UGC-SAP DRS-I (Rs.40 Lakhs) and DST-FIST (Rs.45 Lakhs).

Shaila S. G. is a Professor and Chairperson for the Data Science Program in the Department of Computer Science & Engineering. She earned her PhD in Computer Science from National Institute of Technology, Tiruchirapalli, Tamil Nadu for her thesis on "Multimedia Information Retrieval in Distributed System". She has 17 years of experience in teaching & research in the concerned field. She has worked for Central Power Research Institute as a Trainee Engineer. Later, she worked as a Junior Research Fellow for a project funded by Dept. of Science and Technology (DST), India for a period of 3 years. She has also worked in Indo-US collaborated project in a Student Exchange program for "Obama-Singh Knowledge Initiative Program" in the University of Nevada (UNLV), Las Vegas, United States. She is a certified IBM trainer for the Business Intelligence. Her research areas are Data mining, Information Retrieval, Image Processing and Computational Neuroscience. She has published more than 45 research articles in reputed journals and conferences, books and book chapters. She has 11 Indian patents and 2 Australian patents.

List of Contributors

Balamurugan M
Dept. of Computer Science and Engineering
CHRIST (Deemed to be University), Bengaluru, India

Bhasha P
Dept. of Data Science
Mohan Babu University (Erstwhile Sree Vidyanikethan Engineering College), Tirupati, India

Brighty Ebenezer L
School of Electronics Engineering
Vellore Institute of Technology, Chennai, India

Sameer Chauhan
School of Computer Science and Engineering
VIT University, Vellore, India

Chanthini Baskar
School of Electronics Engineering
Vellore Institute of Technology, Chennai, India

Smita Darandale
Department of Computer Science and Engineering
GITAM School of Technology
GITAM Deemed to be University, Bengaluru, Karnataka, India

Raghavendra M Devadas
Information Technology
Manipal Institute of Technology Bengaluru
Manipal Academy of Higher Education (MAHE)

Deepa S
SRM Institute of Technology, Ramapuram Campus, Chennai, India

Divya P
Department of Electronics Engineering
Madras Institute of Technology, Anna University, Chennai, India

Ganesh Hariharan
School of Computer Science and Engineering
Vellore Institute of Technology, Chennai, India

Vani Hiremani
Symbiosis Institute of Technology
Symbiosis International (Deemed), University, Pune, India

Jenefa J
Department of Computer Science and Engineering
CHRIST (Deemed to be University), Bengaluru, India

Jennie Gratia Franklin
Department of Electronics Engineering
Madras Institute of Technology, Anna University, Chennai, India

Jyothsna V
Department of Data Science
Mohan Babu University (Erstwhile Sree Vidyanikethan Engineering College), Tirupati, India

Kanimozhi S
School of Computer Science Engineering
Vellore Institute of Technology, Chennai, India

Kasthuri P
Department of Electronics Engineering
Madras Institute of Technology, Anna University, Chennai, India

Lubna Ambreen
Management Studies
JAIN (Deemed-to-be University), Bengaluru, India

Luv Sachdeva
School of Computer Science and Engineering
VIT University, Vellore, India

Manickavasagam Suruthi
School of Computer Science and Engineering
Vellore Institute of Technology, Chennai, India

Meena K
Department of Computer Science and Engineering
GITAM School of Technology
GITAM Deemed to be University, Bengaluru, Karnataka, India

Melvin
School of Computer Science Engineering
Vellore Institute of Technology, Chennai Campus, India

Menaka Pushpa A
School of Computer Science and Engineering
Vellore Institute of Technology, Chennai Campus, Tamil Nadu, India

Sandhya Rani Nallola
Department of Computer Science and Engineering
GITAM School of Technology
GITAM Deemed to be University, Bengaluru, Karnataka, India

Naga Swetha N
Dept. of Information Technology
Sree Vidyanikethan Engineering College, Tirupathi, India

Narayanan Ganesh
School of Computer Science and Engineering
Vellore Institute of Technology, Chennai, India

Niranjan L
Department of Electronics and Communication Engineering
CMR Institute of Technology, Bengaluru, Karnataka, India

Prakash P
Madras Institute of Technology, Anna University, MIT Rd, Radha Nagar, Chromepet, Chennai, India

Prasan Mittal
School of Computer Science and Engineering
Vellore Institute of Technology, Chennai, Tamilnadu, India

Praveen Gujjar J
Management Studies
JAIN (Deemed-to-be University), Bengaluru, India

Persiya J
School of Electronics Engineering
Vellore Institute of Technology, Chennai, India.

Poorani M
Department of Information Science & Engineering
CMR Institute of Technology, Bengaluru, India

Rakoth Kandan Sambandam
Department of Computer Science and Engineering
CHRIST (Deemed to be University), Bengaluru, India

Rashmi D
School of Computing and Information Technology
REVA University, Bengaluru, India

Sai Divya Sree T
Department of Information Technology
Sree Vidyanikethan Engineering College, Tirupathi, India

Sairamesh L
Department of Computer Science and Engineering
St.Joseph's Institute of Technology, Chennai, Tamilnadu, India

Sandhya E
Department of CSE (Artificial Intelligence)
Madanapalle Institute of Technology & Science, Madanapalle, India

Sanjith Surya S
School of Computer Science and Engineering
VIT University, Vellore, India

Sahaaya Arul Mary S. A.
School of Computer Science and Engineering
VIT University, Vellore, India

Sasithradevi A
Centre for Advanced Data Science
Vellore Institute of Technology, Chennai, India

Senthil Kumar J. P.
Associate Professor, Department of Management
GITAM School of Business,
GITAM Deemed to be University, Bangalore, Karnataka, India

Sengamali K N
Department of Electronics Engineering
Madras Institute of Technology, Anna University, Chennai, India

Shoba S
Centre for Advanced Data Science
Vellore Institute of Technology, Chennai Campus, India

Srinath Srinivasan
School of Mechanical Engineering
Vellore Institute of Technology, Chennai, India

Srivarshinee S
School of Computer Science and Engineering
Vellore Institute of Technology, Chennai Campus, Tamil Nadu, India

Sumathy P
Department of Computer Science
Bharathidasan University, Tiruchirappalli, Tamil Nadu, India

Suneetha J
Department of Computer Science Engineering
GITAM School of Technology
GITAM Deemed to be University, Bengaluru, Karnataka, India

Thaiyalnayaki S
Department of CSE, SOC
Bharath Institute of Higher Education and Research (Deemed to be University)
Chennai, India

Tanvir H Sardar
Department of Computer Science and Engineering
GITAM School of Technology
GITAM Deemed to be University, Bengaluru, Karnataka, India

Vadivel A
Department of Artificial Intelligence & Data Science
GITAM School of Technology
GITAM Deemed to be University, Bengaluru, Karnataka, India

Divya Vetriveeran
Dept. of Computer Science and Engineering
CHRIST (Deemed to be University), Bengaluru, India

Vijayalakshmi M
School of Electronics Engineering,
Vellore Institute of Technology, Chennai, India

Chapter 1

Bibliometric analysis on visual data analysis and dynamic dashboard tools

A literature review

Sandhya Rani Nallola, Vadivel A, Senthil Kumar J.P., and Meena K
GITAM Deemed to be University, Bangalore, Karnataka, India

Sumathy P
Bharathidasan University, Tiruchirappalli, Tamil Nadu, India

1.1 INTRODUCTION: DATA ANALYSIS

Visual data analysis is a multi-dimensional discipline that merges the realms of data analytics and visual communication to extract meaningful insights through graphical representations of complex data. In recent times, voluminous of data flows in every aspect of our lives, and the ability to decipher, interpret, and communicate it effectively is paramount. Visual data analysis is one of the important tools that provides a dynamic framework to explore, understand, and communicate insights derived from diverse datasets across various domains (Keim 2001).

The visual data analysis process the data quickly to present the result to human for better understanding. The raw data is transformed into visual forms such as charts, graphs, maps, and interactive dashboards. The analysts distill complex datasets into intuitive representations that facilitate comprehension and decision-making. These visualizations serve as a bridge between the intricacies of data and the cognitive processes of human perception. As a result, the analysts to uncover patterns, trends, and relationships that may remain obscured within raw numerical data (Elmqvist, Stasko, and Tsigas 2008).

The visual data analysis has a series of iterative steps, say, data collection and preparation, etc. This phase involves gathering data from various sources, cleaning, and transforming it into a format suitable for analysis. Once the data is ready for exploration, analysts use/apply different visualization techniques to investigate the dataset from various dimensions. This may involve creating simple descriptive visualizations to gain an initial understanding of the structure of data, followed by more sophisticated techniques such as multivariate analysis and dimensionality reduction for deeper

DOI: 10.1201/9781003542735-1

insights. Crucially, visual data analysis is not merely about generating static images but rather fostering an interactive dialogue between analysts and data. Interactive visualizations empower users to dynamically manipulate parameters, filter subsets of data, and drill down into specific details, fostering a more exploratory and iterative approach to analysis. Through this interactive feedback loop, analysts can iteratively refine their hypotheses, validate assumptions, and unearth novel insights that may have remained hidden within the complexity of data (Keim 1997).

Moreover, visual data analysis extends beyond the realm of exploratory analysis, which encompasses explanatory and communicative analysis. Once insights are obtained from the data, analysts he effectively communicates the findings to various stakeholders, ranging from fellow analysts to decision-makers and the general public. Visualizations are considered as the a powerful medium for communication, enabling analysts to distill complex analyses into intuitive narratives that resonate with their audience. Whether through static infographics, interactive dashboards, or immersive data storytelling experiences, visualizations play a pivotal role in conveying insights (De Oliveira, Levkowitz, and Nq 2001). In recent years, the advent of advanced technologies such as machine learning, big data analytics, and data visualization tools has revolutionized the landscape of visual data analysis. These technologies empower analysts to grapple with increasingly large and complex datasets, extract deeper insights, and create more immersive visualizations ever before. The frontier of visual data analysis continues to evolve at a rapid pace, offering boundless opportunities for innovation and discovery, such as identifying patterns from vast datasets using AI, using virtual reality to create immersive data experience, etc. (Stalling, Westerhoff, and Hege 2005).

In conclusion, visual data analysis stands as a cornerstone of modern data science, blending analytical rigor with the communicative power of visualization to unlock the latent insights contained within complex datasets. By harnessing the human capacity for visual perception and leveraging advanced technologies, analysts can navigate the intricacies of data with agility and precision, empowering stakeholders to make informed decisions in an increasingly data-driven world (Elmqvist, Stasko, and Tsigas 2008).

1.1.1 Dynamic dashboard tools

In ever-evolving landscape of data-driven decision-making, dynamic dashboard tools emerge as indispensable assets, empowering organizations to transform raw data into actionable insights with unparalleled speed and agility. These tools represent a fusion of sophisticated data visualization capabilities, interactive functionalities, and intuitive user interfaces, offering a comprehensive solution for harnessing the full potential of data in real time (Urrutia et al. 2016). Dynamic dashboard tools serve as the conduit through which data is transformed into actionable intelligence, driving strategic decision-making and fostering a culture of data-driven innovation (Ji et al. 2014; Garwood, Steingard, and Balduccini 2020). At its essence, a dynamic dashboard is a customizable interface that consolidates diverse data streams into a cohesive and visually appealing presentation. Unlike traditional static reports or spreadsheets, which offer limited interactivity and flexibility, dynamic dashboards provide users with the ability to interact with data in real time, exploring trends, drilling down into details, and uncovering insights with unprecedented fluidity. Whether tracking Key Performance Indicators (KPIs), monitoring operational metrics, or conducting ad-hoc analyses, users can tailor their dashboard experience to suit their unique needs and preferences, empowering them to extract insights that drive informed decision-making (Toasa et al. 2018; Adekoya-Cole 2024).

Dynamic dashboard tools offer a myriad of features designed to enhance the user experience and maximize analytical capabilities. These may include drag-and-drop interfaces for effortless dashboard customization, real-time data connectivity for up-to-the-minute insights, interactive filters and slicers for dynamic data exploration, and intuitive visualization options ranging from charts and graphs to maps and gauges. Additionally, advanced functionalities such as predictive analytics, geospatial analysis, and natural language processing enable users to delve deeper into data and extract actionable insights with ease (Jakubiec et al. 2017; Badgeley et al. 2016).

One of the hallmark features of dynamic dashboard tools is their ability to democratize data access and foster collaboration across organizational silos. By centralizing data from disparate sources within a single, cohesive interface, these tools break down barriers to information sharing and facilitate cross-functional collaboration (Pluto-Kossakowska et al. 2022; Molina-Marin et al. 2016). Decision-makers can gain holistic perspectives on organizational performance, while frontline employees can access relevant insights to drive day-to-day operations. Moreover, collaborative features such as annotation tools, commenting capabilities, and shared dashboards enable teams to collaborate in real-time, facilitating data-driven discussions and aligning stakeholders around common objectives (Antonini, Ganuza, and Castro 2022; Shamsuzzoha et al. 2014). In Table 1.1, various interactive dashboard tools and their feature is presented.

Furthermore, dynamic dashboard tools play a pivotal role in driving data-driven culture within organizations, empowering users at all levels to make data-informed decisions with confidence. By providing intuitive interfaces, actionable insights, and self-service analytics capabilities, these tools reduce reliance on IT or data science teams for routine analysis tasks, enabling business users to take ownership of their data and drive innovation from the ground up. Moreover, by surfacing insights in real-time and fostering a culture of continuous improvement, dynamic dashboards serve as catalysts for organizational agility and responsiveness in rapidly evolving business environment (Abd-Elfattah, Alghamdi, and Amer 2014; Taber et al. 2021).

Table 1.1 Interactive dashboards tools and features

Tool	*Features*	*Subscription options*	*Cost (monthly)*
Tableau	– Interactive Dashboards	– Creator	$70
	– Data Blending	– Explorer	$35
	– Real-time Collaboration	– Viewer	$12
	– Embedding Options		
Power BI	– Customizable Dashboards	– Pro	$9.99
	– Natural Language Query	– Premium	Custom Pricing
	– AI-powered Insights	– Free	Free
	– Integration with Microsoft Products		
Plotly	– Interactive Web-based Visualizations	– Basic	$0
	– Support for Python, R, and Julia	– Pro	$39
	– Dashboards and Reporting	– Enterprise	Custom Pricing
	– Community Support		
Qlik sense	– Associative Data Indexing	– Qlik Sense Cloud	$30
	– Drag-and-Drop Interface	– Qlik Sense Desktop	$30
	– Advanced Analytics	– Qlik Sense Enterprise	Custom Pricing
	– Scalable and Secure		

Source: Self constructed.

In conclusion, dynamic dashboard tools represent a transformative force in the realm of data analytics, empowering organizations to unlock the full potential of their data and drive strategic decision-making at scale. By combining advanced visualization capabilities with interactive functionalities and intuitive user interfaces, these tools democratize data access, foster collaboration, and cultivate a culture of data-driven innovation. As organizations strive to navigate the complexities of the digital age, dynamic dashboard tools emerge as indispensable allies, enabling them to thrive in a landscape defined by data abundance and opportunity (Seong Kam et al. 2012; Ivanov et al. 2019).

This study aims to examine research on visual data analysis and dynamic dashboard tools, as well as their insights and an overview of the topic, using the science mapping review approach. Concerning the following research questions, this study addresses them:

RQ 1: To identify and analyze the key authors, organizations, and scholarly trends in the literature on Visual data analysis and Dynamic Dashboard tools.

RQ 2: How does the social structure of the knowledge base impact Visual Data Analysis and dynamic dashboard tools?

RQ 3: Are there any themes in the literature of Visual Data Analysis and dynamic dashboard tools that have been explored with frequency of occurrence and are currently getting the most remarkable attention?

1.2 METHODOLOGY

The study utilized a bibliometric analysis methodology, which involved implementing documented search strategies and rigorous data-tracking procedures. The research primarily relied on Scopus, a widely recognized platform for scientific literature. Utilizing this platform, a comprehensive analysis is performed. The search is conducted on specific search strategy that involved combining the keywords "Visual data analysis" and "dynamic dashboard tools" or "visualization". The initial search has retrieved 436 documents. The second stage of the process retrieved raw data in CSV format from the relevant publications from the online database. The data is analyzed meticulously to evaluate its relevance to the research topic and confirm its alignment with the predefined search strategy. In order to enhance the understanding of the methodology, visual representation of the process is depicted in Figure 1.1. This visual representation includes the specific keywords and search operators used.

Stage 1: Search Criteria
- The first stage of the methodological framework involves defining the initial search criteria for identifying relevant articles within the Scopus database. This stage sets the foundation for the entire analysis by specifying the keywords and potentially other filters to be used in the search query.

Stage 2: Search with Multiple Criteria
- Uses Boolean operators (AND) to combine keywords related to the research topic. keywords are "visual data analysis" AND "dynamic dashboard tools" OR "visualization"

Stage 3: Refinement
- Refines the initial search results based on additional criteria is shown as refinement by publication year (1990–2024).

Stage 4: Subject Area
- Limits the search to specific subject areas "Computer Science, Engineering and Business".

Stage 5: Document Type and Language
- Defines the document types and language for inclusion clearly specifies document type as Articles and language as English only.

Stage 6: Results
- The final, refined list of articles is then exported using software tools like R programming language and VOSviewer for further analysis. Following the export of the final list of articles, the data

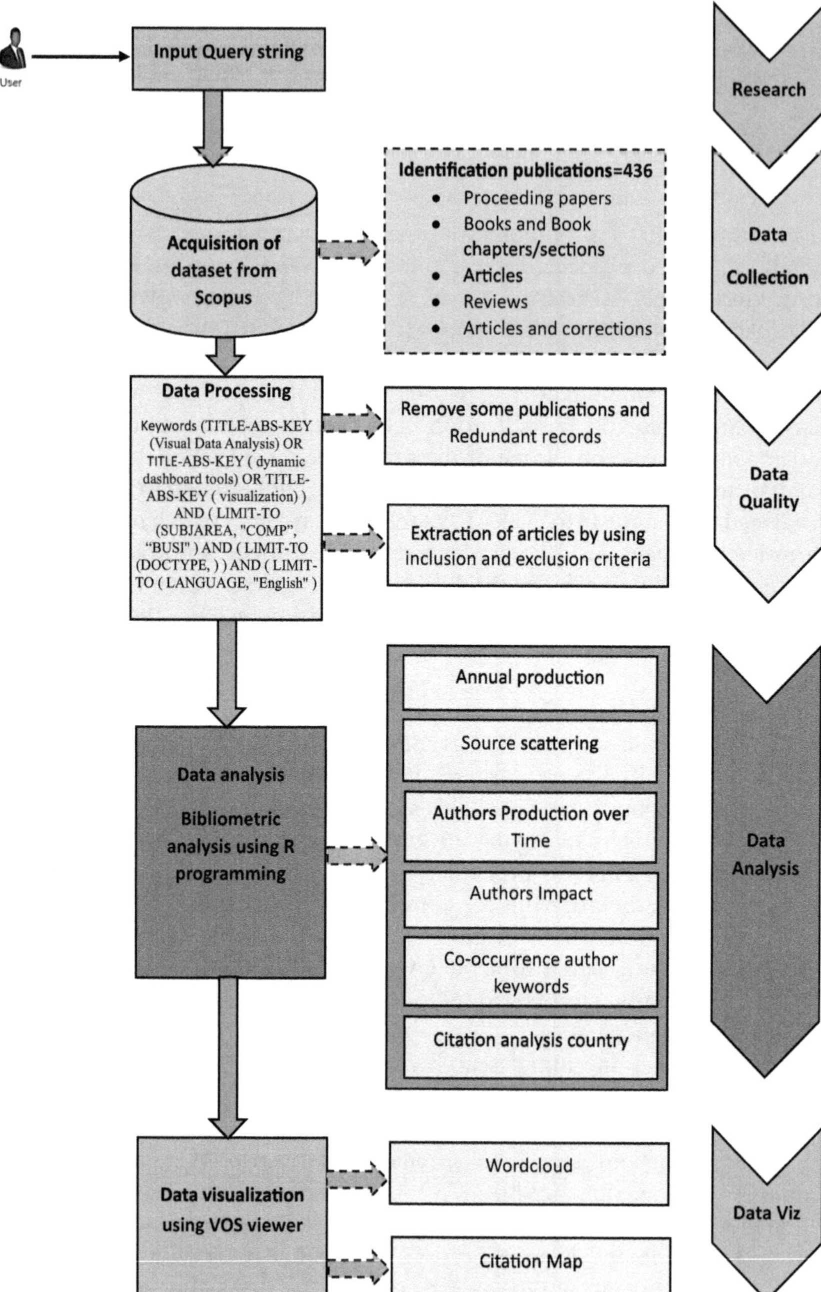

Figure 1.1 Methodological flowchart of bibliometric analysis.

underwent a rigorous analysis and visualization process. The selected manuscripts, matching the predetermined inclusion criteria, were exported into comma-separated values (.csv) format for further manipulation. This data was then transformed into two separate formats:

Interactive tools: Power BI and Tableau

Power BI: Microsoft Power BI is a business intelligence and data visualization application that transforms information from multiple data sources into a variety of business intelligence reports. Utilizing Microsoft Power BI enables organizations to effortlessly identify patterns, monitor progress, and arrive at decisions based on data.

Tableau: Tableau Desktop is an application that facilitates the development of attractive and interactive data visualizations.

VOSviewer Software: The data is prepared for use within VOSviewer software, a powerful tool for creating and analyzing bibliographic networks.

R Programming Data Frame: The data is also converted into a data frame structure compatible with the R programming language. This facilitated the utilization of the open-source R package, bibliometrix, to ensure the accuracy and reliability of the results.

A dimensionality-reduction method focused on most relevant information and eliminated extraneous data. This technique helped to streamline the data analysis and visualization process. Finally, the power of VOSviewer and the bibliometric package within R is leveraged to create comprehensive network maps of the filtered data. These visualizations provided valuable insights into the research landscape, allowing the identification of key themes, trends, and relationships within the scholarly corpus. By carefully filtering parameters and constructing the data's social structures, the visualizations aimed to present a clear and concise picture of the research domain on visual data analysis.

Query Strings:

(TITLE-ABS-KEY (Visual Data Analysis) OR TITLE-ABS-KEY (dynamic dashboard tools) OR TITLE-ABS-KEY (visualization)) AND (LIMIT-TO (SUBJAREA, "COMP", "BUSI")) AND (LIMIT-TO (DOCTYPE,)) AND (LIMIT-TO (LANGUAGE, "English"))

1.3 RESULTS AND DISCUSSION

In the context of bibliometric analysis, annual production/publications refer to the quantification of publications generated on a topic, discipline, journal, institution, or by an individual author within a specified year. A quantitative metric is employed to evaluate the productivity and degree of activity within a scholarly community or a particular study field over a while. The annual production/publications can be determined by tallying the number of publications (such as articles, books, conference papers, etc.) that are listed in bibliographic databases or acquired from other sources for each specific year. This statistic facilitates the comprehension of patterns, detection of new subjects, and assessment of the influence of research endeavors within a specific period for researchers and policymakers. Figure 1.2 illustrates the annual publication of documents from 1990 onwards, specifically focusing on Visual Data Analysis and dynamic dashboard tools. It is evident from Figure 1.2 that the first article published in 1990 in the interested research filed. The number of publications increased and reached the peak in the year 2023. The curve falls substantially after 2023 and this is due to the fact that the number of publications in 2024 is considered till March 2024 (Table 1.2).

It is known that Bradford's law categorized the scientific journals based on their productivity for grouping the articles as core and additional groups. Bradford's Law is a concept in bibliometrics that describes the pattern of distribution of articles on a specific subject within scientific journals. It was first described by Samuel C. Bradford in 1934. The law suggests that if journals are sorted by the number of articles on a subject, they fall into three groups, each containing about one-third of all articles. In Figure 1.3, we depict the source clustering using Bradford's rule to present the relationship between the number of paper published and the name of the journals. Because Bradford's Law explains how articles are dispersed throughout several publications, it is often referred to as the law of scattering.

It is noticed that a large number of articles are published in the IEEE Transactions, Lecture notes on computers, ACM International conference, and so on. Also, the research field that we have considered is being covered by most of the publishers. This shows that the research area considered is vital to the research community.

1.3.1 Production cycle of authors

The Bibliometric analysis helps in understanding the productivity of authors over time by analyzing their publication history for the period of time. This enables us to understand the research interests of authors expansion, and influence. In this paper, the typical production statistics of the author is presented in Figure 1.4 and Table 1.3.

It is observed from Figure 1.4 that, most cited paper in the field is by KEIM DA with 362 citations in the year 2006. Similarly, the next cited paper is also by KEIM DA with 276 citations in the year 2014. The citations for the rest of the authors is presented in Table 1.3 and the citation is ranging from 1 to 79.

Table 1.3 presents the production statistics of authors in terms of total citations, and total citation per year (TCpY). The top five authors with high

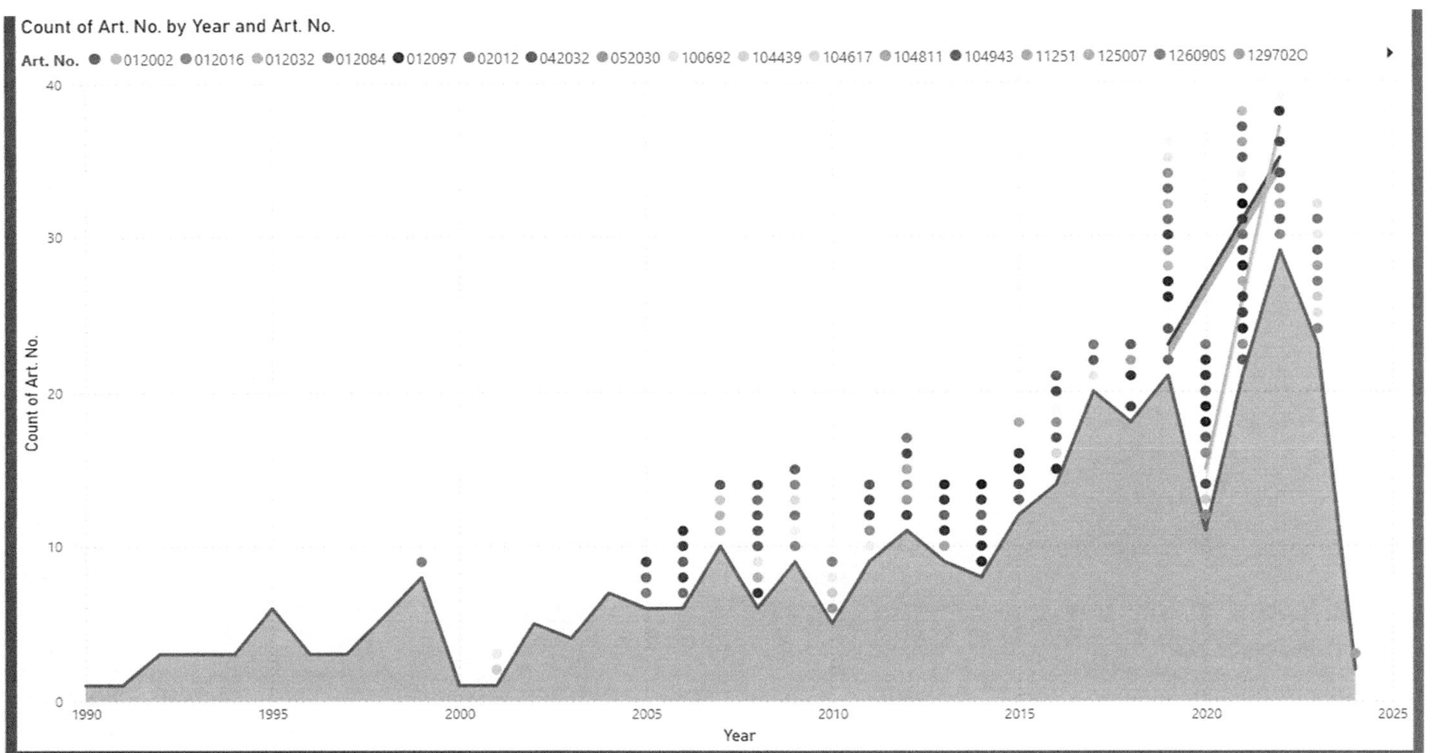

Figure 1.2 Annual production.

Table 1.2 Annual production

Year	*Articles*
1990	1
1991	1
1992	3
1993	0
1994	3
1995	6
1996	3
1997	3
1998	0
1999	9
2000	1
2001	3
2002	5
2003	4
2004	7
2005	9
2006	11
2007	14
2008	14
2009	15
2010	9
2011	14
2012	17
2013	14
2014	14
2015	18
2016	21
2017	23
2018	23
2019	36
2020	23
2021	38
2022	39
2023	32
2024	3

Source: Self constructed.

citations are KEIM DA, BETHEL EW, NARECHANIA A, DACHSELT R and HAGEN H ranging from 35 to 362 times. This shows that these authors are popular in visual data analysis, interactive dashboard and visualization tools and also produced high quality papers that have a significant impact on the field.

Figure 1.3 Relationship between the number of articles and journal titles.

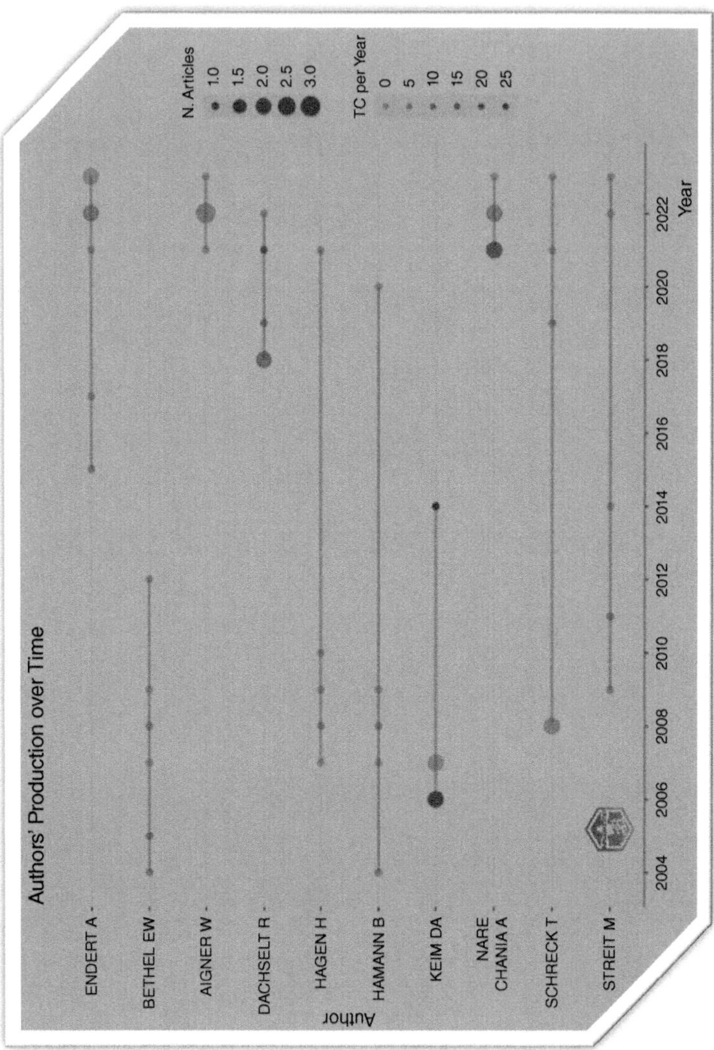

Figure 1.4 Authors' production over time.

Table 1.3 Authors' production over time

Author	*Year*	*Freq*	*TC*	*TCpY*
AIGNER W	2021	1	6	1.5
AIGNER W	2022	3	10	3.333
AIGNER W	2023	1	1	0.5
BETHEL EW	2004	1	34	1.619
BETHEL EW	2005	1	79	3.95
BETHEL EW	2007	1	3	0.167
BETHEL EW	2008	1	35	2.059
BETHEL EW	2009	1	3	0.188
BETHEL EW	2012	1	7	0.538
DACHSELT R	2018	2	55	7.857
DACHSELT R	2019	1	43	7.167
DACHSELT R	2021	1	57	14.25
DACHSELT R	2022	1	4	1.333
ENDERT A	2015	1	28	2.8
ENDERT A	2017	1	1	0.125
ENDERT A	2021	1	1	0.25
ENDERT A	2022	2	11	3.667
ENDERT A	2023	2	1	0.5
HAGEN H	2007	1	5	0.278
HAGEN H	2008	1	35	2.059
HAGEN H	2009	1	3	0.188
HAGEN H	2010	1	17	1.133
HAGEN H	2021	1	8	2
HAMANN B	2004	1	34	1.619
HAMANN B	2007	1	3	0.167
HAMANN B	2008	1	35	2.059
HAMANN B	2009	1	3	0.188
HAMANN B	2020	1	2	0.4
KEIM DA	2006	2	362	19.053
KEIM DA	2007	2	13	0.722
KEIM DA	2014	1	276	25.091
NARECHANIA A	2021	2	59	14.75
NARECHANIA A	2022	2	11	3.667
NARECHANIA A	2023	1	1	0.5
SCHRECK T	2008	2	25	1.471
SCHRECK T	2019	1	15	2.5
SCHRECK T	2021	1	0	0
SCHRECK T	2023	1	0	0

Source: Self constructed.

1.3.2 Properties of query string and keywords

As shown in Figure 1.1, the query string is formed with visual data analysis, dynamic dashboard tools and visualizations and the same is shown in Figure 1.5. As the first step, the frequency of occurrence of these items is used to obtain the retrieval set. In Table 1.4, the frequency of occurrence of each keyword term is presented for better understanding and analysis. The highest frequency of occurrence is for "visual data analysis" with 226 occurrences. "Visualization" has appeared 163 times and so on.

1.3.3 Relationship among terms in keywords

The co-occurrence matrix captures the variations in the number of articles by each author. The cosine similarity or Jaccard index is used for normalization, where the number of publications is divided by the total number of publications to get the result in the range of [0–1]. In addition, it is important to capture the relationship between various keywords used in the query string. In this paper, we have used visual data analysis, visualization, interactive dashboard tools and the same as depicted in Figure 1.6. It is observed from the figure that there is a notable relationship among various other terminologies such as information analysis, data handling, etc. These terms are referring to various technologies that are contributing to the growth of data visualization and interactive dashboards. It is also imperative that all these terminologies must be cohesive among each other such that most of the real-time issues can be solved.

Figure 1.5 Term cloud of terms in keyword/query.

Table 1.4 Frequency of occurrence of key terms

Terms	*Frequency*
Visual data analysis	226
Visualization	163
Data visualization	128
Information analysis	80
Data handling	78
Data reduction	52
Visual analytics	40
Human	37
Article	35
Computer graphics	31
Data analysis	30
Decision making	26
Information systems	24
User interfaces	24
Visual analysis	24
Data mining	22
Information visualization	22
Learning systems	21
Humans	20
Human computer interaction	17
Flow visualization	16
Algorithms	15
Big data	15
Deep learning	15
Software	15
Computer vision	14
Three dimensional computer graphics	14
Artificial intelligence	13
Visualization technique	13
Adult	12
Female	12
Male	12
Semantics	12
Behavioral research	11
Computer simulation	11
Data sets	11
Design	11
Visual languages	11
Classification (of information)	10
Database systems	10
Digital storage	10

(Continued)

Table 1.4 (Continued)

Terms	*Frequency*
Human engineering	10
Machine learning	10
Virtual reality	10
Clustering algorithms	9
Computer program	9
Exploratory analysis	9
Graphical user interfaces	9
Information management	9
Interactive computer systems	9

Source: Self constructed.

Figure 1.6 Co-occurrence author keywords.

1.3.4 Citation analysis based on country of author

The citation analysis is shown in Figure 1.7, and that help us to determine the relationship between the country of authors. The link between various nodes represents the relationship among the authors. The width of the link is considered as weight, representing the number of citations among

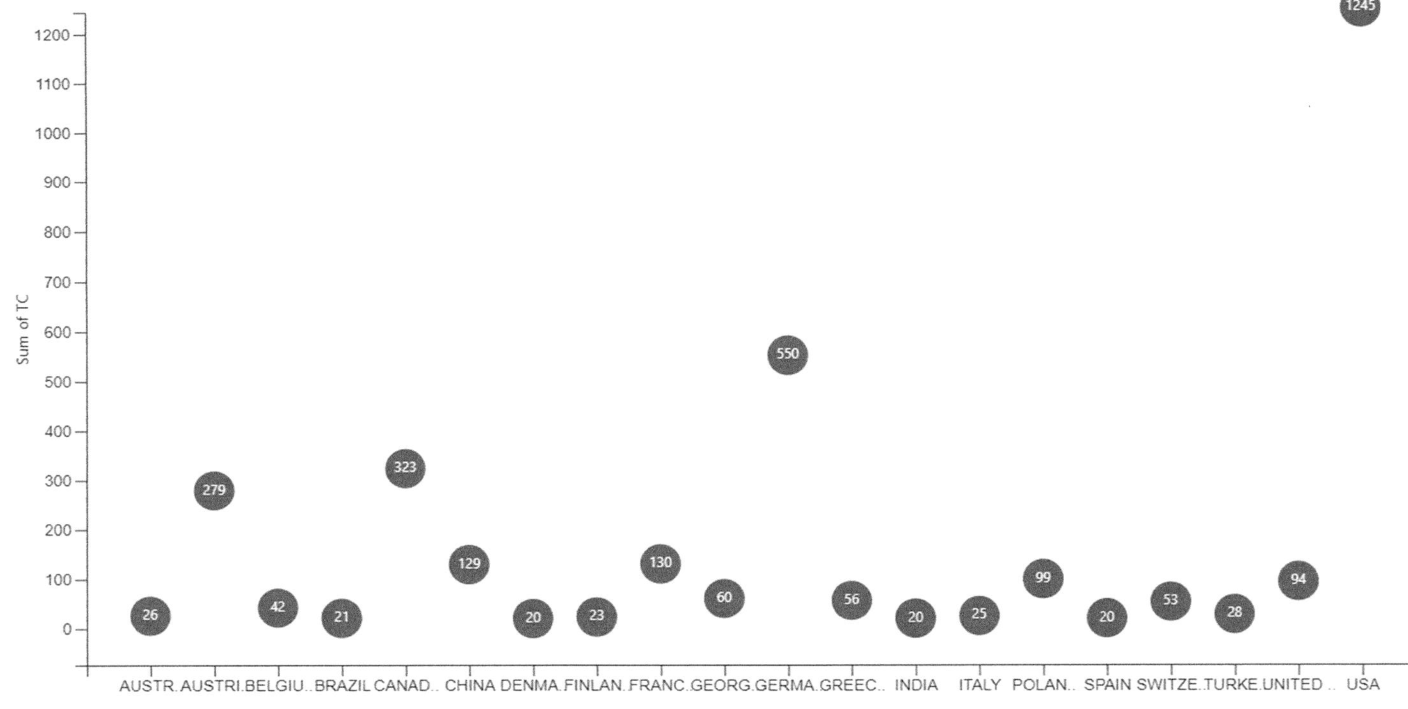

Figure 1.7 Citation analysis.

them. Thus, a high weight denotes a stronger relationship, and a low weight denotes a weak relationship between the authors.

It is observed from Figure 1.7, that the publications are contributed in this field by authors from different countries. However, co-authors belonging to a few countries have larger number of contributions. It is also found that the contributions from the United States and Germany have higher citations on publications in the area of visual data analysis, dashboard tools, and visualization.

1.4 CONCLUSION

In this work, we have examined the research trends in visual data analysis and dynamic dashboard tools using a science mapping review approach. The annual publications from 2009 onwards is focused for analysis. We have applied Brandford's law of categorizing scientific journals based on publications. This analysis has created various groups of publications such as core and additional groups. It has also created a zone having the same number of publications as the core group. We have analyzed bibliometric values to calculate the publications over time based on number of publications to get the efficiency, expansion, and influence of the research interest of authors. The outcome of this analysis are presented as Annual production, relationship between articles and title of the journal, publications trends of authors over time, co-occurrence of keywords, and citation analysis for researchers consumption.

REFERENCES

Abd-Elfattah, Mohamed, Turki Alghamdi, and Eslam Amer. 2014. "Dashboard Technology Based Solution to Decision Making." *International Journal of Computer Science Engineering and Information Technology Research (IJCSEITR) ISSNP* 4 (2): 59–70.

Adekoya-Cole, Temitope. 2024. "From Text to Insights: Building a Dynamic Dashboard for Sentiment Analysis - A Twitter Case Study." 1–26.

Antonini, A. S., M. L. Ganuza, and S. M. Castro. 2022. "VISUEL - A Web Dynamic Dashboard for Data Visualization | VISUEL - Un Tablero Dinámico Web Para La Visualización de Datos." *Journal of Computer Science and Technology (Argentina)* 22 (1): 42–57.

Badgeley, Marcus A., Khader Shameer, Benjamin S. Glicksberg, Max S. Tomlinson, Matthew A. Levin, Patrick J. McCormick, Andrew Kasarskis, David L. Reich, and Joel T. Dudley. 2016. "EHDViz: Clinical Dashboard Development Using Open-Source Technologies." *BMJ Open* 6 (3): 1–11. https://doi.org/10.1136/bmjopen-2015-010579

De Oliveira, Maria Cristina Ferreira, Haim Levkowitz, and A Nq. 2001. "Visual Data Exploration and Mining: A Survey." São Carlos: ICMC-USP. Recuperado de https://repositorio.usp.br/directbitstream/986a148e-d7aa-4281-b19e-0303dd839359/1215743.pdf

Elmqvist, Niklas, John Stasko, and Philippas Tsigas. 2008. "DataMeadow: A Visual Canvas for Analysis of Large-Scale Multivariate Data." *Information Visualization* 7 (1): 18–33. https://doi.org/10.1057/palgrave.ivs.9500170

Garwood, Kathleen Campbell, David Steingard, and Marcello Balduccini. 2020. "Dynamic Collaborative Visualization of the United Nations Sustainable Development Goals (SDGs): Creating an SDG Dashboard for Reporting and Best Practice Sharing." *VISIGRAPP 2020 - Proceedings of the 15th International Joint Conference on Computer Vision, Imaging and Computer Graphics Theory and Applications 3 (Visigrapp 2020)*: 294–300. https://doi.org/10.5220/0009172302940300

Ivanov, Vladimir, Daria Larionova, Dragos Strugar, and Giancarlo Succi. 2019. "Design of a Dashboard of Software Metrics for Adaptable, Energy Efficient Applications." *Proceedings - DMSVIVA 2019: 25th International DMS Conference on Visualization and Visual Languages*, 75–82. https://doi.org/10.18293/jvlc2019-n2-009

Jakubiec, J. Alstan, Max C. Doelling, Oliver Heckmann, Ramkumar Thambiraj, and Vedashree Jathar. 2017. "Dynamic Building Environment Dashboard: Spatial Simulation Data Visualization in Sustainable Design." *Technology Architecture and Design* 1 (1): 27–40. https://doi.org/10.1080/24751448.2017.1292791

Ji, Min, Christine Michel, Élise Lavoué, and Sébastien George. 2014. "*DDART, a Dynamic Dashboard for Collection, Analysis and Visualization of Activity and Reporting Traces.*" *Lecture Notes in Computer Science (Including Subseries Lecture Notes in Artificial Intelligence and Lecture Notes in Bioinformatics)* 8719 LNCS: 440–445. https://doi.org/10.1007/978-3-319-11200-8_39

Keim, Daniel A. 2001. "Visual Exploration of Large Data Sets." *Communications of the ACM* 44 (8): 38–44. https://doi.org/10.1145/381641.381656

Keim, Daniel A. 1997. "Introduction Goals of Visualization Techniques Visual Data Mining." In: *International Conference on Very Large Databases (VLDB'97)*, Vienna, September, 1997.

Molina-Marin, M., E. Granados-Gomez, A. Espinosa-Reza, and H. R. Aguilar-Valenzuela. 2016. "CIM-Based System for Implementing a Dynamic Dashboard and Analysis Tool for Losses Reduction in the Distribution Power Systems in México." *WSEAS Transactions on Computers* 15: 24–33.

Pluto-Kossakowska, Joanna, Anna Fijałkowska, Małgorzata Denis, Joanna Jaroszewicz, and Sylwia Krzysztofowicz. 2022. "Dashboard as a Platform for Community Engagement in a City Development—A Review of Techniques, Tools and Methods." *Sustainability (Switzerland)* 14 (17): 1–33. https://doi.org/10.3390/su141710809

Seong Kam, Tin, Ketan Barshikar, Shaun Jun Hua Tan, Tin Seong, and Shaun Jun Hua. 2012. "DIVAD: A Dynamic and Interactive Visual Analytical Dashboard for Exploring and Analyzing Transport Data." *International Journal of Computer, Electrical, Automation, Control and Information Engineering* 6 (11): 834–1353.

Shamsuzzoha, Ahm, Yuqiuge Hao, Petri Helo, and Khan Khadem. 2014. "Dashboard User Interface for Measuring Performance Metrics: Concept from Virtual Factory Approach." *Proceedings of the 2014 International Conference on Industrial Engineering and Operations Management*, Bali, Indonesia, January 7–9, 2014, 124–133.

Stalling, Detlev, Malte Westerhoff, and Hans Christian Hege. 2005. "Amira: A Highly Interactive System for Visual Data Analysis." *Visualization Handbook* 1: 749–767. https://doi.org/10.1016/B978-012387582-2/50040-X

Taber, Peter, Charlene Weir, Jorie M. Butler, Christopher J. Graber, Makoto M. Jones, Karl Madaras-Kelly, Yue Zhang, et al. 2021. "Social Dynamics of a Population-Level Dashboard for Antimicrobial Stewardship: A Qualitative Analysis." *American Journal of Infection Control* 49 (7): 862–867. https://doi.org/10.1016/j.ajic.2021.01.015

Toasa, Renato, Marisa Maximiano, Catarina Reis, and David Guevara. 2018. "Data Visualization Techniques for Real-Time Information - A Custom and Dynamic Dashboard for Analyzing Surveys' Results." *Iberian Conference on Information Systems and Technologies, CISTI*, Caceres, Spain, 2018-June, 1–7. https://doi.org/10.23919/CISTI.2018.8398641

Urrutia, Manuel Leon, Ruth Cobos, Kate Dickens, Su White, and Hugh Davis. 2016. "Visualising the MOOC Experience: A Dynamic MOOC Dashboard Built through Institutional Collaboration." *EMOOCs 2016*, 1–8.

Chapter 2

Visual data analysis and inference through dimensionality reduction techniques

Jyothsna V
Mohan Babu University (Erstwhile Sree Vidyanikethan Engineering College), Tirupati, India

Sandhya E
Madanapalle Institute of Technology & Science, Madanapalle, India

Bhasha P
Mohan Babu University (Erstwhile Sree Vidyanikethan Engineering College), Tirupati, India

N. Naga Swetha and Sai Divya Sree T
Sree Vidyanikethan Engineering College, Tirupathi, India

2.1 INTRODUCTION

A dashboard is generally referred to as a visual interface that gives consumers a condensed and easily comprehensible summary of important data, metrics, or information. Dashboards are widely utilized in many domains like project management, analytics, business, and finance to oversee operations, monitor developments, and make well-informed choices. Measures such as revenue, sales, customer acquisition rates, and website traffic may be shown in real-time or almost real-time on a dashboard in a commercial setting. Users can rapidly evaluate the current situation and spot trends or abnormalities thanks to the frequent use of charts, graphs, tables, and other visual components in the presentation of these measurements. Dashboards are frequently interactive, enabling users to drill down into certain data points or change parameters to investigate various facets of the information being displayed. They can be tailored to meet the unique requirements and preferences of users. In general, dashboards are useful instruments for enhancing organizational performance and supporting data-driven decision-making.

DOI: 10.1201/9781003542735-2

Dashboard requirements:

Dashboards are used for a wide range of applications in many fields [1]. Some typical use cases are as follows:

a. Business Performance Monitoring: Dashboards are used in businesses to track key performance indicators (KPIs) like revenue, sales, expenses, profit margins, employee productivity, and customer satisfaction ratings. Company executives can monitor these indicators in real time or over predetermined intervals of time to evaluate performance and pinpoint areas in need of development.
b. Data Analytics and Visualization: To view and examine big datasets, data scientists and analysts employ dashboards. Data-driven decision-making is made possible by analysts' ability to spot trends, patterns, and correlations in data by presenting it in charts, graphs, and other visualizations.
c. Project Management: To assign resources, monitor task completion, track project progress, and spot possible hazards or bottlenecks, project managers use dashboards. To give a thorough overview of the state of the project, project dashboards frequently incorporate Gantt charts, task lists, timetables, and resource allocation graphs.
d. Marketing Campaign Analysis: Using dashboards, marketers evaluate how well their campaigns are performing on a variety of platforms, including social media, email, search engines, and traditional advertising. In order to assess the efficacy of a campaign, marketing dashboards might show indicators such as website traffic, conversion rates, click-through rates, and return on investment (ROI).
e. Financial Reporting and Analysis: Finance experts monitor financial parameters like cash flow, profitability, variation from the budget, and accounts payable/receivable using dashboards. Financial dashboards assist stakeholders in making well-informed decisions regarding resource allocation, investments, and budgeting by offering insights into the organization's financial health.
f. Customer Relationship Management (CRM): To watch sales pipelines, measure customer satisfaction metrics, manage customer interactions, and anticipate income, sales teams and customer support agents use dashboards coupled with CRM systems. CRM dashboards give companies the ability to better engage customers, find upselling and cross-selling opportunities, and prioritize prospects.
g. Logistics and Supply Chain Management: Dashboards are used in supply chain management to track shipments, keep an eye on inventory levels, plan routes, and maintain connections with suppliers. Supply chain dashboards assist businesses in streamlining processes, cutting expenses, and guaranteeing on-time delivery of goods and services.

All things considered, dashboards are adaptable instruments that enable users to efficiently display, evaluate, and understand data, promoting educated decision-making and enhancing organizational performance in a variety of roles and sectors.

2.1.1 Interactive dashboard

A dynamic and user-friendly interface that enables people to view and interact with data in real-time is called an interactive dashboard [2]. In contrast to static dashboards that offer fixed views of data, interactive dashboards let users explore and modify the information that is shown to get deeper understanding and help them make better decisions.

Key features of interactive dashboards include:

- *Quick filters*: Users using interactive dashboards can quickly modify the data they see in a dashboard and return it to the original data without affecting other users' ability to use the dashboard. A user may be able to easily filter an interactive dashboard to reveal only sales data for the northeast region, for instance, even though the dashboard displays sales data for all 50 U.S. states.
- *Drill-Down Capabilities*: With this function, users of interactive dashboards can go deeper into the data layers displayed beneath the top layer. Drill down may enable a user to click on a single state and then view the sales figures per city inside that state, using our sales-by-state example from earlier.
- *Interactive Visualization*: Charts, graphs, maps, and tables are just a few examples of the interactive visualizations that are frequently used in dashboards. Hovering over data points, clicking on components to reveal more details, and changing display parameters are all ways that users can engage with these visualizations.
- *Real-Time Updates*: A few interactive dashboards offer updates in real-time or very near-real-time, enabling users to keep an eye on shifting data trends and act fast on new information.
- *Customization Options*: To fit their needs and tastes, users can alter the colour schemes, dashboard layout, visualization kinds, and other display choices. This adaptability makes data analysis more efficient and improves user experience.
- *Export and Sharing Functions*: Data exporting and insight sharing are common capabilities found in interactive dashboards. For cooperation and decision-making, users can share links to interactive dashboards with stakeholders or colleagues, or export data to a variety of formats (such as CSV or Excel).
- *Hide cells*: Users of interactive dashboards can choose the specific data they wish to view or share with other users by hiding all other data, making the selected data stand out and be emphasized.

Across all industries, interactive dashboards are extensively used for a variety of tasks, such as project management, data analysis and visualization, company performance monitoring, marketing campaign analysis, and more. They enable users to make smarter decisions based on facts and trends, unearth actionable insights, and examine data dynamically.

2.1.2 Dynamic interactive dashboard

A dynamic interactive dashboard is a vital resource in a variety of fields, such as business, finance, healthcare, and education, since it enables users to interact with data in real-time [3].

The following are some essential features and advantages of dynamic interactive dashboards:

- *Real-Time Data Updates*: With the ability to update data in real-time or on a regular basis, dynamic interactive dashboards guarantee that users always have access to the most recent information. This function comes in very handy for tracking indicators that change quickly, such stock prices, website traffic, or social media interaction.
- *Interactive Visualizations*: These dashboards frequently include interactive visualizations that let users explore data in a more natural and interesting way, like graphs, charts, maps, and gauges. By dragging their cursor over data points, clicking on items to reveal more information, or changing display parameters, users can interact with these visualizations.
- *Customizable Filters and Parameters*: Users can customize the parameters and filters on dynamic interactive dashboards to make the data provided more relevant to their individual requirements and tastes. In order to undertake more focused analysis, users can use filters to focus on particular time periods, geographic locations, product categories, or other pertinent dimensions.
- *Drill-Down and Hierarchical Navigation*: By clicking on particular data points or categories to expose more comprehensive information, users can explore deeper into the data on these dashboards, which frequently offer drill-down and hierarchical navigation. Users can do more in-depth analysis and more successfully spot trends, patterns, and outliers with the aid of this hierarchical navigation.
- *Responsive Design*: The flexible design of dynamic interactive dashboards allows them to adjust to many screen sizes and devices, including smartphones, tablets, and desktop PCs. This makes it possible for users to view and engage with the dashboard from anywhere at any time, while also guaranteeing a uniform and smooth user experience across all platforms.
- *Collaboration and Sharing Features*: Collaboration and sharing tools can be incorporated into dynamic interactive dashboards, enabling users to work together with coworkers or stakeholders by exchanging

insights, annotations, or comments right within the dashboard experience. Additionally, users can share links to interactive dashboards and export data to help in decision-making and communication.

All things considered, dynamic interactive dashboards enable users to dynamically examine data, unearth relevant insights, and make better decisions in real time. These dashboards help firms remain competitive and adaptable in the fast-paced business world of today by giving them access to current data and simple visualization tools.

2.2 CLASSIFICATION OF DYNAMIC INTERACTIVE DASHBOARD TOOLS

2.2.1 Tableau: powerful data visualization tool

Tableau, a top platform for business intelligence and data visualization, users can build interactive dashboards with a variety of analytics and visualization options. With the help of the well-known data visualization tool Tableau, users can create dynamic, informative dashboards and reports from their data [4].

Key Features:

- *Drag-and-drop interface*: Even without significant technical knowledge, users may easily build visualizations with Tableau's user-friendly interface. Users can create charts, graphs, and maps that effectively express their findings by just dragging and dropping data pieces as shown in Figure 2.1.

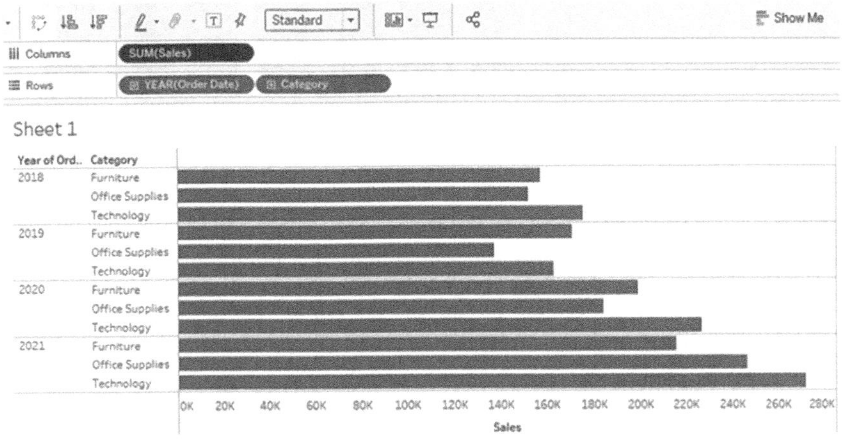

Figure 2.1 Drag and drop interface.

- *Broad variety of visualizations*: Bar charts, line charts, pie charts, scatter plots, heat maps, and more are just a few of the many chart kinds that Tableau provides. Because of its adaptability, users can select the best visualization format for precisely and successfully representing their data as shown in Figure 2.2.

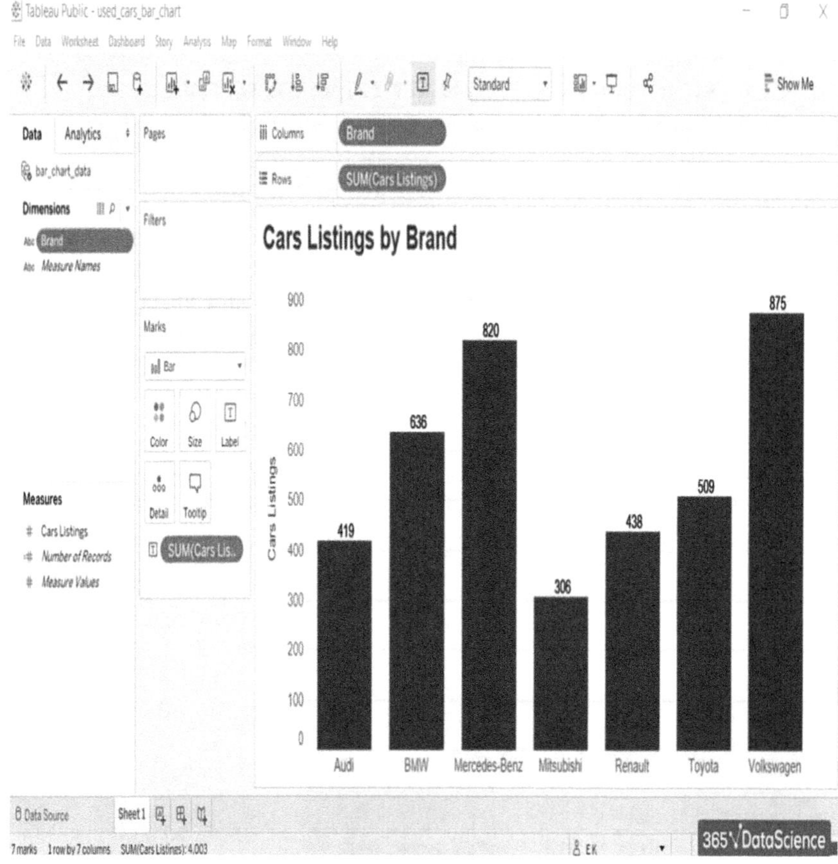

Figure 2.2 Broad variety of visualizations.

- *Data blending and cleaning*: Tableau allows users to effortlessly merge data from several sources, including spreadsheets, databases, and cloud apps as shown in Figure 2.3. In order to guarantee the accuracy and consistency of the data used in visualizations, the software also has built-in capabilities for data cleaning and manipulation.

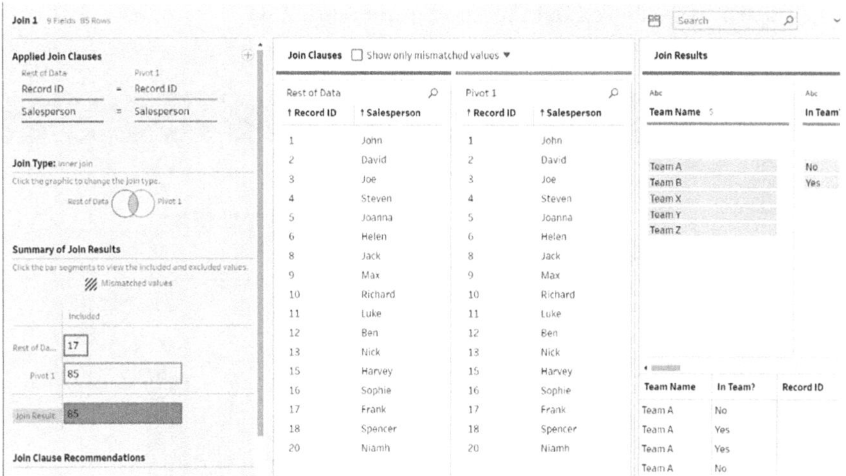

Figure 2.3 Data blending and cleaning.

- *Interactive stories and dashboards*: Tableau users can construct interactive dashboards that let visitors filter, delve deeper, and examine data in real time. Users can also create data tales, which are narratives or analyses that incorporate text, graphics, and other aspects to guide viewers through.
- *Sharing and collaboration*: Tableau helps users collaborate by enabling them to safely share reports and dashboards with other users. Members of the team can ask questions, leave comments on visualizations, and learn from shared data as shown in Figure 2.4.

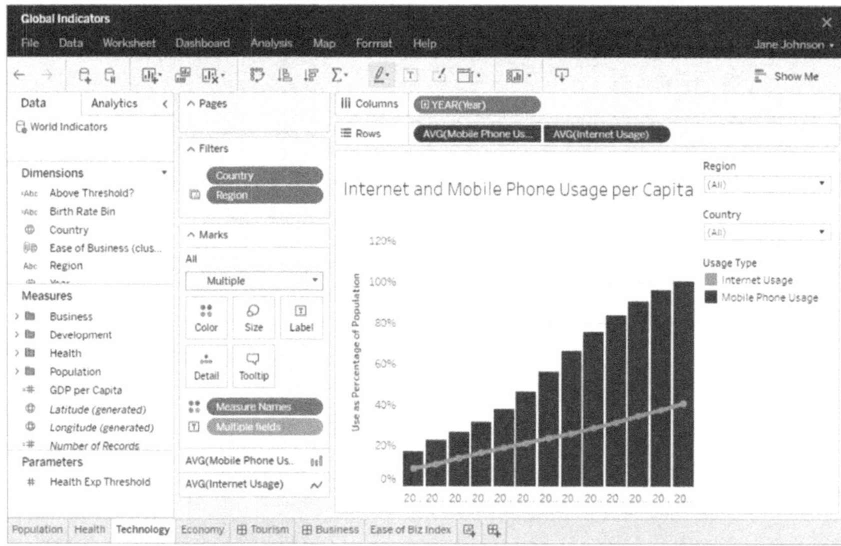

Figure 2.4 Sharing and collaboration.

Benefits:

Tableau's visual representations as shown in Figure 2.5 make data easier to grasp and interpret, even for non-technical audiences. This leads to improved data understanding and better decision-making and problem-solving.

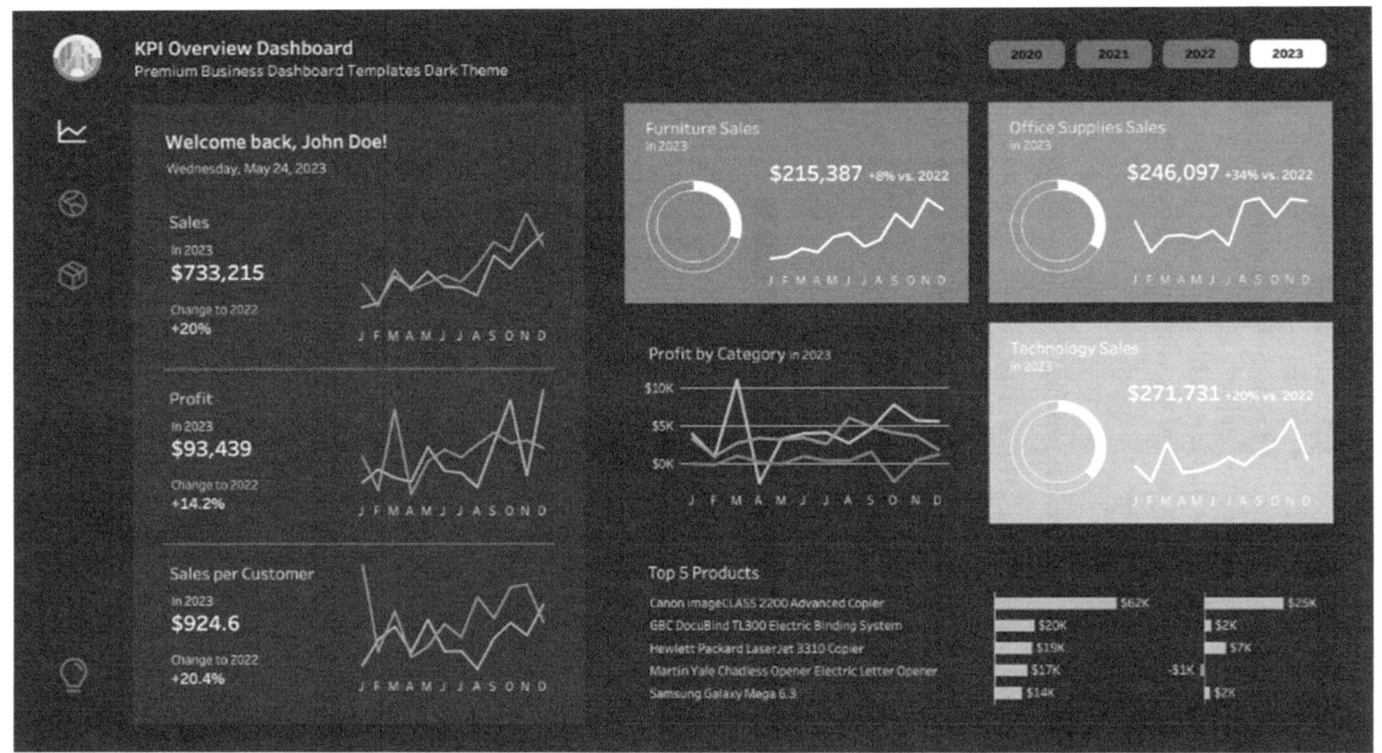

Figure 2.5 Tableau dashboard.

- *Improved teamwork and communication*: Interactive dashboards facilitate teamwork and communication by giving groups a common platform to examine data and exchange insights.
- *Enhanced productivity and efficiency*: Tableau expedites data analysis procedures, saving time and money in comparison to more conventional data exploration techniques.
- *Data-driven decision making*: Tableau enables users to make well-informed judgments devoid of guesswork or intuition by offering concise and useful insights.

2.2.2 Microsoft power BI: unleashing data insights

A robust business intelligence (BI) and data visualization solution, Microsoft Power BI [5, 6] enables users to turn data into informative dashboards and reports as shown in Figure 2.6. Below is a summary of its main attributes, advantages, and an example dashboard image:

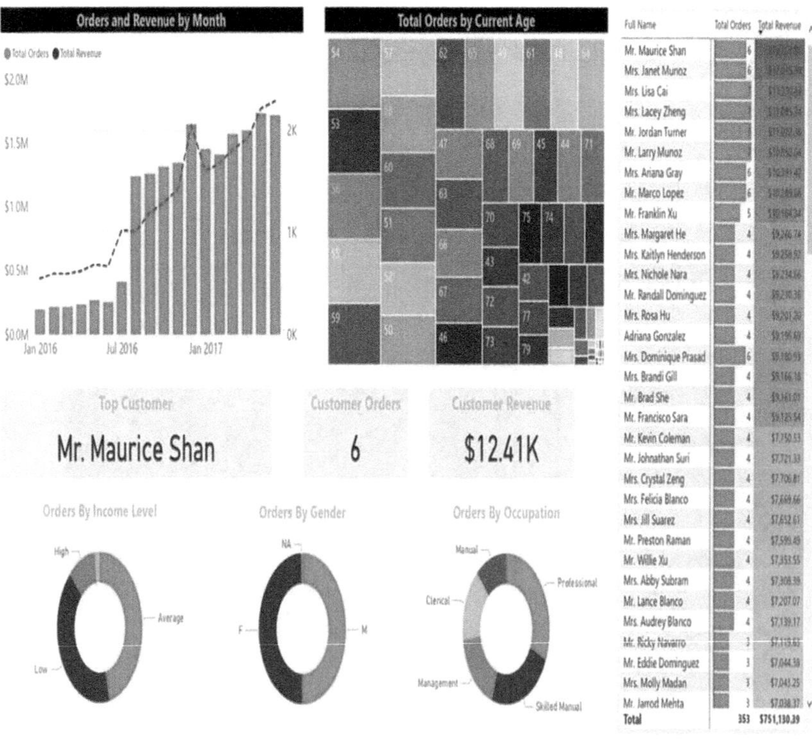

Figure 2.6 Simple interface.

Key Features:

- *Simple interface*: Users of various technical skill levels may easily navigate Power BI's intuitive interface, which features drag-and-drop functionality. Figure 2.6 show simple interface of Power BI.
- *Data connectivity*: To get a comprehensive picture of your data, connect to a variety of data sources, such as databases, cloud storage services, Excel spreadsheets, and APIs as shown in Figure 2.7.

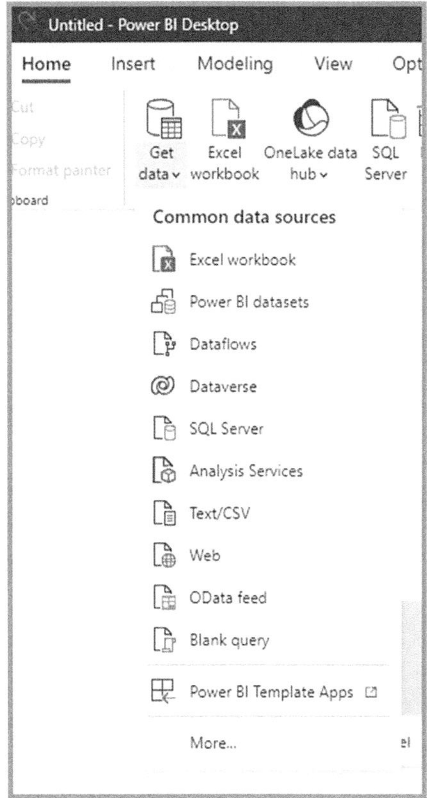

Figure 2.7 Data connectivity.

- *Data modelling and transformation*: Before making visualizations, make sure your data is accurate and consistent by cleaning, shaping, and transforming it. Figure 2.8 displays data modelling and transformation.

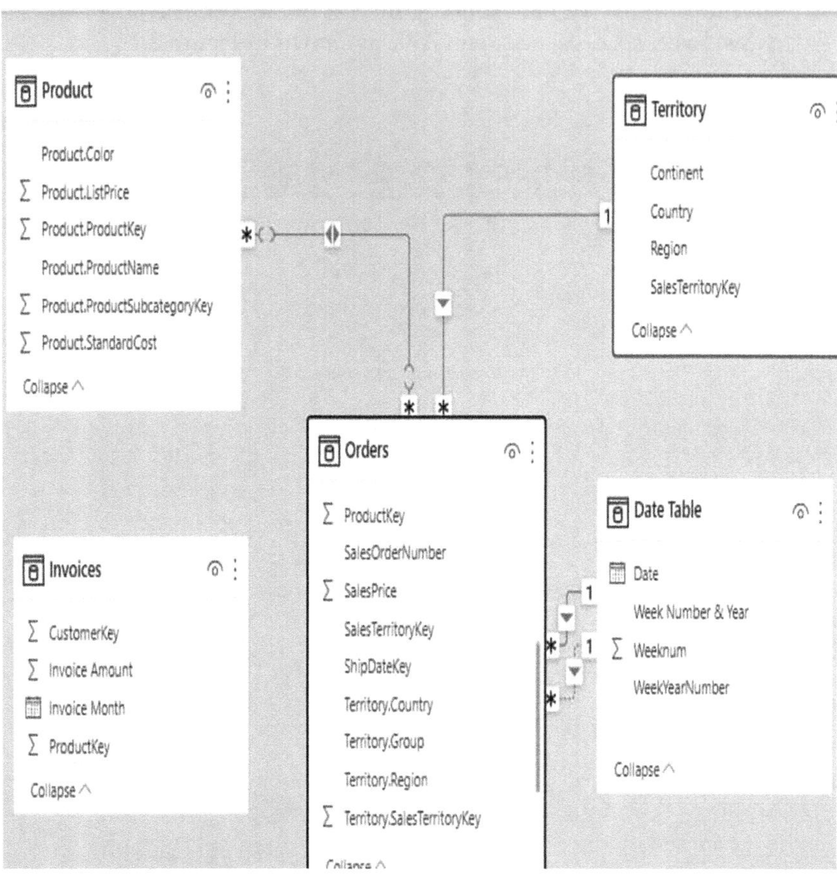

Figure 2.8 Data modelling and transformation.

- *Comprehensive visualizations*: To effectively convey your data, select from a variety of chart formats such as bar charts, line charts, maps, and custom visuals as shown in Figure 2.9.

Figure 2.9 Comprehensive visualizations.

- *Interactive dashboards and reports*: Figure 2.10 provide dynamic dashboards and reports that stimulate deeper knowledge by enabling users to filter, delve down, and examine data in real-time.

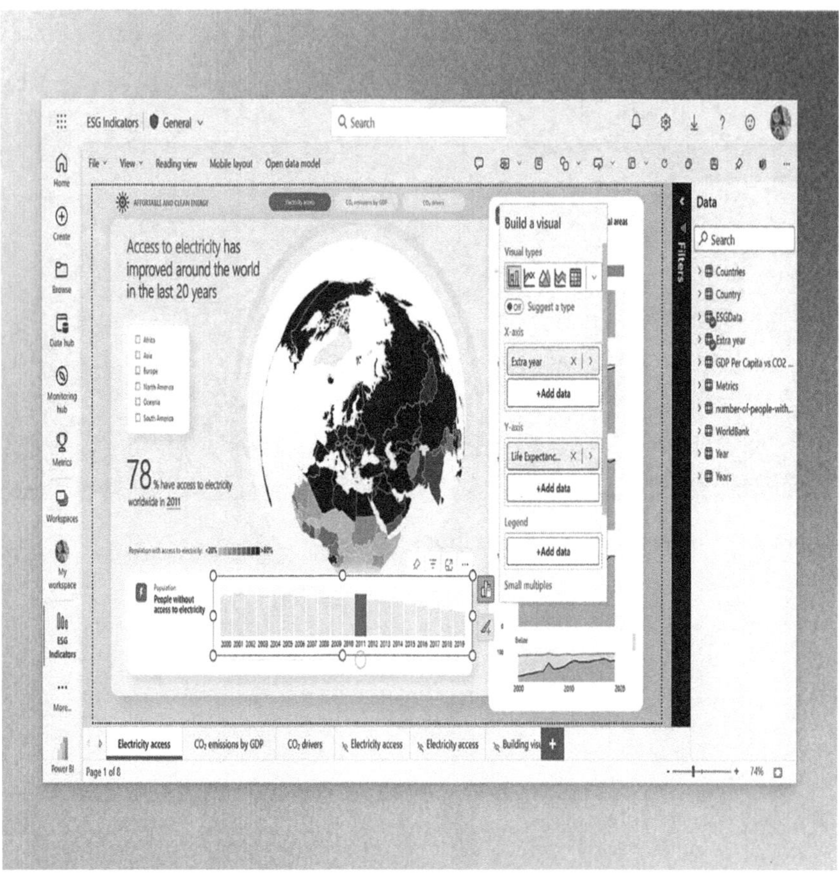

Figure 2.10 Interactive dashboards and reports.

- *Governance and Security*: As shown in Figure 2.11 data security and governance are made easier with the help of Power BI's many features. Data encryption, auditing, and role-based access control are some of these characteristics.

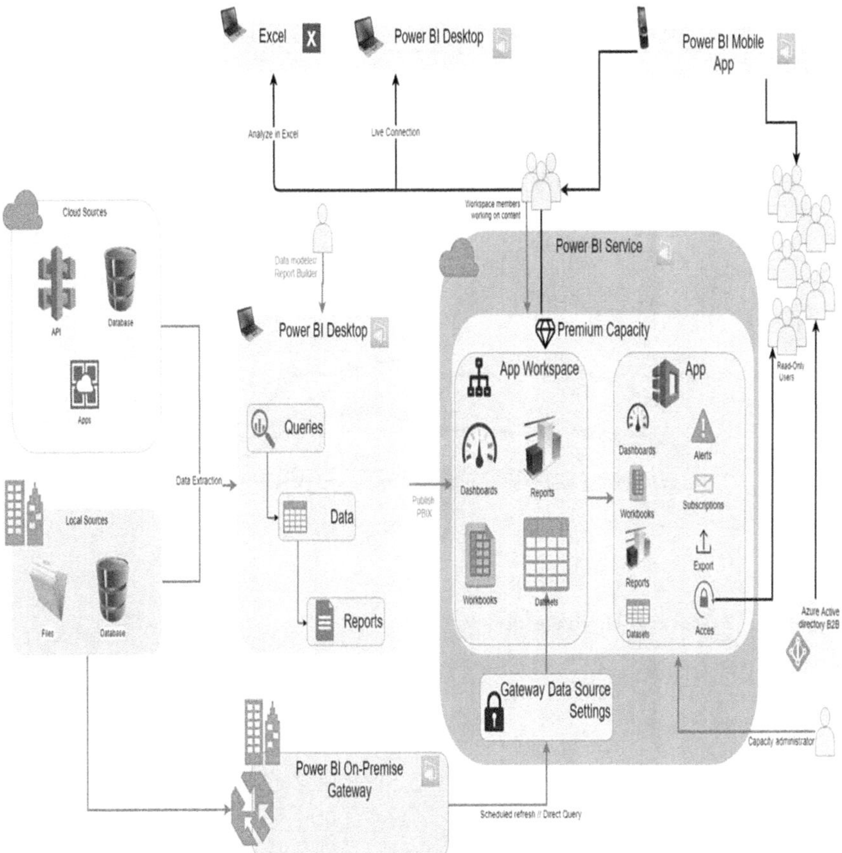

Figure 2.11 Governance and security.

- *Natural language query*: In Power BI as shown in Figure 2.12, you can ask queries about your data using natural language. For instance, you can ask Power BI, "What were my sales last month?" and it will display the results.

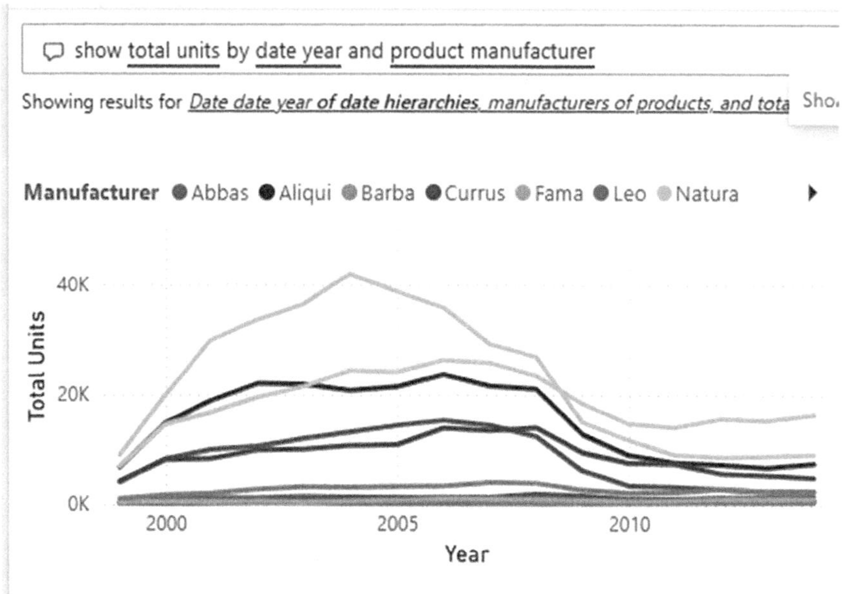

Figure 2.12 Natural language query.

- *Mobile access*: Use the Power BI mobile app as shown in Figure 2.13 to view and interact with your dashboards and reports while on the go.

Figure 2.13 Mobile access.

Benefits:

- *Better decision-making*: At every level of your company, use data-driven insights to make well-informed and strategic decisions.
- *Improved cooperation and communication*: With interactive dashboards and reports, participants can exchange ideas and promote data-driven conversations.
- *Enhanced productivity and efficiency*: Use self-service BI features to save time and streamline data analysis procedures.
- *Security and scalability*: Power BI prioritizes data security with strong governance features and grows to meet the demands of enterprises of all sizes. Figure 2.14 displays Power BI dashboard.

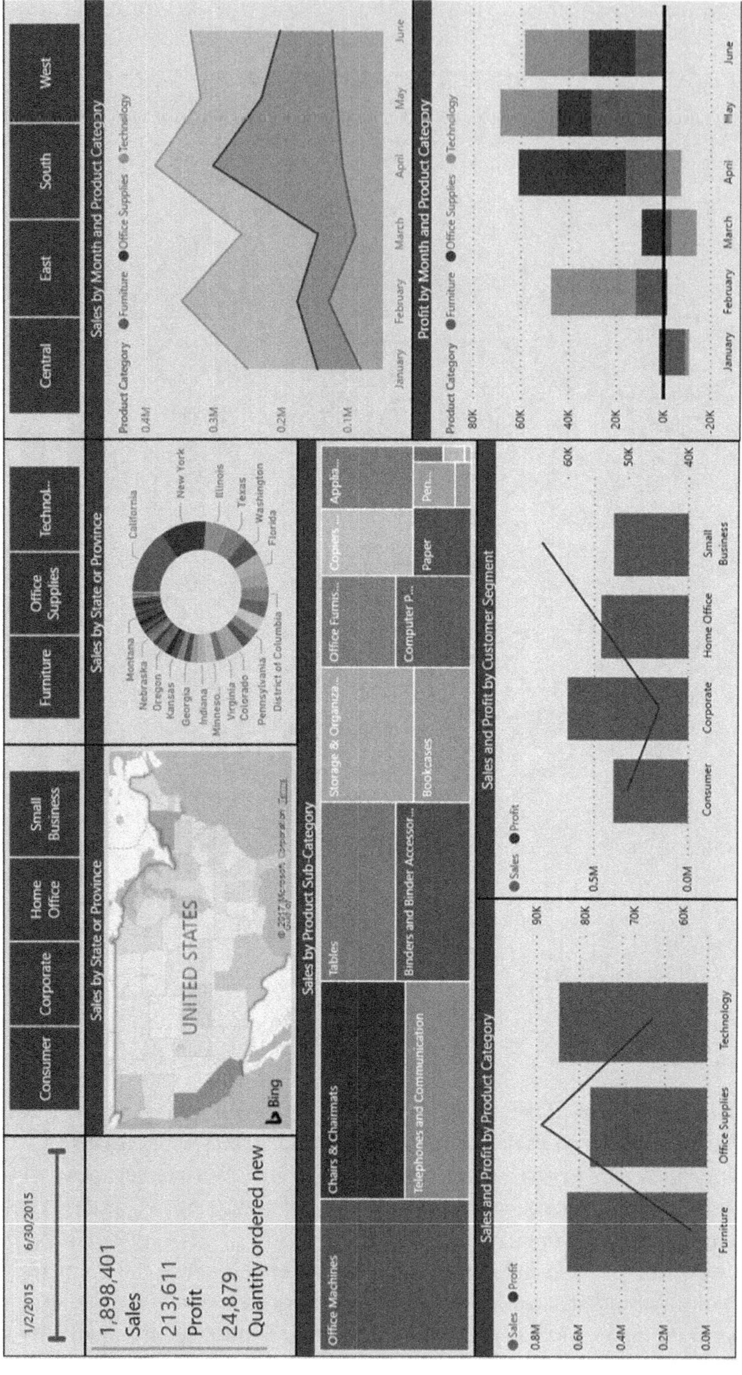

Figure 2.14 Power BI dashboard.

2.2.3 QlikView and Qlik Sense: powerful tools for associative exploration

Strong Resources for Associative Investigation Qlik has developed two robust systems for data analytics: QlikView and Qlik Sense [7]. Qlik Sense provides choices for both on-premises and cloud deployment, whereas QlikView is a desktop-only solution. Although they serve somewhat different user demands, both platforms have comparable basic features and advantages.

Key Features:

- *Associative engine*: Without the need to write intricate queries, users may freely explore data, make choices, and discover insights thanks to this special feature.
- *Data integration*: For a complete picture of your data, connect to several data sources, such as databases, cloud apps, and flat files.
- *Self-service analytics*: Encouraging data democratization within enterprises by enabling users of various skill levels to independently explore data.
- *Advanced visuals*: To effectively convey your data, select from a variety of chart formats, such as interactive dashboards, maps, and custom visualizations.
- *Sharing and collaboration*: Using safe dashboards and reports, participants may exchange insights and work together on data analysis.
- *Governance and security*: Use strong governance elements to preserve user access control and data security.

Benefits:

- *Quicker time to insights*: Users can make decisions more rapidly by using the associative engine to find hidden patterns and trends in their data.
- *Enhanced data accessibility*: Self-service analytics encourage data literacy by enabling consumers to examine data on their own and lessening need on IT teams.
- *Improved teamwork and communication*: Data-driven conversations and knowledge exchange across teams are made easier by interactive dashboards and reports.
- *Flexibility and scalability*: QlikView and Qlik Sense are flexible and scalable solutions that can be tailored to fit the needs of expanding businesses. They offer both on-premise and cloud deployment choices.

How to See a Dashboard:

But, there are lots of public examples of QlikView and Qlik Sense dashboards as shown in Figure 2.15 that demonstrate the software's flexible visualization options and intuitive user interface [8].

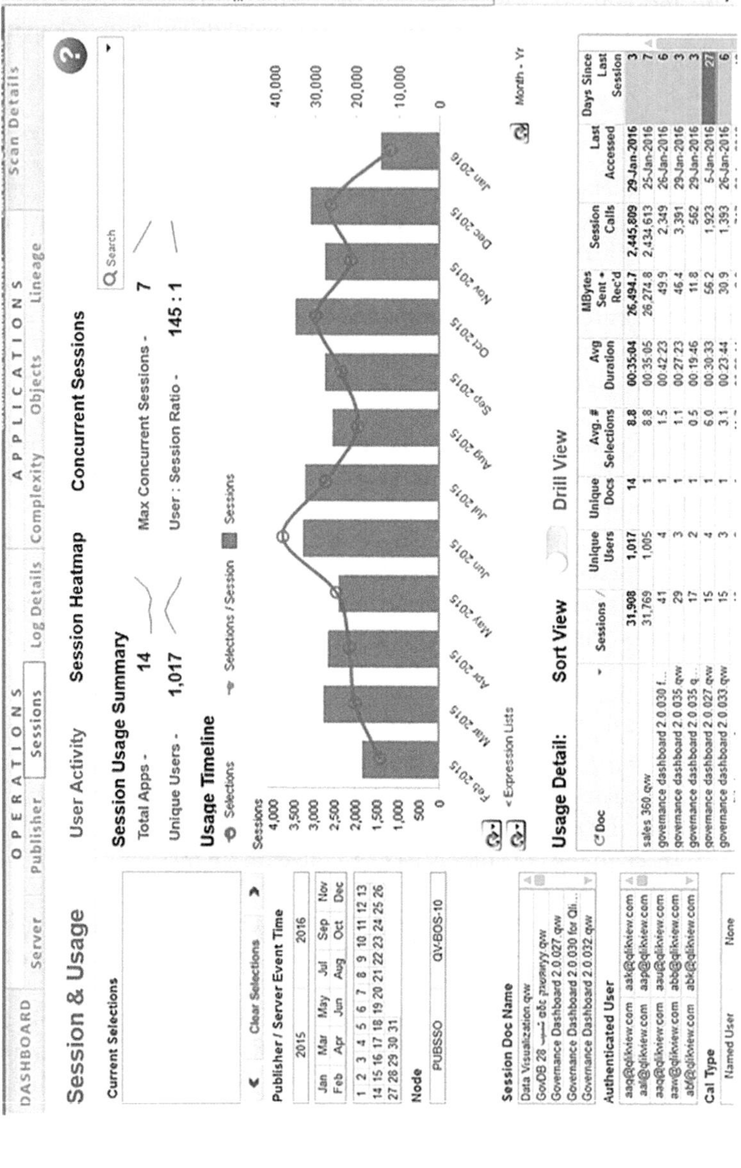

Figure 2.15 QlikView and Qlik Sense dashboard.

Important Variations:

- *Deployment*: Qlik Sense provides choices for both on-premises and cloud deployment, whereas QlikView is solely a desktop solution.
- *Target user*: Qlik Sense targets a wider audience with its user-friendly interface, whereas QlikView caters to more technical users and developers.
- *Cost*: While Qlik Sense offers a freemium model with premium tiers for extra capabilities, QlikView usually requires a licensing charge.

2.2.4 Domo: a single platform for analytics and data management

Domo is a cloud-based platform for data management and analysis that enables companies to manage and centralize data from multiple sources. It provides an extensive feature set aimed at optimizing data operations, promoting data-driven decision-making, and enhancing intra-organizational cooperation.

Key Features:

- *Data integration*: As shown in Figure 2.16, it allows you to combine all of your data onto a single platform by connecting to a variety of data sources, such as file systems, databases, and cloud apps.

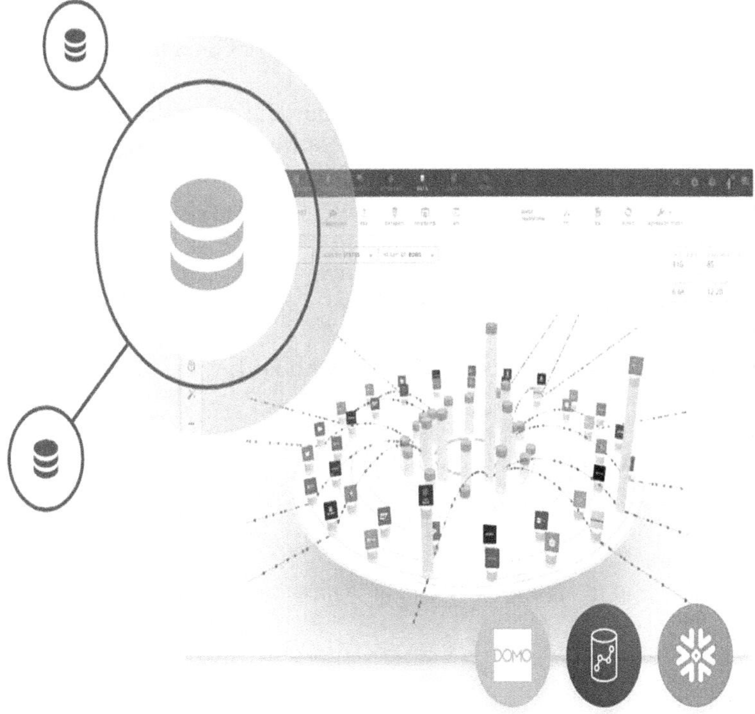

Figure 2.16 Data integration.

- *Data transformation and preparation*: Use visual processes and user-friendly tools to clean, transform, and get your data ready for analysis.
- *Self-service analytics*: With drag-and-drop capability and pre-made visuals, this tool enables users of all skill levels to independently examine data.
- *Advanced analytics*: For deeper insights, anomaly detection, and predictive modeling, take advantage of machine learning and artificial intelligence capabilities.
- *Interactive reports and dashboards*: Produce visually stunning, interactive reports and dashboards that narrate interesting data tales.
- *Mobile access*: Use the Domo mobile app to access and engage with your data and insights while on the go.
- *Cooperation and information sharing*: Promote cooperation and knowledge sharing by securely sharing dashboards, insights, and data with stakeholders and coworkers.
- *Governance and security*: Use strong governance elements to preserve user access control and data security.

Benefits:

- *Enhanced data accessibility and governance*: Domo offers a single platform for data management and access, guaranteeing data quality and consistency.
- *Quicker time to insights*: Self-service analytics enable users to independently examine data and find insights.
- *Data-driven decision making*: Use your data to get deeper insights that will guide strategic choices at all organizational levels.
- *Improved cooperation and communication*: Collaborate and align by efficiently sharing data and insights via interactive dashboards and reports.
- *Enhanced productivity and efficiency*: Use self-service features to streamline data workflows and lessen dependency on IT professionals.

Figure 2.17 gives a glance on Domo dashboard.

Figure 2.17 Domo dashboard.

2.2.5 Looker: powerful data visualization and business intelligence

Looker is business intelligence and data visualization platform that enables users to explore, analyse, and share insights from their data [9]. Looker Studio is the new name for the product. Below is a summary of its main attributes, advantages, and an example dashboard image.

Key Features:

- *Broad data source connectivity*: For a unified picture of your data, connect to a variety of data sources, such as databases, cloud apps, marketing platforms, and more.
- *Data analysis and exploration*: To find hidden patterns and trends in your data, examine it using interactive dashboards, filters, and drill-down features.

- *Variety of visualizations*: To effectively convey your data, pick from a wide selection of chart formats, such as bar charts, line charts, pie charts, maps, and tables.
- *Easy-to-use interface*: Looker Studio's drag-and-drop interface as shown in Figure 2.18 makes it suitable for users with varying technical proficiency levels.

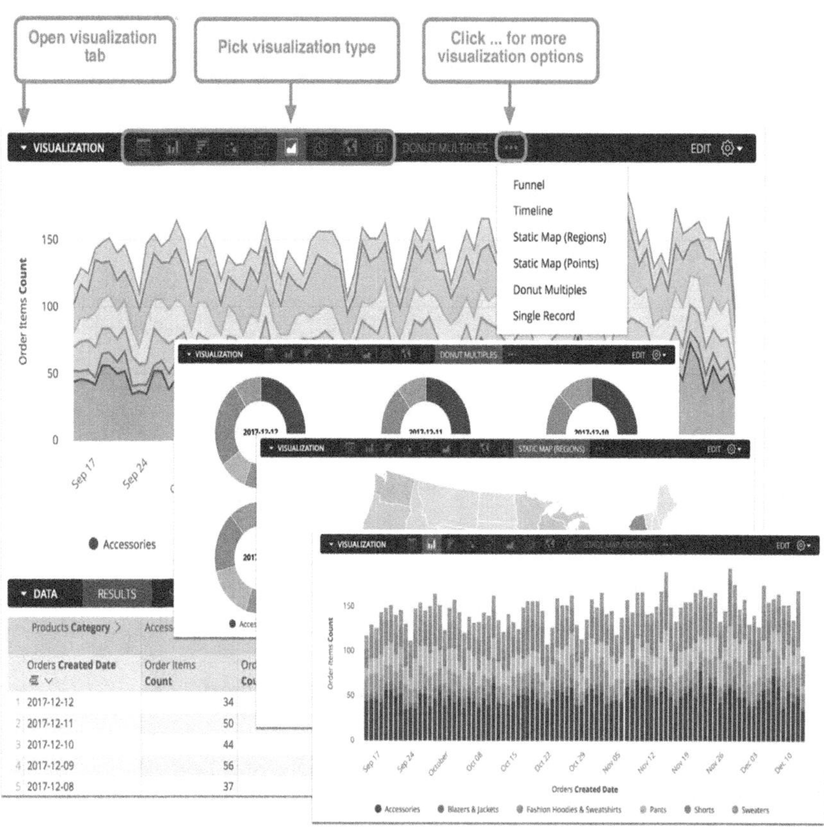

Figure 2.18 Easy-to-use interface.

- *Customization choices*: A wide range of customization options as mentioned in Figure 2.19, including as themes, fonts, and colors, let you personalize the appearance and feel of your dashboards and reports.

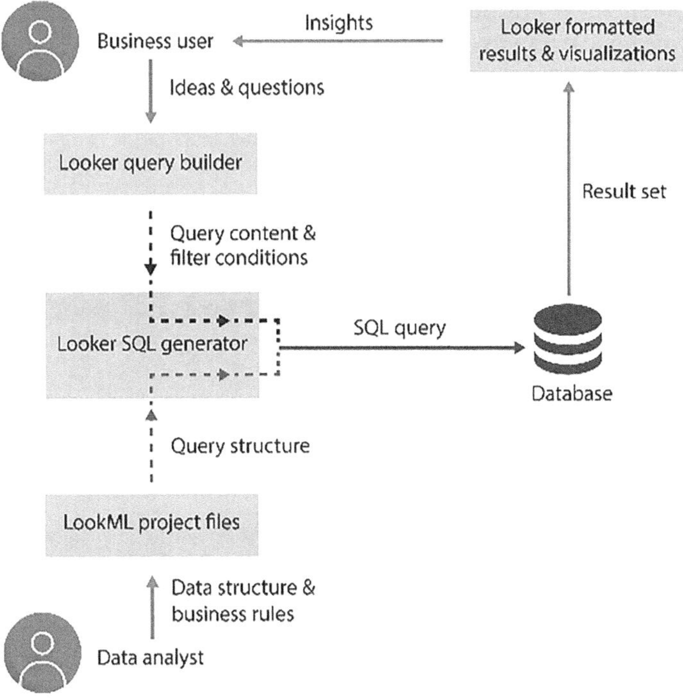

Figure 2.19 Customization choices.

- *Sharing and collaboration*: Using safe dashboards and reports, participants may exchange insights and work together on data analysis. They can be distributed more widely by being integrated into webpages [10].
- *Real-time data updates*: To guarantee that your reports and dashboards always show the most recent information, schedule automated data refreshes.
- *Advanced features*: Looker Studio may be integrated with Looker, a more sophisticated platform that provides embedded analytics, data modeling, and governance.

Benefits:

- *Better comprehension of data*: Looker Studio's visual aids facilitate data interpretation and comprehension, even for non-technical audiences.
- *Data-driven decision making*: Use data insights to guide strategic choices at all organizational levels.
- *Improved cooperation and communication*: Using interactive dashboards and reports, exchange insights and promote data-driven conversations.

- *Enhanced productivity and efficiency*: Use self-service BI features to expedite data analysis procedures and save time.
- *Scalability and flexibility*: Looker Studio's extensive array of data sources, visualizations, and customization choices allow it to accommodate a wide range of requirements and tastes.

Figure 2.20 displays Looker dashboard interface.

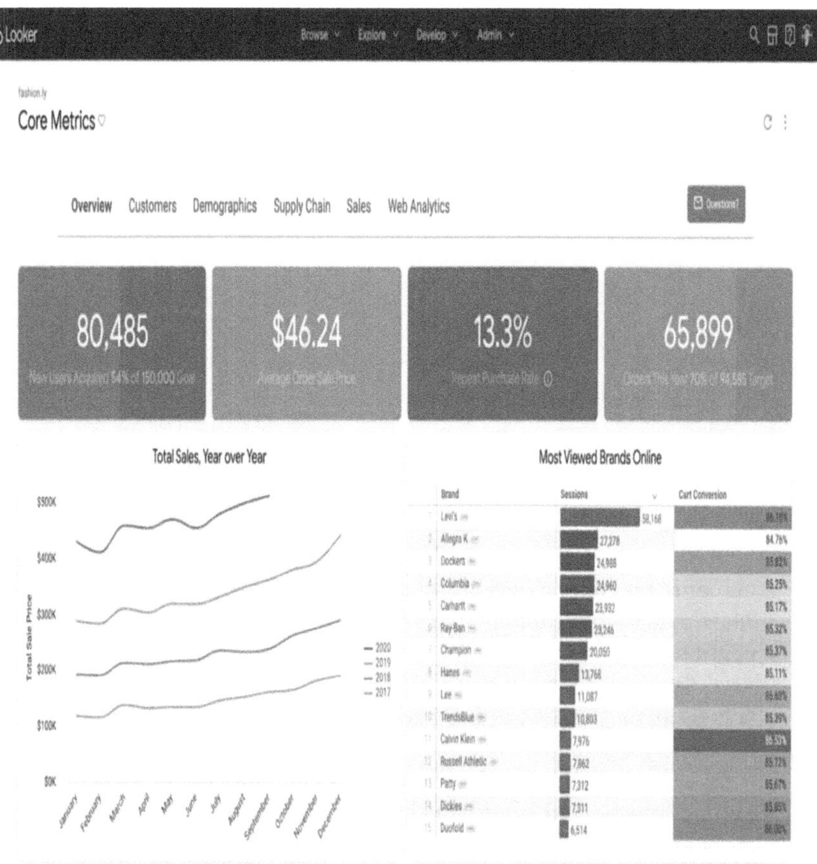

Figure 2.20 Looker dashboard.

2.2.6 Sisense: unlocking data insights with ease and flexibility

Sisense is a platform for data analytics and business intelligence (BI) that is renowned for its flexible deployment options, robust functionality, and easy-to-use interface [11]. By enabling users to examine, evaluate, and present data from several sources, it promotes data-driven decision-making throughout enterprises.

Key Features:

- *Drag-and-drop interface*: As shown in Figure 2.21, Sisense's user-friendly interface, which includes drag-and-drop capabilities, makes it suitable for users with varying technical backgrounds.

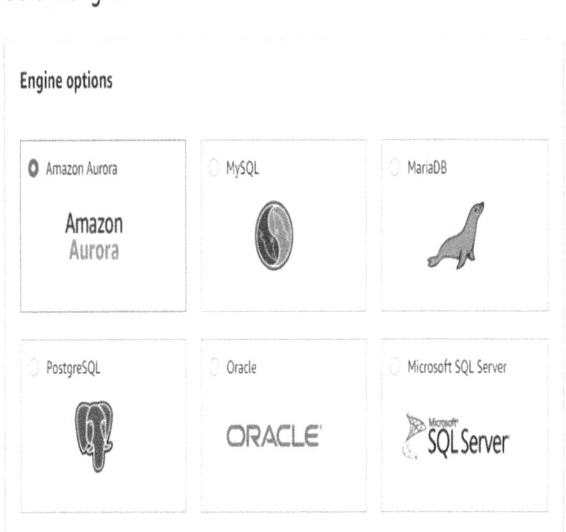

Figure 2.21 Drag-and-drop interface.

- *Broad data connectivity*: To get a complete picture of your data, connect to a variety of data sources, such as databases, cloud apps, flat files, and APIs.
- *In-memory analytics*: Even with enormous datasets, take use of in-memory technology for quick data exploration and real-time insights.
- *Advanced analytics*: To extract more meaning from your data, make use of tools like machine learning, forecasting, and data modeling.
- *Interactive dashboards and reports*: With a variety of chart kinds and customization options, create visually appealing and interactive dashboards and reports as shown in Figure 2.22.

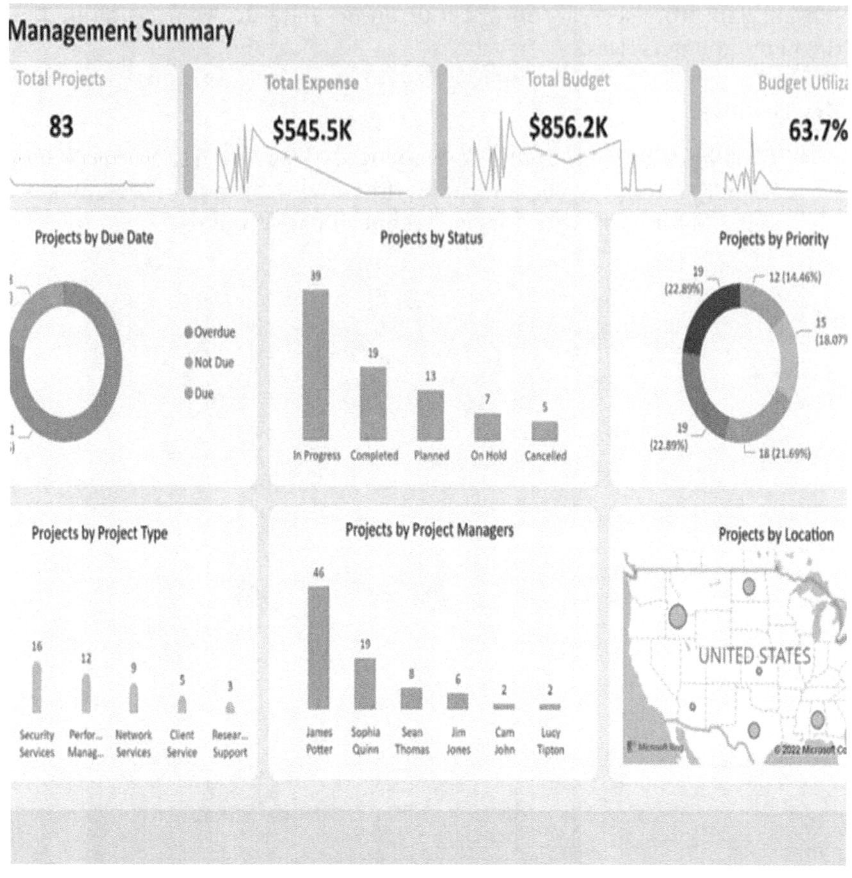

Figure 2.22 Interactive dashboards and reports.

- *Embedded analytics*: Provide data insights in relation to user behaviours by integrating analytics straight into workflows and applications.
- *Mobile access*: Use the Sisense mobile app to access and engage with your data and insights while on the road.
- *Cooperation and information sharing*: Promote cooperation and knowledge sharing by securely sharing dashboards and reports with stakeholders and coworkers.
- *Flexibility in deployment*: Select on-premise, cloud, or hybrid deployment solutions based on your unique requirements and infrastructure.

Benefits:

- *Quicker time to insights*: Sisense's in-memory technology and intuitive interface facilitate rapid data exploration and insight discovery for users.

- *Enhanced governance and accessibility of data*: centralize data, provide users with self-service analytics, and uphold strict governance rules.
- *Data-driven decision making*: Use your data to extract actionable insights that will guide strategic choices at all organizational levels.
- *Improved cooperation and communication: Promote* data-driven conversations and alignment by efficiently sharing insights via interactive dashboards and reports.
- *Enhanced productivity and efficiency*: Use self-service features and embedded analytics to streamline data workflows and lessen dependency on IT teams. Figure 2.23 represents Sisense dashboard.

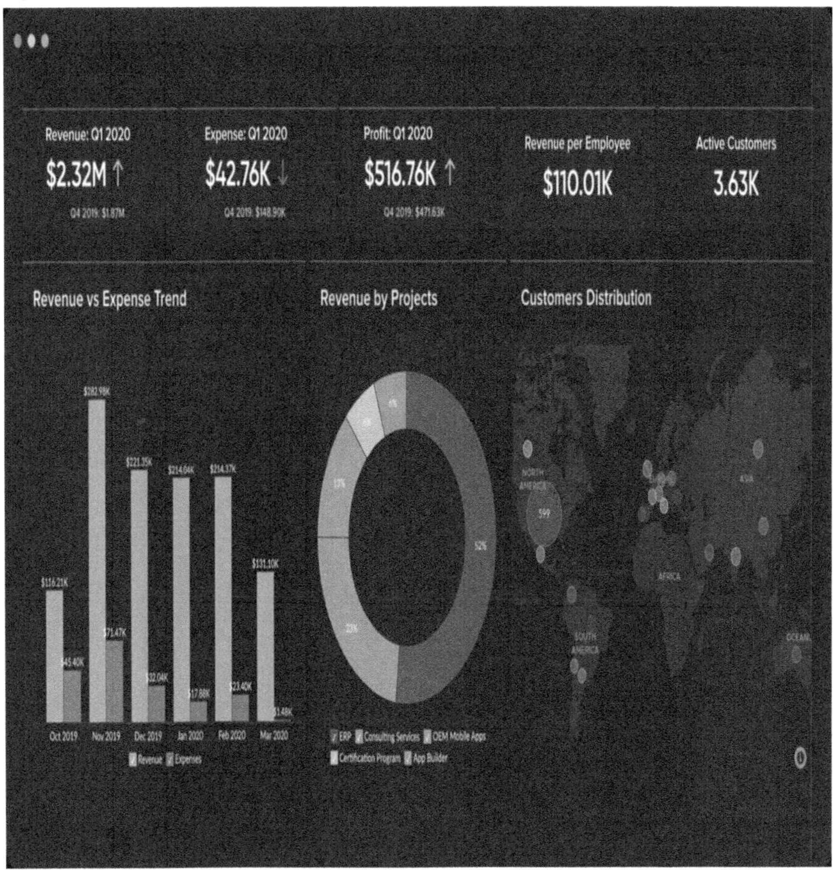

Figure 2.23 Sisense dashboard.

2.2.7 Plotly dash: interactive data visualization for python developers

A well-liked Python framework called Plotly Dash was created especially for creating interactive dashboards and data visualizations that are accessible online. It makes use of Plotly.js's capabilities to produce aesthetically pleasing graphs and charts with a great degree of customization and interactivity [12].

Key Features:

- *Python-based*: Dash is designed to accommodate developers who are acquainted with the Python programming language as shown in Figure 2.24. This allows developers to quickly prototype and implements interactive dashboards.
- *Interactive components*: You may allow users to examine data dynamically by integrating different interactive elements into your dashboards, such as buttons, sliders, and dropdown menus.
- *Reusable components*: To improve code efficiency and maintainability, create reusable components for standard layouts and visualizations.

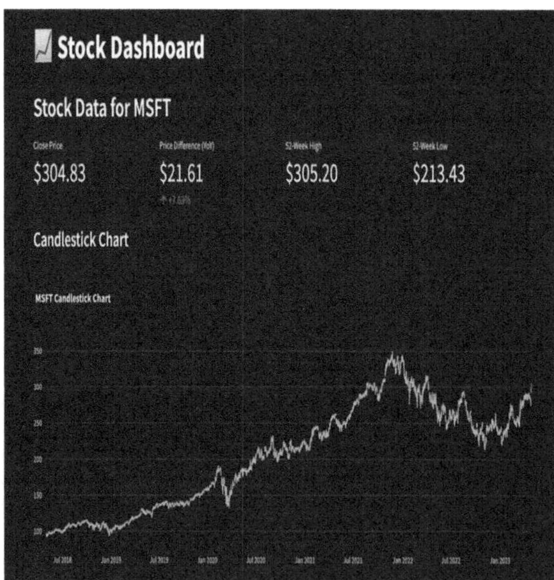

Figure 2.24 Python-based dashboard.

- *Callback system*: An adaptable callback system that starts operations based on user input allows you to quickly manage user interactions and data updates as shown in Figure 2.25.

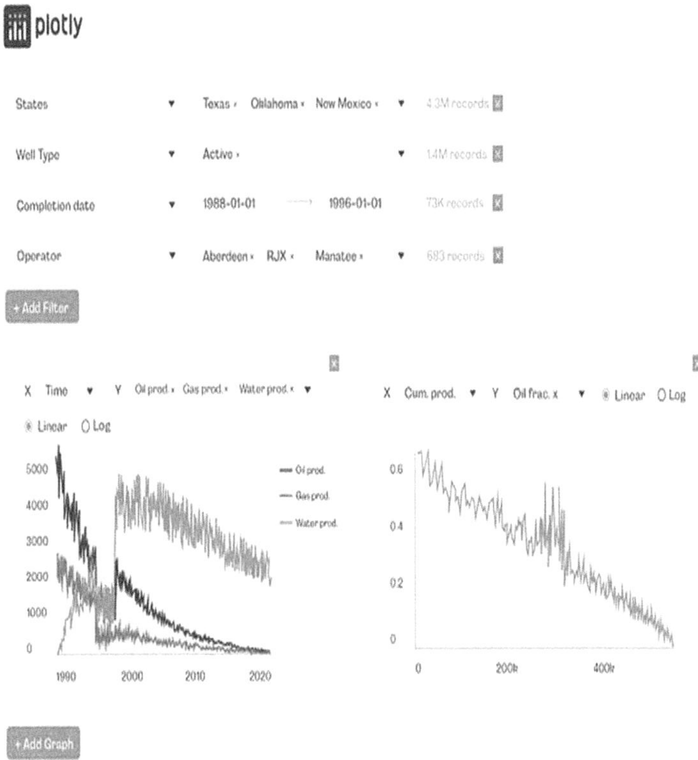

Figure 2.25 Callback system.

- *Styling and customization*: You have a plethora of options to personalize the way your dashboards look, including the ability to integrate CSS for more precise control.
- *Flexibility in deployment*: You may install Dash apps on a range of platforms, such as local computers, cloud servers, and containerized environments.

Benefits:

- *Fast prototyping*: Utilizing Python code, quickly create interactive dashboards and visualizations, hence expediting the development process.
- *Improved user experience*: Involve users with interactive components that let them customize how they examine the data, leading to a deeper comprehension.
- *Customization and control*: Well-made dashboards with a wide range of stylistic options and reusable parts that are tailored to match certain requirements and preferences.
- *Using the Python ecosystem integration*: Use the extensive Python library and tool ecosystem in your Dash apps to manipulate, analyze, and distribute data. Figure 2.26 represents Plotly dashboard

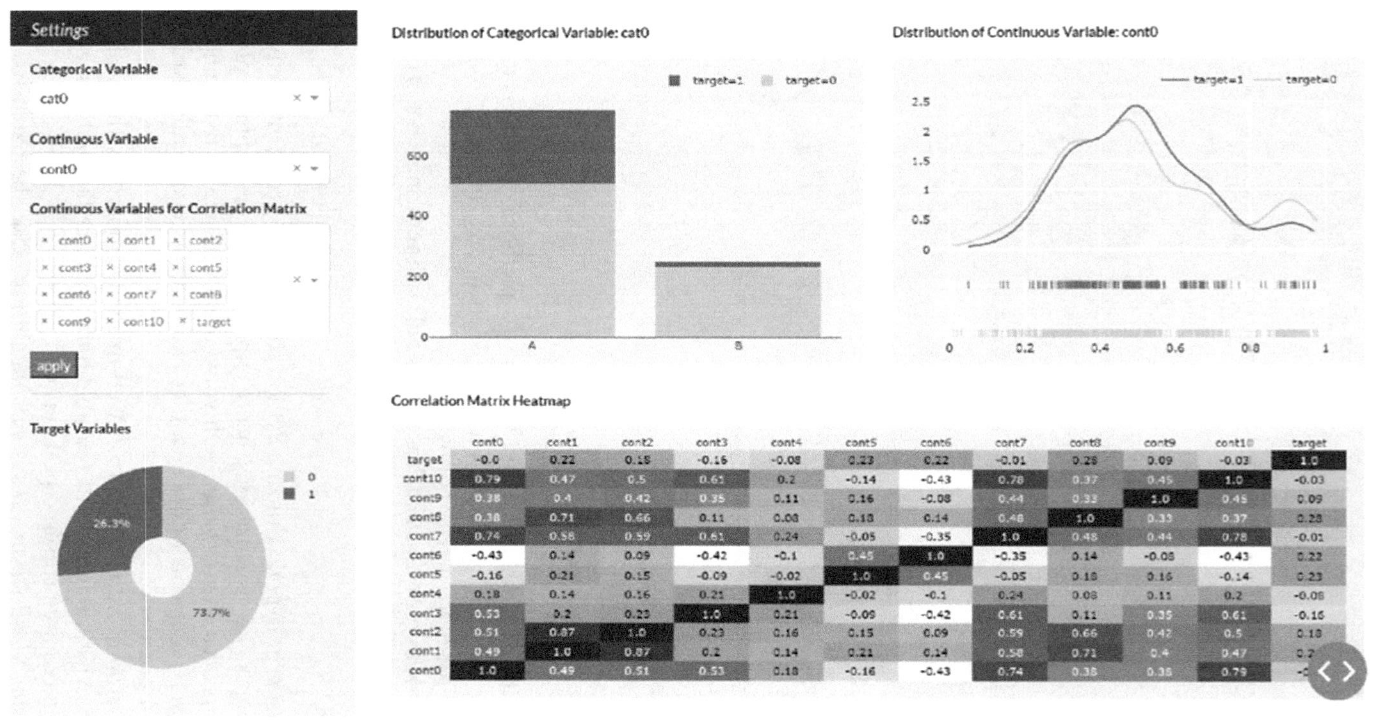

Figure 2.26 Plotly dashboard.

- *Community-driven and open-source*: Take use of a sizable and vibrant community that offers assistance, materials, and framework contributions.

2.2.8 Dundas BI: feature-rich platform for data visualization and analytics

Users may create interactive dashboards, reports, and data visualizations with Dundas BI, a comprehensive platform for business intelligence (BI) and data visualization. It offers a variety of capabilities for data exploration, analysis, and communication, making it suitable for both technical and non-technical users.

Key Features:

- *Drag-and-drop interface*: Figure 2.27 shows user-friendly interface that allows users of all ability levels to create dashboards and reports visually.

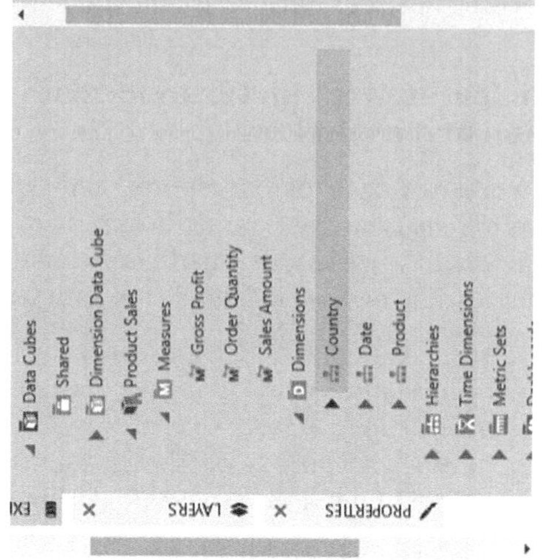

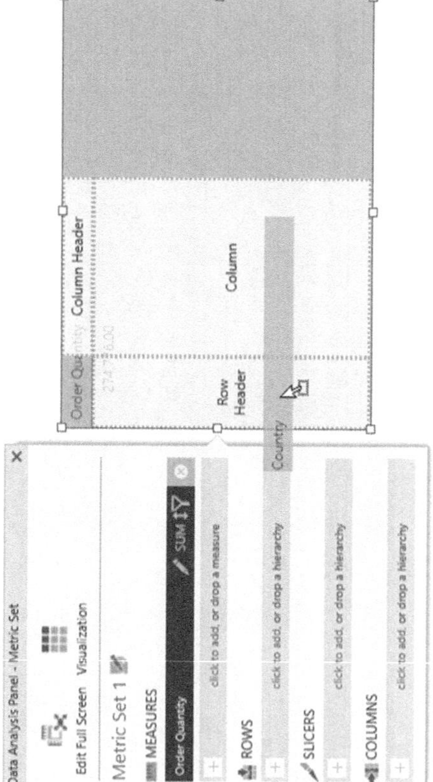

Figure 2.27 Drag-and-drop interface.

- *Broad data source connectivity*: For a unified view of your data, connect to a variety of data sources, including as databases, cloud apps, flat files, and APIs.
- *Advanced visualizations*: To properly depict your data, select from a wide range of chart styles, such as interactive dashboards, maps, heatmaps, and custom visualizations as shown in Figure 2.28.

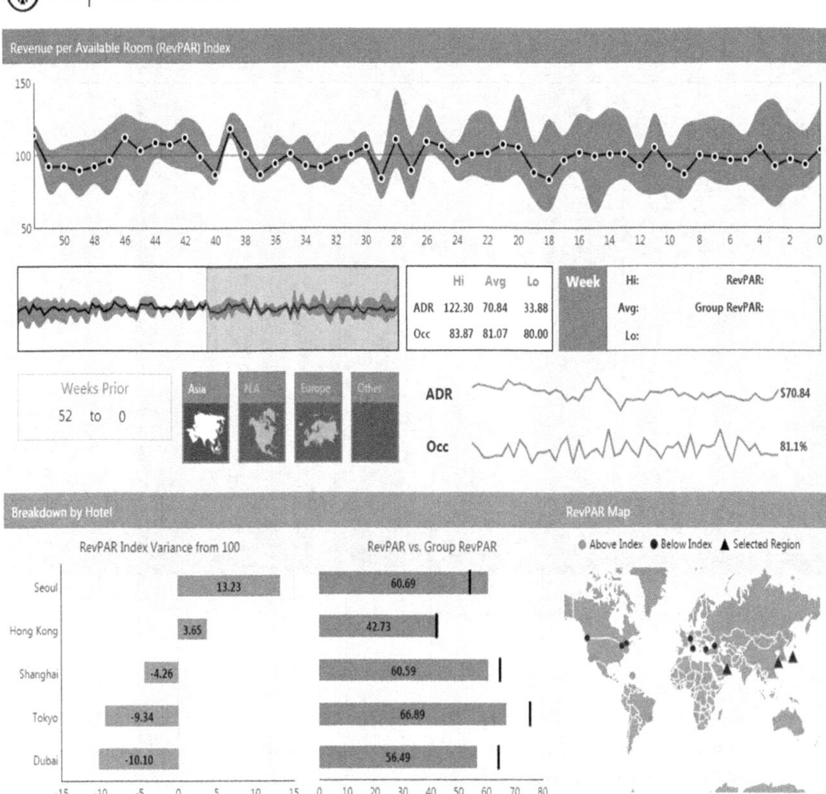

Figure 2.28 Advanced visualizations.

- *Data analysis and exploration*: Analyze your data in-depth, apply filters based on certain standards, and run computations to find hidden trends and patterns as shown in Figure 2.29.

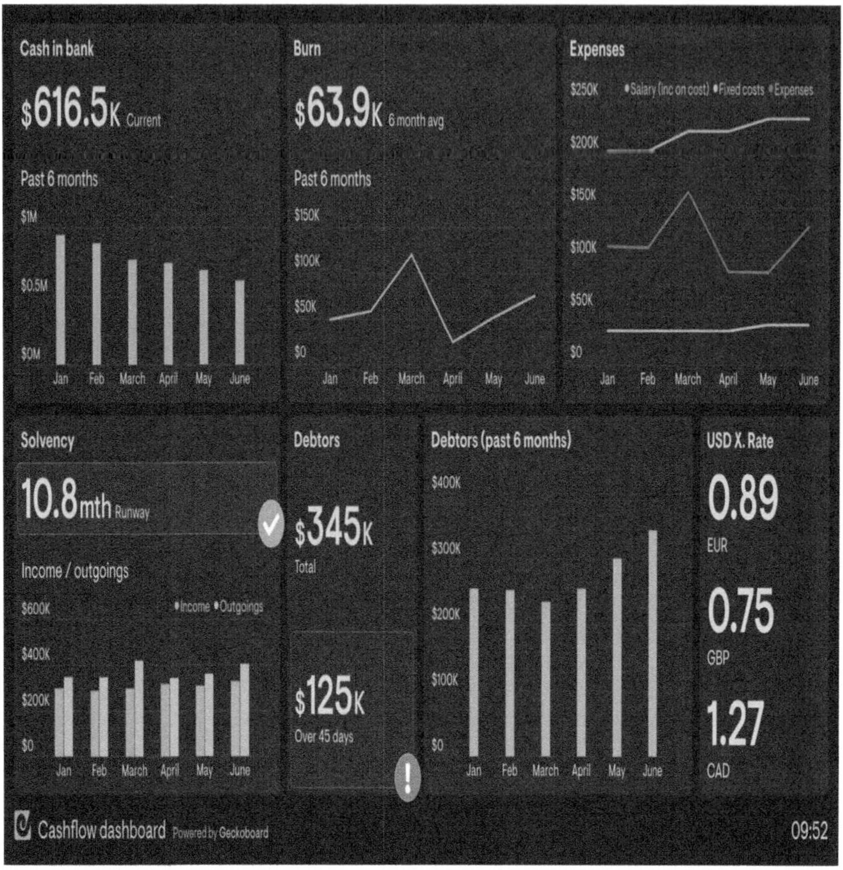

Figure 2.29 Data analysis and exploration.

- *Cooperation and information sharing*: Promote cooperation and knowledge sharing by securely sharing dashboards and reports with stakeholders and co-workers.
- *Embedded analytics*: Provide data insights in relation to user behaviors by integrating analytics straight into workflows and applications [13].
- *Mobile access*: Figure 2.30 represents the use of Dundas BI mobile app to access and interact with your data and insights while on the go.

Figure 2.30 Mobile access.

- *Security and governance*: Use strong governance elements to maintain user access control and data security.
- *Customization choices*: A wide range of customization options, including as themes, fonts, and colors, let you personalize the appearance and feel of your dashboards and reports. Figure 2.31 represents Dundas BI dashboard

Benefits:

- *Enhanced comprehension of data*: The visual representations provided by Dundas BI facilitate the understanding and interpretation of data, especially for non-technical audiences.
- *Data driven decision making*: Use your data to extract actionable insights that will guide strategic choices at all organizational levels.
- *Improved cooperation and communication: Promote* data-driven conversations and alignment by efficiently sharing insights via interactive dashboards and reports.
- *Enhanced productivity and efficiency: Use* embedded analytics and self-service BI features to streamline data analysis procedures and lessen dependency on IT professionals.
- *Scalability and flexibility*: Dundas BI's many features, cloud, on-premise, and hybrid deployment choices, and embeddable design make it suitable for companies of all sizes and sectors.

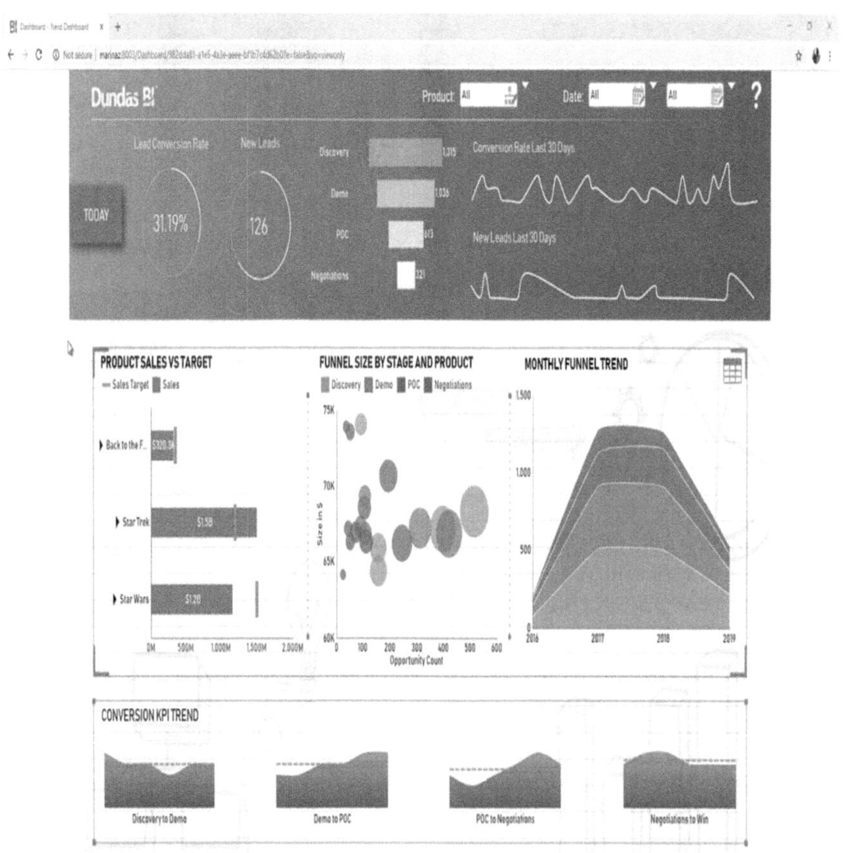

Figure 2.31 Dundas BI dashboard.

2.2.9 Zoho analytics: feature-rich and user-friendly data analytics platform

With the help of the all-inclusive business intelligence (BI) and data visualization platform Zoho Analytics, customers can turn their data into insights that are useful for their daily lives. Because of its comprehensive features, user-friendly interface, and range of functionalities, it's a popular option for companies of all kinds. Figure 2.35 represents Zoho dashboard.

Key Features:

- *Data Integration*: To compile your data in one place, connect to a variety of data sources, such as file systems, databases, cloud apps, and APIs. Figure 2.32 show how data can be integrated.

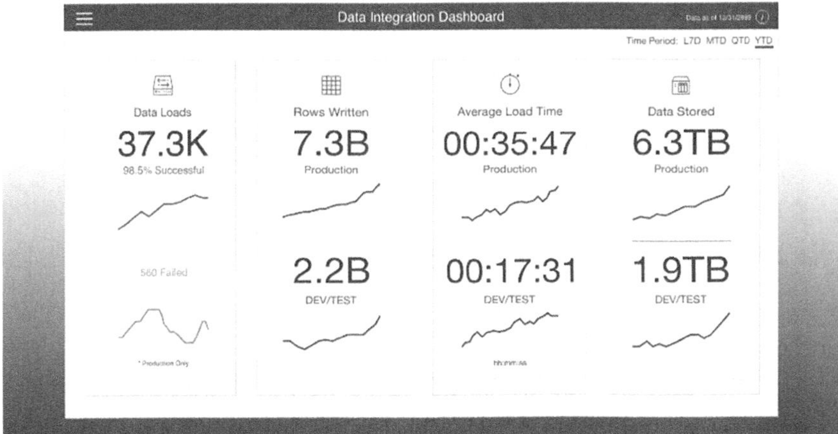

Figure 2.32 Data integration.

- *Data Preparation and Management*: Use user-friendly tools to clean, transform, and enrich your data so that it is accurate and consistent in advance of analysis.
- *Visual Analysis*: With a wide range of chart types and customization choices, explore your data through interactive dashboards, reports, and visualizations as shown in Figure 2.33.

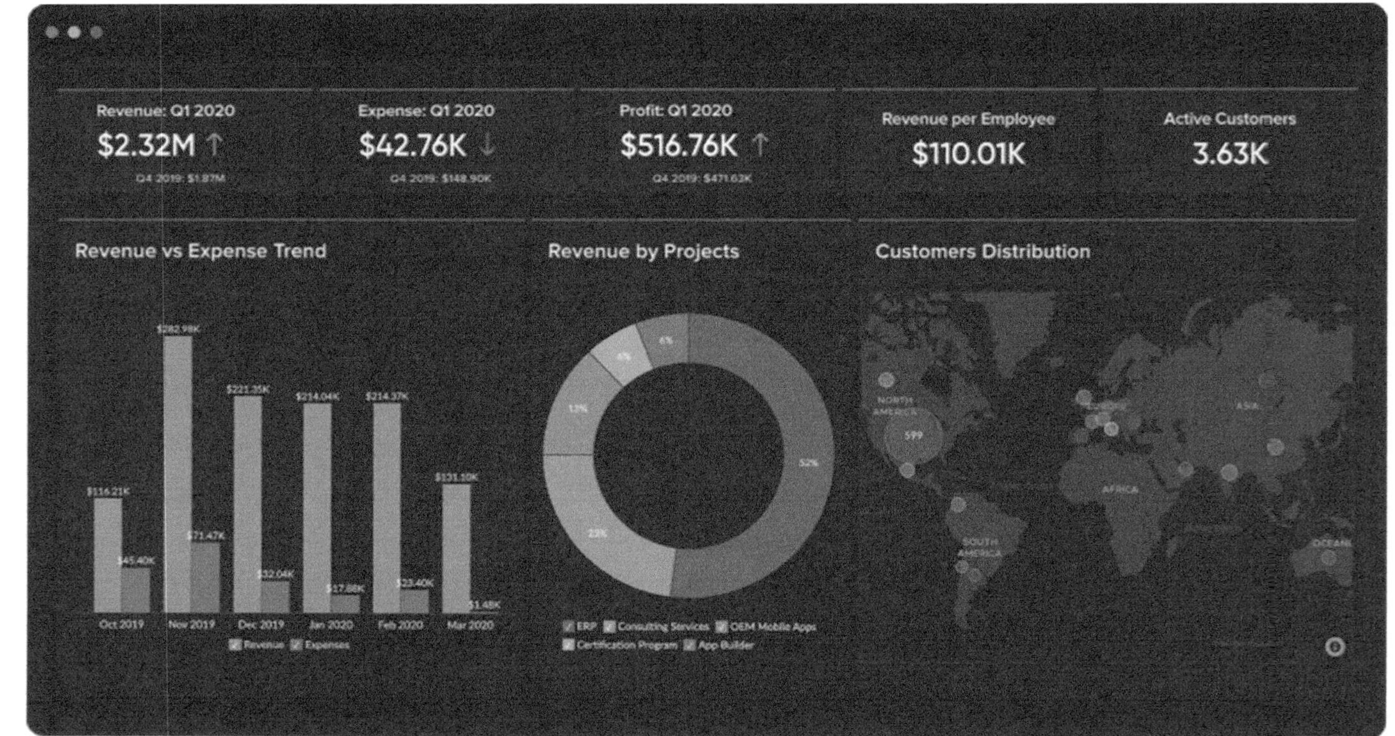

Figure 2.33 Visual analysis.

- *Augmented Analytics*: To obtain deeper insights and automate tasks, take advantage of AI-powered capabilities like anomaly detection, smart warnings, and predictive analytics.
- *Unified Business Insights*: Figure 2.34 shows about combining information from several sources to see your company as a whole and spot trends and patterns across functional boundaries.

Figure 2.34 Unified business insights.

- *Collaborative Analytics*: Through protected dashboards, reports, and conversations, share insights and work together with peers.
- *Data Storytelling*: To successfully convey insights to stakeholders, create captivating data tales by fusing text, visuals, and other elements.
- *Mobility*: Use the Zoho Analytics mobile app to access and interact with your data and insights while on the go.
- *Embedded BI*: For smooth data integration into current operations, integrate analytics straight into workflows and applications.

- *Security and Governance*: Strong governance features help maintain user access control and data security.

Benefits:

- *Better comprehension of data*: Users with varying technical skill levels may easily access and comprehend data thanks to Zoho Analytics' user-friendly interface and visual representations.
- *Data-driven decision making*: Use your data to extract actionable insights that will guide strategic choices at all organizational levels.
- *Improved cooperation and communication*: Use interactive dashboards, reports, and collaborative tools to promote data-driven conversations and knowledge exchange.
- *Enhanced productivity and efficiency*: Automate data analysis tasks with AI-powered automation, self-service BI features, and embedded analytics.
- *Scalability and flexibility*: With its wide range of functionality, cloud-based deployment choices, and reasonably priced pricing levels, Zoho Analytics meets a variety of needs.

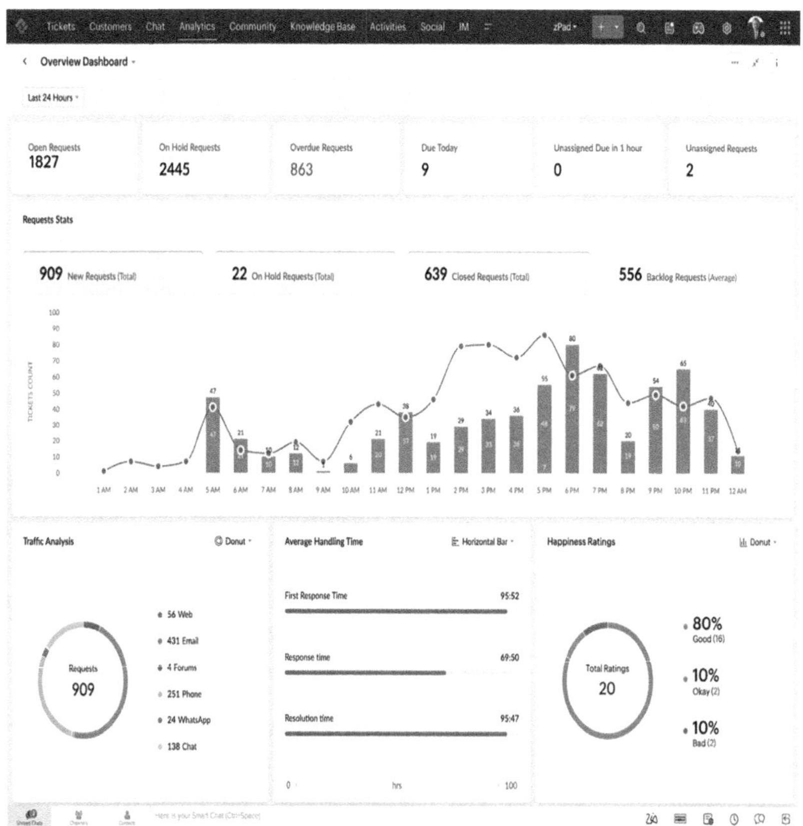

Figure 2.35 Zoho dashboard.

2.3 DIMENSIONALITY REDUCTION

One popular technique for evaluating and displaying high-dimensional data is dimensionality reduction. But considering the outcomes of a dimensionality reduction in a dynamic way is challenging. Dimensionality-reduction methods employ intricate optimizations to minimize the number of dimensions in a dataset; nevertheless, these additional dimensions frequently have no obvious connection to the original data dimensions, making them challenging to understand [14–17].

2.3.1 Principal component analysis

One method of unsupervised machine learning is principal component analysis, or PCA. Dimensionality reduction is possibly the most common application of principal component analysis. with addition to being a useful method for preparing data, PCA may also be applied to aid with data visualization [18]. A picture speaks a thousand words. We can more easily gain some understanding and choose the next course of action for our machine learning models when the data is visualized.

ALGORITHM 1: Principal Component Analysis

1. Get image data: Suppose $a_1, a_2 \ldots a_M$ are represented as $N \times 1$ vectors
2. Compute the average of vector: $\bar{a} = \frac{1}{M}\sum_{i=1}^{M} a_i$
3. Subtract the Mean: $\Phi i = a_i - \bar{a}$
4. Evaluate the covariance matrix: Matrix $P = [\Phi 1, \Phi 2 \ldots . \Phi M]$ ($N \times M$ matrix) from this compute

$$C = \frac{1}{M}\sum_{n=1}^{M} \Phi n \Phi n^T = PP$$

5. Compute eigenvalues and eigenvectors of the covariance matrix

$$C = \lambda_1 > \lambda_1 > \cdots > \lambda_N \text{ (eigenvalues)}$$
$$C = u_1, u_1 \ldots u_N \text{ (eigenvectors)}$$

6. When creating a feature vector, eigenvectors are arranged from highest to lowest eigenvalue. The components are now listed in order of significance. The primary component of the data set is the eigenvector with the highest eigenvalue. The greatest eigenvalue is selected to create the feature vector.

7. Creating a new dataset: After deciding which of the principal components (eigenvectors) in our data to retain, we created a feature vector. To create the new dataset, we just multiply the feature vector by the transpose of the original data set.

Row feature vector * row data adjust equals final data.

2.3.2 t-Distributed stochastic neighbour embedding

A potent non-linear dimensionality reduction method known as t-distributed Stochastic Neighbor Embedding, or t-SNE, is very helpful for visualizing high-dimensional data in a low-dimensional domain, such 2D or 3D. Because it preserves local associations between points, it's particularly suited for building interactive dashboards that help illustrate the underlying structure of the data [19].

- *Similarity Calculation*: Using a Gaussian distribution, t-SNE first determines the pairwise similarities between each data point in the high-dimensional space.
- *Dimensionality Reduction*: In order to maintain the pairwise commonalities, these points are then mapped to a lower-dimensional space.
- *Optimization*: Using a t-distribution, which has fatter tails and aids in more uniformly dispersing points, the algorithm iteratively optimizes the distribution of points in the lower-dimensional space

Perplexity, the primary t-SNE parameter, functions as a variable to balance the attention given to local versus global features in your data. A lower perplexity value indicates that the algorithm concentrates more on the local structure, whereas a greater value considers a wider range of data.

t-SNE is a useful tool for interactive dashboards where you want to allow users to explore and make sense of the data visually since it excels at producing aesthetically pleasing and educational representations of complex datasets.

ALGORITHM 2: t-Distributed Stochastic Neighbour Embedding

Data: data set $A = \{a_1, a_2, \ldots, a_n\}$,

cost function parameters: perplexity Perp,

optimization parameters: number of iterations T, learning rate η, momentum $\alpha(t)$.

Result: low-dimensional data representation $\mathbf{B}^{(T)} = \{b_1, b_2, \ldots, b_n\}$.

begin
compute pairwise affinities $p_{j|i}$ with perplexity Perp
set $p_{ij} = \dfrac{p_{j|i} + p_{i|j}}{2n}$
sample initial solution $B^{(0)} = \{b_1, b_2, \ldots, b_n\}$ from $\mathcal{N}\left(0, 10^{-4} I\right)$
for $t = 1$ to T do
compute low-dimensional affinities q_{ij}
compute gradient $\dfrac{\delta C}{\delta b}$ (using Equation 5)
set $B^{(t)} = B^{(t-1)} + \eta \dfrac{\delta C}{\delta B} + \alpha(t)\left(B^{(t-1)} - B^{(t-2)}\right)$
end
end

2.3.3 Uniform manifold approximation and projection

ALGORITHM 3: Uniform Manifold Approximation and Projection

Input: Training data $\{\boldsymbol{a}_k\}_{k=1}^{n}$ and build kNN graph
Laplacian eigenmap is used to initialize $\{\boldsymbol{b}_k\}_{k=1}^{n}$
Evaluate c_{kl} and d_{kl} for $\forall k, l \in \{1, \ldots, n\}$ by

$\eta \leftarrow 1, v \leftarrow 0$

while not converged do

$v \leftarrow v + 1$

for k from 1 to n do
for l from 1 to n do

$u \sim U(0,1)$

if $u \leq p_{kl}$ then

$\boldsymbol{b}_k \leftarrow \boldsymbol{b}_k - \eta \dfrac{\partial c_{k,l}^{a}}{\partial \boldsymbol{b}_k}$ $\boldsymbol{b}_j \leftarrow \boldsymbol{b}_j - \eta \dfrac{\partial c_{k,l}}{\partial \boldsymbol{b}_l}$

for m iterations do

$l \sim U\{1, \ldots, n\}$

$\boldsymbol{b}_k \leftarrow \boldsymbol{b}_k - \eta \dfrac{\partial c^{*}k,}{\partial \boldsymbol{b}_k}$

//The next line does not exist
in original UMAP:

$$\boldsymbol{b}_l \leftarrow \boldsymbol{b}_l \quad \eta \frac{\partial c^r_{k,l}}{\partial \boldsymbol{b}_l}$$

$$\eta \leftarrow 1 - \frac{v}{v_{\max}}$$

Return $\{\boldsymbol{b}_k\}_{k=1}^{n}$

The dimensionality reduction method known as UMAP, or Uniform Manifold Approximation and Projection, works especially well for presenting high-dimensional data on interactive dashboards. Its theoretical foundations in algebraic topology and Riemannian geometry yield a scalable and useful algorithm. Although UMAP and t-SNE have comparable visualization quality, UMAP is frequently chosen due to its better runtime performance and capacity to maintain a larger amount of global structure. UMAP can be used to project high-dimensional data into a lower-dimensional space for interactive dashboards, which facilitates the exploration and discovery of patterns or clusters in the data.

2.4 CONCLUSION

To sum up, this study explores the vital role that dimensionality reduction approaches play in improving the interpretation of visual data while dealing with the difficulties presented by high-dimensional and increasingly complicated datasets. By means of a methodical investigation and assessment of techniques like PCA, t-SNE, and UMAP, our goal is to offer a thorough grasp of their suitability, advantages, and disadvantages in various contexts. Our work emphasizes the significance of taking into account the trade-offs between representation integrity and computing efficiency, as well as the effects of dimensionality reduction on information preservation and interpretability. Through empirical evaluations on a range of datasets, we evaluate how well these strategies capture intrinsic structures and patterns, offering important insights for practitioners and scholars. Moreover, we highlight the usefulness of dimensionality reduction in enabling more efficient inference and decision-making by integrating it into more comprehensive data analysis pipelines. Examining pragmatic aspects like interpretability, scalability, and adaptability to various data kinds highlights the applicability of our results in a variety of fields, such as image analysis, finance, and biology.

In the end, by directing the best choice and implementation of dimensionality reduction approaches, this research advances visual data analysis strategies. Our goal is to promote progress across several domains and facilitate better data-driven decision-making by enabling users to derive significant insights, boost interpretability, and improve decision-making processes.

REFERENCES

[1] G. Sedrakyan, E. Mannens, and K. Verbert, "Guiding the choice of learning dashboard visualizations: Linking dashboard design and data visualization concepts," *Journal of Computer Languages*, vol. 50, pp. 19–38, Feb. 2019, https://doi.org/10.1016/j.jvlc.2018.11.002

[2] E. Dong, H. Du, and L. Gardner, "An interactive web-based dashboard to track COVID-19 in real time," *The Lancet Infectious Diseases*, vol. 20, no. 5, pp. 533–534, May 2020, https://doi.org/10.1016/s1473-3099(20)30120-1

[3] R. Toasa, M. Maximiano, C. Reis, and D. Guevara, "Data visualization techniques for real-time information – A custom and dynamic dashboard for analyzing surveys' results," *2018 13th Iberian Conference on Information Systems and Technologies (CISTI), Caceres, Spain*, 2018, pp. 1–7, https://doi.org/10.23919/CISTI.2018.8398641

[4] J. N. Milligan, B. Hutchinson, M. Tossell, and R. Andreoli, *Learning Tableau 2022*. Packt Publishing Ltd, 2022. (TextBook).

[5] M. G. Bhargava, K. T. Phani Surya Kiran, and D. R. Rao, "Analysis and design of visualization of educational institution database using power BI tool", *Global Journal of Computer Science and Technology*, vol. 18, no. C4, pp. 1–8, 2018.

[6] S. Mahajan, "Data visualization using a powerful tool – Power BI," *Journal of Data Processing and Business Analytics*, pp. 1–4, June 2023, https://doi.org/10.48001/jodpba.2023.111-4

[7] S. Rovena, C. Thomas Melathil, I. Priya Dr., and S. Mary, "A comparative analysis between data visualization tools for effective communication: Power BI and Qlik Sense," *International Research Journal of Modernization in Engineering Technology and Science*, January 2024, Published, https://doi.org/10.56726/irjmets47851

[8] A. Schreiber, L. von Kurnatowski, A. Meinecke, and C. de Boer, "An interactive dashboard for visualizing the provenance of software development processes," *2021 Working Conference on Software Visualization (VISSOFT), Luxembourg*, 2021, pp. 100–104, https://doi.org/10.1109/VISSOFT52517.2021.00019

[9] V. D. Kolychev andA. A. Shebotinov, "Application of business intelligence instrumental tools for visualization of key performance indicators of an enterprise in telecommunications," *Scientific Visualization*, vol. 11, no. 1, 2019, https://doi.org/10.26583/sv.11.1.03

[10] M. Holjevac and T. Jakopec, "Web application dashboards as a tool for data visualization and enrichment," *2020 43rd International Convention on Information, Communication and Electronic Technology (MIPRO), Opatija, Croatia*, 2020, pp. 1740–1745, https://doi.org/10.23919/MIPRO48935.2020.9245289

[11] I. Syafeâ, D. R. Wibowo, and V. Yordan, "Penggunaan Aplikasi Sisense Untuk Pengolahan Data & Visualisasi Business Intelligence," *INTECOMS: Journal of Information Technology and Computer Science*, vol. 6, no. 1, pp. 463–469, June 2023, https://doi.org/10.31539/intecoms.v6i1.6086
[12] G. D. Méo, "Creating interactive visualizations with plotly," *Programming Historian*, no. 12, December 2023, https://doi.org/10.46430/phen0115
[13] M. Nadj, A. Maedche, and C. Schieder, "The effect of interactive analytical dashboard features on situation awareness and task performance," *Decision Support Systems*, vol. 135, p. 113322, August 2020, https://doi.org/10.1016/j.dss.2020.113322
[14] J. C. Morales-Vega, L. Raya, M. Rubio-Sánchez, and A. Sanchez, "A virtual reality data visualization tool for dimensionality reduction methods," *Virtual Reality*, vol. 28, no. 1, February 2024, https://doi.org/10.1007/s10055-024-00939-8
[15] T. Fujiwara, J.-K. Chou, S. Shilpika, P. Xu, L. Ren, and K.-L. Ma, "An incremental dimensionality reduction method for visualizing streaming multidimensional data," *IEEE Transactions on Visualization and Computer Graphics*, vol. 26, no. 1, pp. 418–428, January 2020, https://doi.org/10.1109/tvcg.2019.2934433
[16] J. An, J. X. Yu, C. A. Ratanamahatana, and Y.-P. Phoebe Chen, "A dimensionality reduction algorithm and its application for interactive visualization," *Journal of Visual Languages & Computing*, vol. 18, no. 1, pp. 48–70, February 2007, https://doi.org/10.1016/j.jvlc.2006.03.001
[17] V. Jyothsna and V. V. Rama Prasad, "FCAAIS: Anomaly based network intrusion detection through feature correlation analysis and association impact scale," *ICT Express*, vol. 2, no. 3, 2016, pp. 103–116, ISSN 2405-9595, https://doi.org/10.1016/j.icte.2016.08.003
[18] R. M. Terol, A. R. Reina, S. Ziaei, and D. Gil, "A machine learning approach to reduce dimensional space in large datasets," *IEEE Access*, vol. 8, pp. 148181–148192, 2020, https://doi.org/10.1109/ACCESS.2020.3012836
[19] K. Pal and M. Sharma, "Performance evaluation of non-linear techniques UMAP and t-SNE for data in higher dimensional topological space," *2020 Fourth International Conference on I-SMAC (IoT in Social, Mobile, Analytics and Cloud) (I-SMAC), Palladam, India*, 2020, pp. 1106–1110, https://doi.org/10.1109/I-SMAC49090.2020.9243502

Chapter 3

Visual data analysis of temperature, ground water level, precipitation for climate-driven socio-economic prediction

S. Shoba, Melvin, and Sasithradevi A
Vellore Institute of Technology, Chennai, India

P. Prakash
Madras Institute of Technology, Anna University, Chennai, India

Deepa S
SRM Institute of Technology, Chennai, India

3.1 INTRODUCTION

In this comprehensive book chapter, we delve into the intricate relationship between rainfall, environmental variables, and the overarching impact on ecosystems and agriculture. Recognizing the inherent complexity of rainfall prediction and its critical role in sustaining life, we embark on a journey to explore the various environmental phenomena that influence this essential climatic factor on a global scale.

Our quest for accurate rainfall prediction begins with the acquisition of relevant datasets, meticulously sourced from official government portals [1–3]. These datasets, serving as the foundation for our analyses, undergo careful filtration and classification to align with the specific requirements of diverse Machine Learning (ML) models. The ensuing chapters unfold a tapestry of analyses, each contributing to a nuanced understanding of the factors driving climatic and rainfall pattern changes.

The application of cutting-edge ML algorithms takes center stage in our exploration. We employ a vector regression model to scrutinize water-related variables such as groundwater and precipitation, offering insights into their influence on temperature dynamics [4]. The K-Nearest Neighbor (KNN) method facilitates a statistical examination through graphs and charts [5], providing a visual narrative of the intricate relationships within environmental variables.

The journey into predictive modeling extends further with the implementation of Feed Forward Neural Network (FFNN) and Adaptive Neuro-Fuzzy Inference System (ANFIS) techniques. These models stand poised to

DOI: 10.1201/9781003542735-3

forecast not only rainfall but also wind patterns and temperature variations [6–8]. The Gaussian model emerges as a powerful tool for global climate modeling [9], providing a holistic perspective on the intricate web of climatic factors.

As we delve deeper, we pivot towards more specialized applications of ML, employing Long Short-Term Memory (LSTM) networks for groundwater monitoring [10]. The integration of Deep Learning (DL) techniques takes precedence in the latter sections of the chapter, encompassing statistical weather prediction [11], numerical weather prediction [12], and synoptic weather prediction [13].

Our exploration extends to hybrid models that amalgamate ML techniques with optimization tools such as genetics and particle swarm optimization, enhancing prediction accuracy through fine-tuned hyperparameters. Additionally, we uncover studies combining multiple ML techniques and integrating ML with established models like Autoregressive Integrated Moving Average (ARIMA) [14].

Recognizing the temporal nature of meteorological data, efforts are made to adapt time series models, such as LSTMs, to 1D data for improved rainfall prediction accuracy. For 2D data, innovative models like ConvLSTMs, integrating Convolutional Neural Network (CNN)s with LSTMs, emerge as promising tools to capture intricate climate patterns [15].

Amidst this exploration of sophisticated models, our chapter does not shy away from the critical analysis of associated issues. We confront challenges and limitations head-on, fostering a transparent evaluation of the strengths and weaknesses inherent in each model. A pivotal aspect of our endeavor involves a meticulous comparison of results across various models, facilitating a nuanced understanding of their respective efficacy.

In sum, this book chapter aspires to be a comprehensive guide and reference for researchers, climatologists, and enthusiasts alike, offering a deep dive into the world of predictive modeling for rainfall and climatic variables. Through the integration of diverse methodologies and the critical examination of their outcomes, we aim to contribute meaningfully to the ongoing discourse surrounding climate science and its profound implications for our planet.

The important variables are as follows for analyzing the climatic variations:

- Precipitation: Understanding and predicting the amount of rainfall, a crucial factor in climate dynamics.
- Temperature: Analyzing temperature patterns to gain insights into climate variations.
- Ground-Water Level: Investigating the fluctuations in the groundwater level, which can be indicative of broader climate changes.

The rest of the paper is organized as AI models and its subfield in Section 2. The detailed view of the collection of datasets and its observations in Section 3.

Section 4 includes the analysis of climatic variations in visual representation and finally concludes this chapter in Section 5.

3.2 ARTIFICIAL INTELLIGENCE MODELS

Artificial Intelligence (AI) model is a program that contains a huge amount of data that trains the data and identifies specific patterns or makes judgements on its own without any human intervention. The judgement made by model is implemented in various device applications to develop an automation system. The subset of artificial intelligence are ML, DL, Generative AI. The ML models are subdivided into supervised ML, Unsupervised ML, semi supervised ML. In general, the types of AI models are linear regression models, decision tree, deep neural networks, Naïve bayes, Linear discriminative analysis, etc.

3.2.1 Machine learning models

Traditional ML models, such as multivariate linear regression (MLR), KNN, Artificial Neural Networks (ANNs), Support Vector Machines (SVMs), and Random Forests (RF), have been effectively utilized to classify with notable performance, particularly when dealing with 1D data obtained from meteorological stations. Their application extends to both short-term and long-term datasets. Prediciting the climatic conditions is necessary not only for agriculture planning but also for industries, aerospace navigation, defense, shipping, and mountaineering. It is frequently used as a warning that sudden changes in the weather can result in natural disasters. The prediction analysis can be done through different regression models. Simple Linear regression model is the most popular statistical method for simulating a relationship between two sets of variables. A linear regression equation is the product which can be utilized to forecast the data. A straightforward linear regression technique [16] to forecast the daily maximum temperatures for Nashville, Tennessee.

MLR makes use of multiple explanatory variables to forecast the value of a response variable [17]. It is building a model of relation between response variables and explanatory variables.

3.2.1.1 Vector regression model

This model is particularly suited for predicting multiple output variables simultaneously, making it valuable for understanding the interdependence of precipitation, temperature, and ground-water level [4]. Predicting the joint behavior of these variables, considering their collective impact on climate dynamics as shown in Figure 3.1.

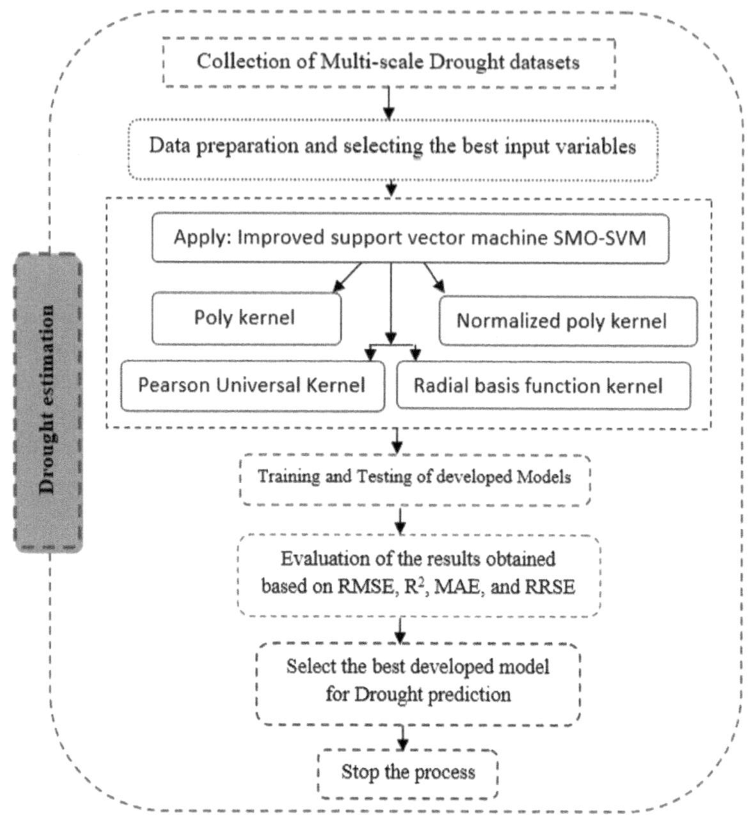

Figure 3.1 Vector regression model having relationship between agriculture and climatic variability [4].

3.2.1.2 Random forest (rf) model creation

The RF model can analyze the intricate relationship between classification and noise with strong robustness in data containing missing values or noise. RF model has been extensively employed as a feature selection technique to determine the importance of variables in high-dimensional data found that a number of earlier research that used RF to predict Daily Global Solar Radiation (DGSR) were primarily concerned with solar radiation at a single site or in a case study within a specific region of China. Conversely, there are few high-density DGSR estimates using surface meteorological measurements for the entirety of China [5]. Figure 3.2 depicts the flow chart for estimating DGSR using the RF algorithm. It consists of the following three steps: Variable selection and data matching. Data quality control and spatiotemporal matching are used to filter the model training and testing

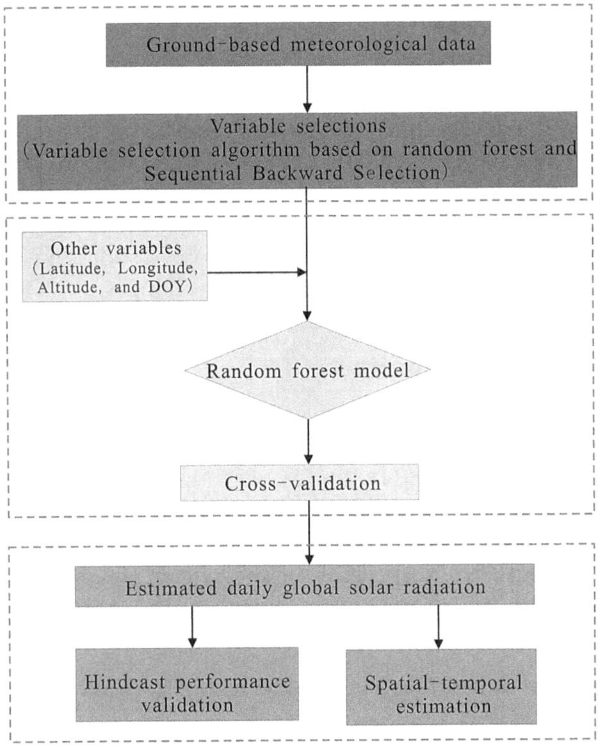

Figure 3.2 Estimating DGSR using the RF algorithm for specific region of China [5].

data pairings. Spatial matching does not need to be taken into consideration because every solar radiation station has overlapped with its matching meteorological station. The backward selection approach is used in the RF model creation, training, and testing to pick and decide the input variables based on the variable important features of the RF. When entering the geoinformation parameters, RF offers functions for spatial interpolation, extrapolation, and interpolation [5].

3.2.1.3 K-nearest-neighbor (KNN) method

Originating as a non-parametric statistical pattern identification process, the KNN method seeks to differentiate between various patterns based on predetermined criteria. It is the most intuitive non-parametric technique among the others, yet it still has strong statistical qualities because no theoretical or analytical relationship between the inputs and the outputs is known or assumed [18].

3.2.1.4 *ML model significance*

The combination of these diverse ML models allows for a comprehensive analysis of the complex interactions between precipitation, temperature, and ground-water levels in the context of climate prediction. The models cater to different aspects of climate dynamics, ranging from capturing non-linear relationships to modeling temporal dependencies, providing a holistic understanding of the variables involved. The insights gained from this study contribute to improved climate prediction models, aiding in better understanding and preparation for potential climate changes.

3.2.2 Deep learning models

DL models is the subset of AI models that consists of convolutional neural networks (CNNs), Long-Short-Term-Memory (LSTM) networks, Deep Neural Network (DNN), ANFIS, FFNN offer a powerful approach to handle the complexities of predicting rainfall and understanding climate change. The capability of DL models to manage vast datasets has led to their prominence, especially in short-term datasets characterized by 2D data.

3.2.2.1 *Adaptive neuro fuzzy inference system (ANFIS)*

ANFIS combines fuzzy logic with neural networks to model complex relationships and uncertainties. The occurrence of meteorological anomaly phenomena might have certain detrimental effects, such as flooding, which can destroy public infrastructure and impede local business and transportation operations. To rapidly and reliably give information on future weather conditions, this research uses weather factors as a variable to predict the amount of rainfall using the ANFIS approach and Support Vector Regression (SVR). Individuals can become ready and set up the necessary tools to deal with it [6, 7]. Predicted rainfall is based on synopsis data, including temperature, wind, and relative humidity. Rainfall prediction is based on the results of each parameter that has been forecast using ANFIS. MSE and RMSE are used to calculate accurate prediction.

3.2.2.2 *Feedforward neural network (FFNN)*

FFNN is versatile and effective for capturing non-linear relationships in data. Application: The occurrence of meteorological anomaly phenomena might have certain detrimental effects, such as flooding, which can destroy public infrastructure and impede local business and transportation operations. In order to rapidly and reliably give information on future weather conditions, this research uses weather factors as a variable to predict the amount of rainfall using the ANFIS approach and Support Vector Regression (SVR). Individuals can become ready and set up the necessary tools to deal with it. Predicted rainfall is based on synopsis data, including temperature, wind, and relative humidity. Rainfall prediction is based on the results of each parameter that has been forecast using ANFIS. MSE and RMSE are used to calculate accurate prediction [8].

3.2.2.3 Gaussian model (global climate modeling)

Gaussian models are often employed in global climate modeling to represent the probability distribution of climate variables. In this study, it might be used to model the distribution of precipitation and temperature, providing probabilistic insights into climate trends at a global scale. Gaussian variable, which makes it easier to model rainfall in conjunction with other weather variables. Zero rainfall data corresponds to suppressed values below a threshold. Here, we apply this technique to the joint modeling of daily precipitation and other meteorological parameters, including wind speed, temperature, radiation, and relative humidity [9].

3.2.2.4 Long short-term memory (LSTM) (ground water modeling)

Long-sequence temporal dependencies in sequential data are captured by recurrent neural networks, or LSTMs. An Recurrent Neural Network (RNN) type called LSTM neural networks was created to get over the problems with standard RNNs' vanishing and exploding gradients [38]. By requiring continual error flow between hidden cell states without going via an activation function, the LSTM architecture reduces gradient concerns. An LSTM cell has three multiplicative units called gates in addition to this constant error path: the forget gate, the input gate, and the output gate. By feeding inputs forward, backpropagating the error, and modifying the weights, a gate—which functions as a neuron—can learn which cell states and inputs are critical for forecasting the output [10, 19].

3.2.3 Hybrid models

Researchers have embraced hybrid models that blend ML techniques with optimization tools like genetics and particle swarm optimization. These hybrids are designed to fine-tune hyperparameters and enhance prediction accuracy. Other studies have explored combining multiple ML techniques, while some integrate ML with time-tested models like ARIMA. ARIMA model is to forecast the weather that shows better visibility for the parameter used in grid technique [14].

3.2.3.1 ConvLSTM

Recognizing the temporal nature of meteorological data, efforts have been made to adapt and practice the time series ML models such as LSTMs to 1D data for improved accuracy in rainfall prediction. For 2D data, innovative models like ConvLSTMs, which integrate CNNs with LSTMs, have shown promise since their inception [15]. Various iterations of these models have since been developed to capture the intricate patterns in climate data.

3.2.4 Challenges in AI models

3.2.4.1 Statistical reasoning

Challenge: Insufficient training data often leads to the extraction of weak hypotheses by learning algorithms.

Solution: Combining multiple classifiers helps in leveraging the tendencies of diverse hypotheses, resulting in a more robust and accurate hypothesis. Ensemble methods exploit the collective strength of multiple classifiers to enhance statistical reasoning.

3.2.4.2 Computational reasoning

Challenge: A hypothesis generated by a single classifier may be complex and time-consuming.

Solution: To address computational challenges, aggregating multiple classifiers with appropriate parameterization becomes crucial. By leveraging the strengths of each classifier, the ensemble can provide a more efficient and accurate hypothesis while mitigating computational time concerns. This involves careful consideration of parameters such as speed, efficiency, and accuracy.

3.2.4.3 Representational reasoning

Challenge: A correct hypothesis is often not fully represented by a single classifier in the hypothesis space.

Solution: Ensemble methods, through the formation of a weighted sum of hypotheses from the hypothesis space, extend the representational capacity. This leads to a more comprehensive and presentable hypothesis, overcoming the limitations of individual classifiers in representing the complexity of the underlying patterns.

3.3 DATASET

The dataset required for this prediction of climate is taken for the government website [1–3], is displays the data of climate and other environmental variables in a tabulated manner.

3.3.1 Dataset collection

The knowledge-based climate analysis system for united states was created that estimates the GIS data daily, monthly, annually, and event-based climatic elements using point climate data, a digital elevation model, and other spatial data sets [20]. The data collection was mainly focused on 30 years from 1961-1990 of parameters of temperature, precipitation, humidity, heat, cool degree days, growing degree days, frozen period, freeze season and others for predicting the climatic variations in day wise, month and year wise manner. Datasets are also collected from government websites which displays the data and other environmental variables for predicting the values [1–3]. Dataset collected for Australia contains the parameters of over the environmental variables like rainfall, temperature (maximum and minimum), and vapour pressure at daily and monthly timeframes [21]. The purpose of these analyses is to enhance the definition of historical climatic variability and change over the Earth and to improve modern climate estimations. Though the datasets are available in plenty for finding the prediction of climatic variations, the suitability of datasets to the model is much needed to obtain the prediction accurately. This work explores the connection between scale and spatial climatic elements. Terrain and water bodies have the greatest influence on spatial climate patterns; these impacts are mainly due to direct effects of terrain-induced climate transitions, elevation, cold air drainage and inversions, and coastal effects [22]. Datamining places an important role for extracting the suitable data required for the accurate prediction.

3.3.2 Data mining

Data mining serves as a valuable technique for extracting meaningful patterns from extensive datasets. It involves the extraction of implicit, previously unknown, and practical information from vast amounts of unpredictable, noisy and incomplete data [23]. This technology proves promising for businesses by enabling a focus on crucial data within their databases. The fundamental approach in data mining is prediction, wherein a model is constructed using pre-classified samples to categorize data and establish connections between independent and dependent variables.

Ensemble methods represent a significant advancement in data mining, offering impactful developments. Weather conditions, with their wide-ranging effects on human civilization, pose challenges addressed by data mining techniques. Remote sensing satellites play a crucial role in collecting relevant data, and as weather factors change over time, prediction models can be created statistically or using various data mining methods such as artificial NN, regression, decision tree and clustering [23].

Data mining techniques encompass classification, prediction, clustering, outlier identification, and regression. Notable algorithms for categorization and prediction include ANN, Decision Trees, Bayesian Classification, SVM, and Regression. The performance of these algorithms is evaluated using various criteria [24].

3.4 RESULTS AND ANALYSIS

The dataset collected provides a valuable insight into the diverse array of ML techniques applied for the detection of rainfall and the study of climate change. Various metrics are used to assess the performance of these models based on the nature of the problem being addressed. In classification problems, common metrics include precision, recall, and accuracy. However, in instances where data imbalance is a concern, the f1-score is preferred over accuracy to account for class imbalances. For sequence classification prediction, the critical success index (CSI) is employed, while continuous outputs are evaluated using metrics such as mean absolute error and root mean squared error.

The challenge of directly comparing results across different studies is acknowledged due to differences in models, pre-processing methods, metrics, datasets, and parameters. Despite this, certain ML algorithms consistently stand out as high performers within individual studies. ANNs and DL models are frequently cited as top-performing models, demonstrating their effectiveness in both long-term and short-term rainfall predictions. SVMs also receive notable recognition across various studies. Additionally, ensemble methods, logistic regression, and KNNs emerge as effective choices in specific contexts. The ML and DL models have challenges in direct comparisons for addressing rainfall detection and climate change studies. This suggests promising avenues for further research and application in understanding and mitigating the impacts of climate change. The weather is predicted in different ways depending on the dependent and independent variables of data.

3.4.1 Ways of weather prediction

3.4.1.1 Synoptic weather prediction

Traditional in nature, synoptic weather prediction involves the systematic observation of various meteorological phenomena over a specified duration. Meteorological departments generate a series of synoptic charts to document the evolving patterns of weather, utilizing a vast array of observed data from numerous weather stations [13]. These charts serve as foundational elements for subsequent weather forecasting activities.

3.4.1.2 *Statistical weather prediction*

In conjunction with numerical methods, statistical weather prediction depends on historical climate data to project future weather conditions. This approach assumes that forthcoming weather patterns will resemble past conditions. The primary objective is to identify which components of historical climate data yield the most accurate predictions for future weather. Consequently, this method enables an overarching anticipation of climate conditions [11].

3.4.1.3 *Numerical weather prediction*

Leveraging computational power, numerical weather prediction relies on sophisticated computer programs executed on supercomputers to forecast various climate parameters. While the equations guiding this process may not achieve absolute precision, the initial stages of weather conditions are partially known. Consequently, the accuracy of weather forecasting using numerical methods is subject to the inherent complexity of the atmospheric dynamics [12, 25].

3.4.2 Visual representation

The analysis of the climatic variations can be visually represented in various ways using line graphs, scatter plots, residual plots, variable plots, and matrix representations. The visual way of analysis shows a clear idea of their advantages and disadvantages.

3.4.2.1 *Line graph*

Regression analysis can be applied by computing the coefficient, slope, and the analyzed climate data set on a daily, monthly, or annual basis as shown in Figure 3.3. Additionally, the effectiveness of forecasting future values can be computed through the application of a basic linear regression technique [26].

3.4.2.2 *Scatter plot*

The estimated DGSR samples use a tenfold cross validation from the RF model on the prediction of DGSR at 97 ground sites during 2013–2014 [5]. The error analysis of RMSE, R and MB are analyzed and represented in a visual form as depicted in Figure 3.4.

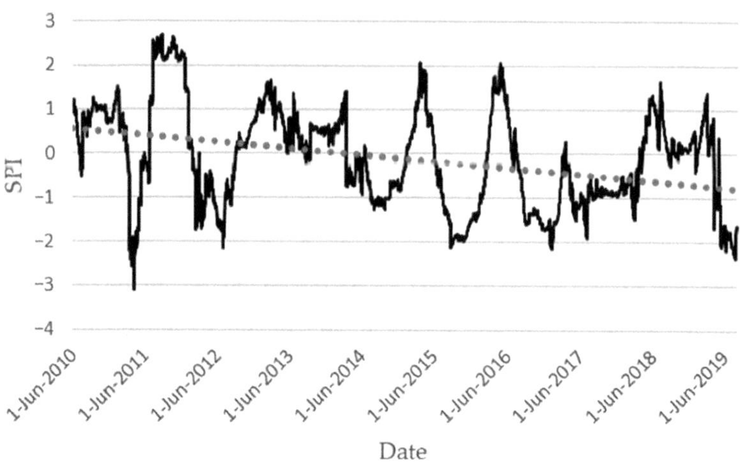

Figure 3.3 Linear Regression analysis of climate data on monthly basis [26].

3.4.2.3 Bar charts

The canadial dataset used in [13] predicts the fire weather from synoptic pattern automatically. The graph in Figure 3.5. shows the experimental parameter distribution. The predictions are taken for 4 months from June to August. Three index are taken for consideration for predicting the climatic change. The highest AUC Area Under Curve for The Fire weather index (FWI)is 0,802 in May, The Initial Spread Index ISI is 0.802 in August, Fine Fuel Moisture Code FFMC is 0.837 in August. When the comparison is made with FFMC and ISI, the performance was good during august month. The fire season in Canada is May and August than June and July which shows a bad performance.

3.4.2.4 Residual scatter plots

When compared to baseline-all and baseline-30 approaches, ARIMA shows somewhat better point forecasts and much better interval forecasts for 20-year estimates of the annual average temperature as shown in Figure 3.6. The negative correlation shown in the forecast scores of the ARIMA, and linear trend approaches suggests that one approach can provide more accurate forecasts when the other is not up to par. For the annual average temperature, the linear trend approach shows more outliers with high forecast scores. The estimates of the maximum annual 1-day precipitation between the ARIMA and the other four statistical methods [14].

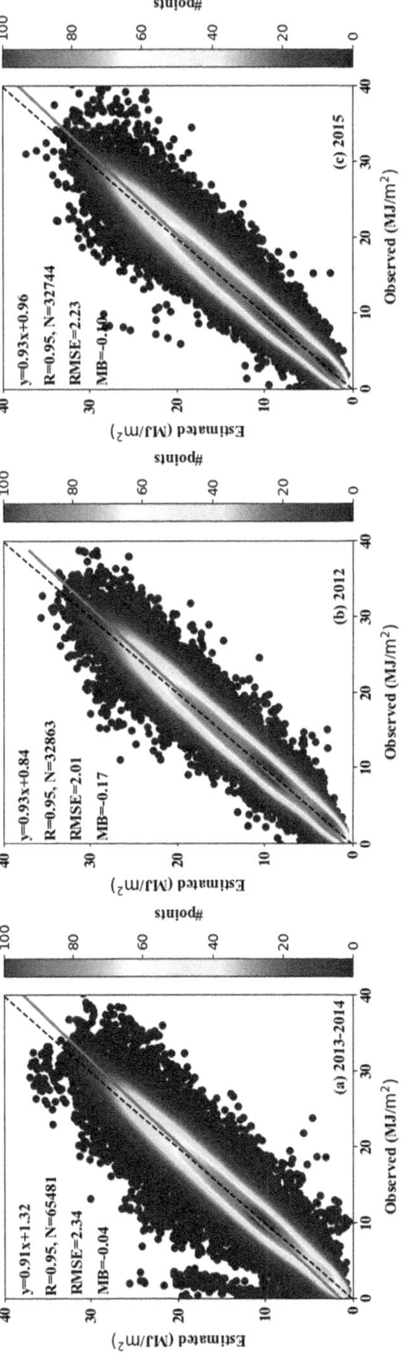

Figure 3.4 Analysis of error for DGSR samples using RF model for 2012–2015 [5].

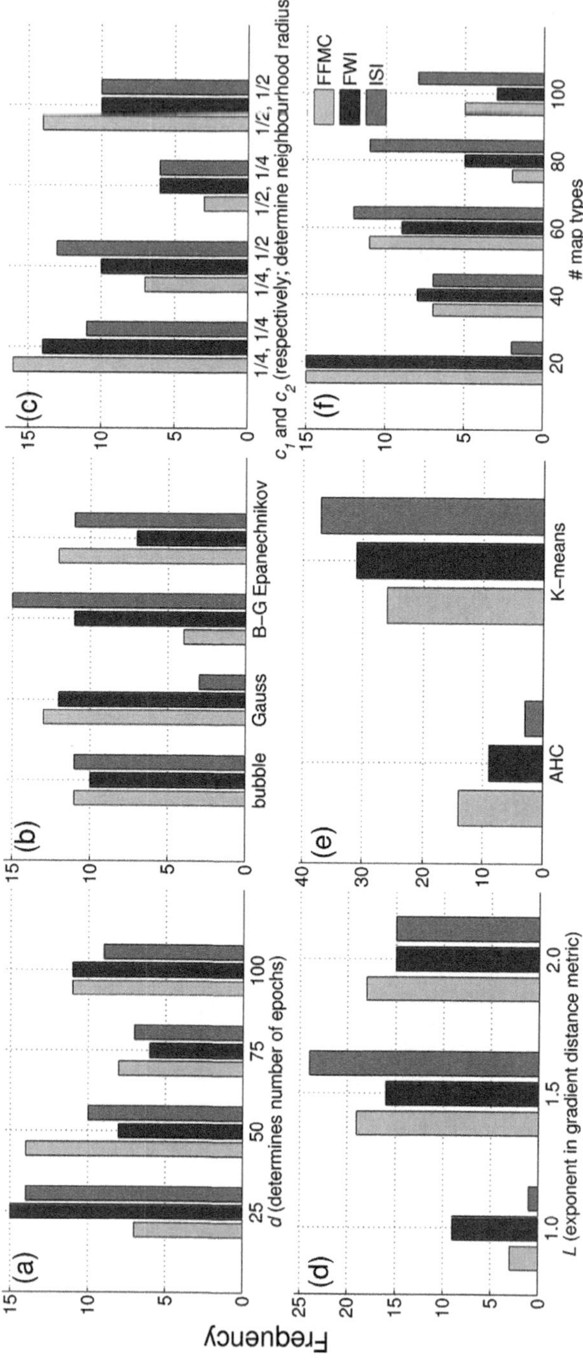

Figure 3.5 Fire season (Canada) FWI, FFMC and ISI index measurement data [13].

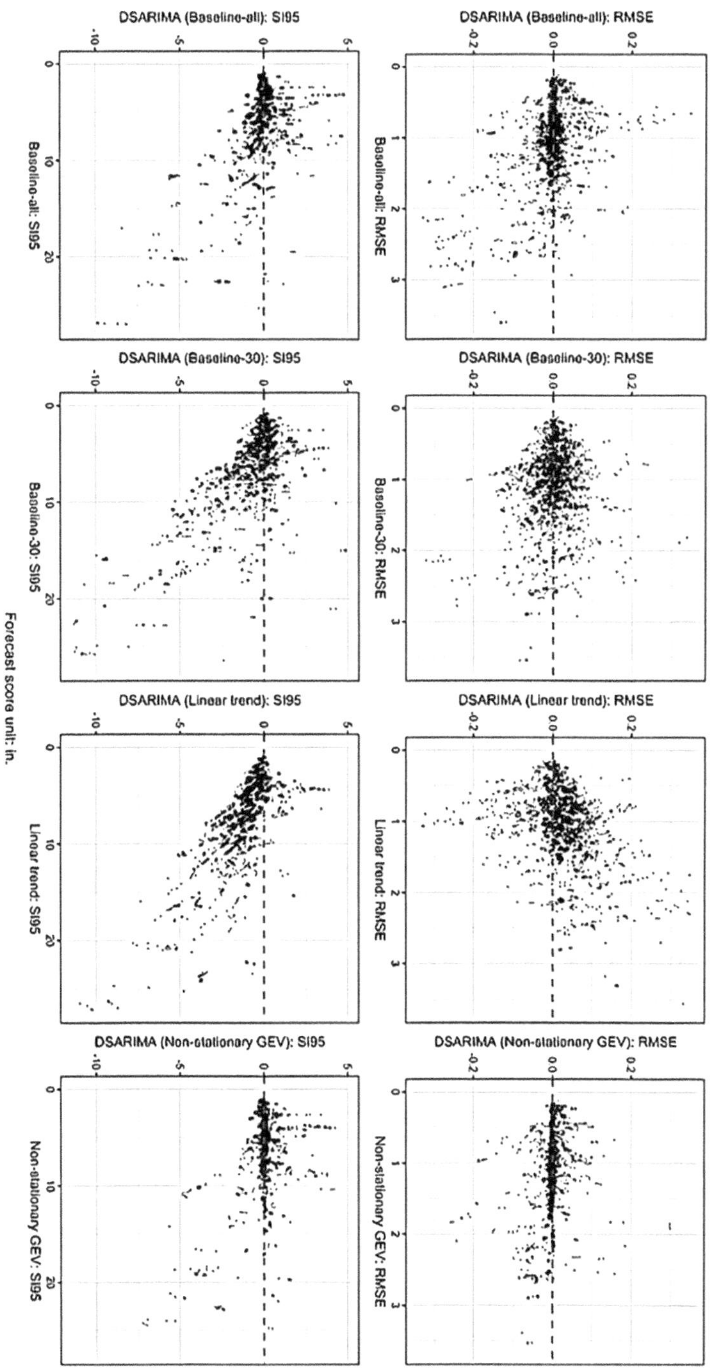

Figure 3.6 ARIMA comparison of baseline-all and baseline-30 approaches [14].

3.5 CONCLUSION

The focus of this paper is on the application of ML models for the prediction of rainfall, a topic that has garnered considerable attention in past research due to its inherent challenges. To enhance the accuracy of rainfall predictions, it is crucial to appropriately analyze and train the collected data, followed by testing with ensemble learning techniques. These techniques are known for their efficiency in yielding predicted outcomes with minimal error compared to the standard set. The critical challenges addressed in this study include the identification of the most suitable ensemble learning technique, addressing the sensitivities of factual procedures, and managing the reliance on element treatment within the prediction process. The primary objective is to scrutinize the data to improve the anticipation of rainfall beyond the capabilities of current models. The proposed approach is designed for simplicity, allowing for easy testing and verification. Additionally, the paper aims to enhance the accuracy of the proposed rainfall prediction model by integrating model predictions with ML methods. Through this research, the goal is to contribute to the advancement of rainfall prediction models, minimizing errors and fostering a more reliable understanding of precipitation patterns.

REFERENCES

1. https://data.gov.in/catalog/all-india-seasonal-and-annual-temperature-series
2. https://data.opencity.in/dataset/daily-temperature-70-years-data-for-major-indian-cities
3. https://ourworldindata.org/co2/country/india
4. Pande, C.B., Kushwaha, N.L., Orimoloye, I.R., Kumar, R., Abdo, H.G., Tolche, A.D. and Elbeltagi, A., 2023. Comparative assessment of improved SVM method under different kernel functions for predicting multi-scale drought index. *Water Resources Management*, 37(3), pp. 1367–1399.
5. Zeng, Z., Wang, Z., Gui, K., Yan, X., Gao, M., Luo, M., Geng, H., Liao, T., Li, X., An, J. and Liu, H., 2020. Daily global solar radiation in China estimated from high-density meteorological observations: a random forest model framework. *Earth and Space Science*, 7(2), p. e2019EA001058.
6. Bushara, N.O. and Abraham, A., 2015. Using adaptive neuro-fuzzy inference system (anfis) to improve the long-term rainfall forecasting. *Journal of Network and Innovative Computing*, 3(2015), pp. 146–158.
7. Novitasari, D.C.R., Rohayani, H., Junaidi, R., Setyowati, R.D., Pramulya, R. and Setiawan, F., 2020, March. Weather parameters forecasting as variables for rainfall prediction using adaptive neuro fuzzy inference system (ANFIS) and support vector regression (SVR). *Journal of Physics: Conference Series*, 1501(1), p. 012012. IOP Publishing.
8. Tran Anh, D., Duc Dang, T. and Pham Van, S., 2019. Improved rainfall prediction using combined pre-processing methods and feed-forward neural networks. *J*, 2(1), pp. 65–83.
9. Durbán, M. and Glasbey, C.A., 2001. Weather modelling using a multivariate latent Gaussian model. *Agricultural and Forest Meteorology*, 109(3), pp. 187–201.

10. Wu, C., Zhang, X., Wang, W., Lu, C., Zhang, Y., Qin, W., Tick, G.R., Liu, B. and Shu, L., 2021. Groundwater level modeling framework by combining the wavelet transform with a long short-term memory data-driven model. *Science of the Total Environment*, 783, p. 146948.
11. Deloncle, A., Berk, R., d'Andrea, F. and Ghil, M., 2007. Weather regime prediction using statistical learning. *Journal of the Atmospheric Sciences*, 64(5), pp. 1619–1635.
12. Coiffier, J., 2011. *Fundamentals of numerical weather prediction*. Cambridge University Press.
13. Lagerquist, R., Flannigan, M.D., Wang, X. and Marshall, G.A., 2017. Automated prediction of extreme fire weather from synoptic patterns in northern Alberta, Canada. *Canadian Journal of Forest Research*, 47(9), pp. 1175–1183.
14. Salman, A.G. and Kanigoro, B., 2021. Visibility forecasting using autoregressive integrated moving average (ARIMA) models. *Procedia Computer Science*, 179, pp. 252–259.
15. Kim, T. and Kim, H.Y., 2019. Forecasting stock prices with a feature fusion LSTM-CNN model using different representations of the same data. *PloS one*, 14(2), p. e0212320.
16. Massie, D.R. and Rose, M.A., 1997. Predicting daily maximum temperatures using linear regression and Eta geopotential thickness forecasts. *Weather and Forecasting*, 12(4), pp. 799–807.
17. Agrawal, A., Kumar, V., Pandey, A. and Khan, I., 2012. An application of time series analysis for weather forecasting. *International Journal of Engineering Research and Applications*, 2(2), pp. 974–980.
18. Toth, E., Brath, A. and Montanari, A., 2000. Comparison of short-term rainfall prediction models for real-time flood forecasting. *Journal of Hydrology*, 239(1-4), pp. 132–147.
19. Bowes, B.D., Sadler, J.M., Morsy, M.M., Behl, M. and Goodall, J.L., 2019. Forecasting groundwater table in a flood prone coastal city with long short-term memory and recurrent neural networks. *Water*, 11(5), p. 1098.
20. Daly, C., Taylor, G.H., Gibson, W.P., Parzybok, T.W., Johnson, G.L. and Pasteris, P.A., 2000. High-quality spatial climate data sets for the United States and beyond. *Transactions of the ASAE*, 43(6), pp. 1957–1962.
21. Jones, D.A., Wang, W. and Fawcett, R., 2009. High-quality spatial climate data-sets for Australia. *Australian Meteorological and Oceanographic Journal*, 58(4), p. 233.
22. Daly, C., 2006. Guidelines for assessing the suitability of spatial climate data sets. *International Journal of Climatology: A Journal of the Royal Meteorological Society*, 26(6), pp. 707–721.
23. Zainudin, S., Jasim, D.S. and Bakar, A.A., 2016. Comparative analysis of data mining techniques for Malaysian rainfall prediction. *International Journal on Advanced Science, Engineering and Information Technology*, 6(6), pp. 1148–1153.
24. Kumar, R.S. and Ramesh, C., 2016, August. A study on prediction of rainfall using datamining technique. In *2016 International conference on inventive computation technologies (ICICT)* (Vol. 3, pp. 1–9). IEEE.
25. Kimura, R., 2002. Numerical weather prediction. *Journal of Wind Engineering and Industrial Aerodynamics*, 90(12-15), pp. 1403–1414.
26. Kim, S.W., Jung, D. and Choung, Y.J., 2020. Development of a multiple linear regression model for meteorological drought index estimation based on landsat satellite imagery. *Water*, 12(12), p. 3393.

Chapter 4

AI-based online interview bot with an interactive dashboard

Rakoth Kandan Sambandam, Divya Vetriveeran, J. Jenefa, and Balamurugan M
CHRIST (Deemed to be University), Bengaluru, India

Thaiyalnayaki S
Bharath Institute of Higher Education and Research (Deemed to be University), Chennai, India

4.1 INTRODUCTION

The use of video interviews in the recruitment process has become increasingly prevalent in recent years. While this method offers convenience and efficiency, it presents new challenges for employers and hiring managers in evaluating candidates' communication skills and perceived personality traits. In many cases, the subjective nature of human evaluation may lead to inconsistent and unreliable results. An intelligent video interview agent is proposed to address this issue. This agent utilizes natural language processing and computer vision techniques to analyze candidates' verbal and nonverbal behavior during the interview. By focusing on linguistic features such as fluency, grammar, and vocabulary, as well as nonverbal cues such as facial expressions and body language, the agent predicts the candidate's communication skills and perceived personality traits. By combining advanced technologies with traditional recruitment methods, this agent may improve the efficiency and effectiveness of the recruitment process, allowing employers to make better-informed decisions when selecting candidates for various positions.

The aim is to validate the effectiveness of the intelligent video interview agent through a series of experiments and evaluations. The results may offer valuable insights into the potential applications and benefits of advanced technologies in recruitment and human resource management. The use of video interviews in the recruitment process has become increasingly prevalent in recent years due to its convenience and efficiency. However, evaluating candidates' communication skills and perceived personality traits from a video interview can be challenging, subjective, and inconsistent. This subjectivity can lead to missed opportunities for highly qualified candidates or hiring individuals who do not fit the company culture. To address this issue,

 DOI: 10.1201/9781003542735-4

an intelligent video interview agent is proposed to provide more accurate and objective evaluations of candidates' communication skills and perceived personality traits.

This study aims to validate the effectiveness of the intelligent video interview agent by comparing the agent's predictions with the ratings given by human evaluators. By combining advanced technologies with traditional recruitment methods, this study aims to improve the efficiency and effectiveness of the recruitment process. The use of video interviews in the recruitment process has become increasingly prevalent in recent years due to its convenience and efficiency. However, evaluating candidates' communication skills and perceived personality traits from a video interview can be challenging, subjective, and inconsistent. The main objective of this study is to develop and validate the effectiveness of an intelligent video interview agent that can provide accurate and objective evaluations of candidates' communication skills and perceived personality traits.

4.2 RELATED WORKS

The existing system for evaluating candidates' communication skills and perceived personality traits during video interviews relies heavily on human evaluators. Human evaluators are responsible for reviewing the video interviews and providing ratings for each communication skill and perceived personality trait. However, the evaluation process can be subjective and inconsistent due to factors such as human biases, misinterpretation of nonverbal cues, and missing important aspects of the candidate's communication and personality traits.

The article [1] People versus machines: introducing the "HIRE Framework" proposes a framework for balancing the use of human judgment and machine learning in the hiring process. The framework is called HIRE, which stands for Human Intervention Required and Expected. The authors argue that while machine learning can provide many benefits in terms of efficiency and objectivity in hiring, it is important to maintain a human element in the process to ensure fairness, accountability, and ethical decision-making.

The article [2] "vRecruit: An Automated Smart Recruitment Web app using Machine Learning" presents a web application called vRecruit that uses machine learning to automate and streamline the recruitment process. The authors first provide an overview of the challenges and benefits of using machine learning in recruitment, including improved efficiency, reduced bias, and increased objectivity.

The article [3] "Examining perceptions towards hiring algorithms" explores the attitudes and perceptions of job seekers and hiring managers towards the use of hiring algorithms in the recruitment process. The authors first provide an overview of the benefits and challenges of using hiring algorithms, such as reducing bias and increasing efficiency, but also potential ethical concerns and the need for transparency and accountability.

The article [4] "Artificial intelligence video interviewing for employment: perspectives from applicants, companies, developer and academicians" discusses the use of artificial intelligence (AI) in video interviewing for employment and presents perspectives from various stakeholders including applicants, companies, developers, and academicians. The authors first provide an overview of the benefits and challenges of using AI in video interviewing, including improved efficiency, reduced bias, and potential ethical concerns.

The authors [5] discuss the potential benefits and drawbacks of using AI in video inter- views, such as improved efficiency and reduced bias, but also concerns around privacy, discrimination, and the lack of transparency in decision-making. They also provide an overview of relevant laws and regulations, such as anti-discrimination laws and data protection laws, and highlight gaps in current legal frameworks that leave job applicants vulnerable to potential abuses.

The article [6] "Automated Video Interviewing as the New Phrenology" argues that the use of automated video interviewing (AVI) in recruitment may be perpetuating discriminatory practices similar to those of phrenology, a discredited practice from the 19th century that claimed to determine personality traits and intelligence based on physical characteristics.

Few researches have been done in the area of trainings too, and some of the notable works like the authors [7] have utilized TensorFlow for automatic personality recognition during asynchronous video interviews, improving the accuracy and efficiency of candidate evaluation which plays a great role in the use of AI. The work by the authors [8] has focused on how AI and synchrony in video interviews affect the ratings of the interviews and applicant attitudes, focusing on how technology impacts the interviewing process. The paper [9] investigates the effectiveness of Virtual Interactive Training Agents (ViTA) in improving job interview skills for adults with autism and developmental disabilities. There has also been the use of AI for sentiment analysis of tweets and participatory research method with people with intellectual disabilities, addressing the challenges and potential benefits of photovoice.

4.3 PROPOSED WORK

An intelligent video interview agent is a software program that is designed to assist in the recruitment process by conducting video interviews with candidates. The agent can analyze the candidate's responses to interview questions and provide feedback to the hiring team, helping them to make informed decisions. There have been several studies and works done on the development of intelligent video interview agents.

There has been a significant amount of experimental and analytical work completed in the field of intelligent video interview agents. Here are some examples:

Experimental Work: Researchers have conducted experiments to evaluate the effective- ness of intelligent video interview agents in various contexts. For example, they have tested the agents' ability to accurately assess candidates' nonverbal behavior, to detect deception, and to predict job performance.

Analytical Work: Researchers have also conducted analytical work to understand the limitations and potential biases of intelligent video interview agents. For example, they have analyzed the impact of cultural differences on the agents' ability to accurately assess candidates, and they have examined the potential for the agents to discriminate against certain groups of candidates based on factors such as age, gender, or ethnicity.

Development Work: In addition to experimental and analytical work, researchers and developers have also worked to improve the technology behind intelligent video inter- view agents. They have developed new algorithms and models to enhance the agents' ability to assess candidates' behavior and performance, and they have incorporated ma- chine learning techniques to enable the agents to learn from their interactions with candidates and improve over time.

Overall, the experimental and analytical work completed in the field of intelligent video interview agents has helped to shed light on the potential benefits and limitations of the technology, and has helped to guide the development of more effective and unbiased agents in the future as given in Figure 4.1.

The modelling, analysis, and design for an intelligent video interview agent using Tensor Flow model can be done by following block diagram.

Support Vector Machines (SVMs): SVMs are a popular machine learning algorithm for classification tasks. They work by finding a hyper plane that separates the data into different classes. While SVMs have been used in the development of intelligent video interview agents, they may not be as effective as Tensor Flow models for complex tasks such as emotion recognition and behavior analysis. Random Forest: Random Forest is an ensemble learning algorithm that uses decision trees to classify data. It works by constructing multiple decision trees and then aggregating the results to make a final prediction. While Random Forest is effective for many classification tasks, it may not be as effective as Tensor Flow models for video analysis tasks due to its limited ability to capture temporal dependencies in the data.

Convolutional Neural Networks (CNNs): CNNs are a deep learning model commonly used for image and video analysis tasks. They work by extracting features from the input data using convolutional filters and then

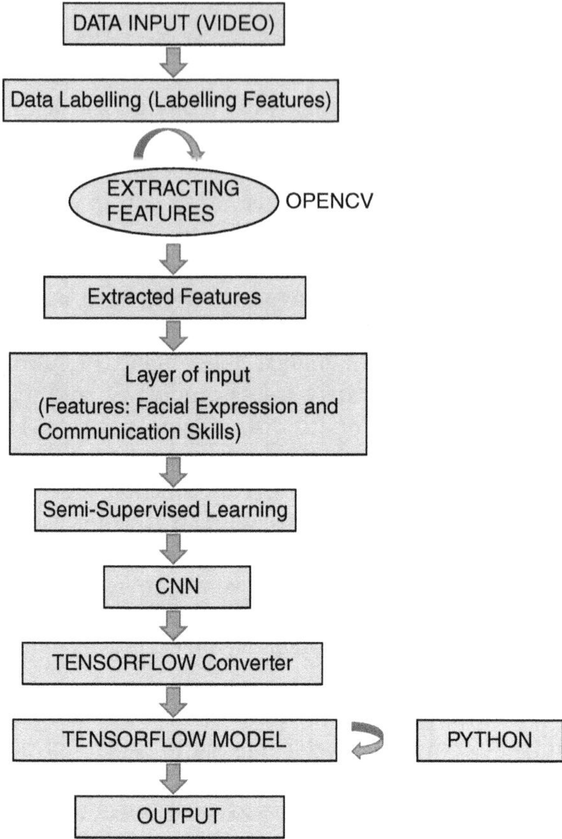

Figure 4.1 Block diagram.

using these features to classify the data. CNNs have been used in the development of intelligent video interview agents with good results, but they can be more computationally expensive than TensorFlow models and require large amounts of training data. The comparison of the two modes of implementation is given in Figure 4.2.

To develop and test a prototype for an intelligent video interview agent using Tensor- Flow, the following steps can be taken: Data Collection: Collect a dataset of video interviews with job candidates. The dataset should be diverse, including candidates from different backgrounds, genders, and ethnicities. The video data should also include a range of emotions and communication styles. Data Pre-processing: Pre-process the video data by segmenting the videos into short clips, labelling the data with relevant metadata, and extracting features from the video using pre-trained models. Model Training: Train a deep neural network using Tensor Flow to predict a candidate's suit- ability for the job based on their video data. The neural network

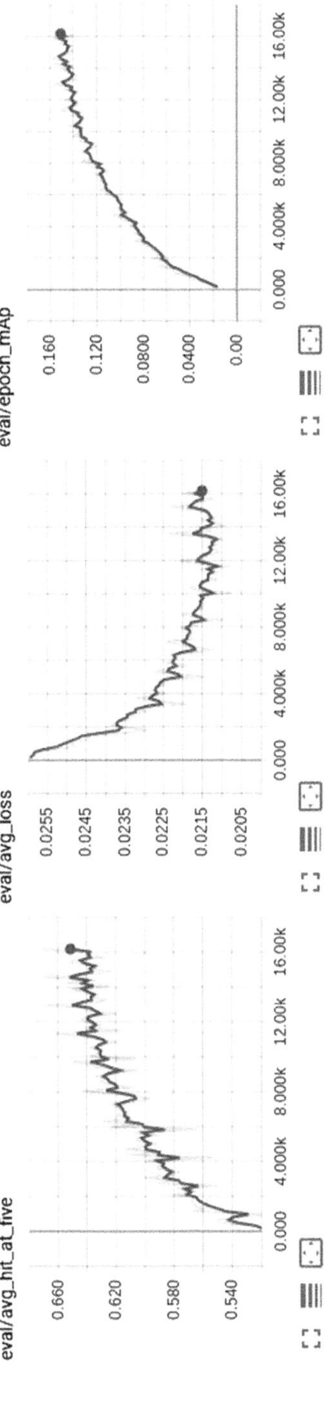

Figure 4.2 CNN vs tensorflow.

should be trained using supervised learning techniques and validated using a separate dataset of video data

Model Optimization: Optimize the neural network by fine-tuning the hyper parameters and testing different architectures to improve the model's accuracy and performance.

Prototype Development: Develop a prototype of the intelligent video interview agent using the optimized neural network. The prototype should be able to analyze video data in real-time and provide a prediction of a candidate's suitability for the job.

Testing and Evaluation: Test the prototype using a separate dataset of video data and evaluate its accuracy and performance. The prototype should be tested for biases and errors to ensure that it is fair and accurate.

4.3.1 Proposed algorithms

The frontend of the application is an interactive HTML/CSS page with the script to connect to the backend Django. The intelligent model is taken care by OpenCV and neural networks. The HTML page consists of the provisions for the student to register and login to their profile thus keeping track of their mock interviews. The user interface interactions in the dashboard is given in Figure 4.3.

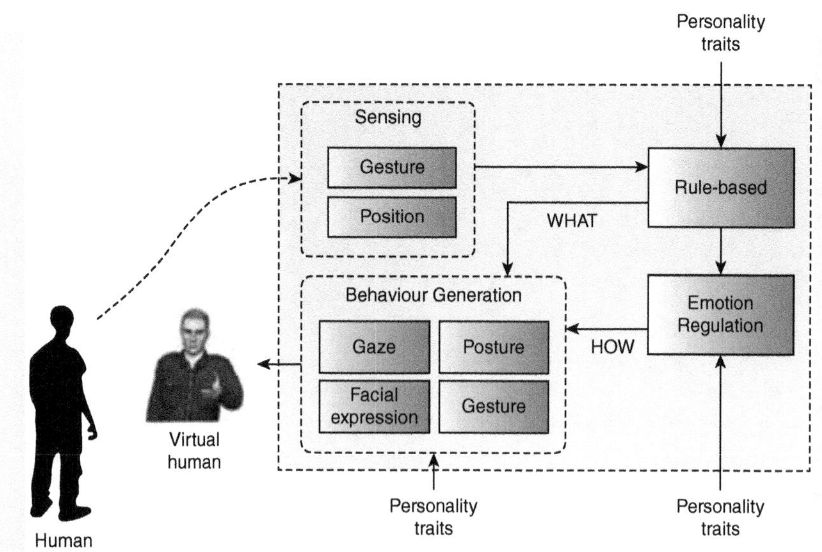

Figure 4.3 User interface model.

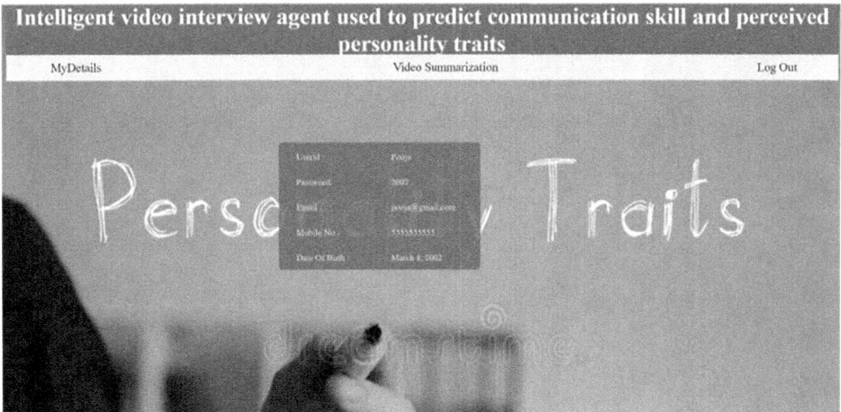

Figure 4.4 Dashboard page.

a. ***Dashboard Setup*:**
   The dash board basic page is set with the following steps. The dashboard setup is given in Figure 4.4.
   1. Set the background of the body to an image.
   2. Define the style for the menu table, including text alignment, width, and margin.
   3. Define the style for the table cells, including background color and padding.
   4. Define the style for the header, including color, background color, font size, and margin.
   5. Define the style for the table cells within the header, including color, text-decoration, font size, and margin.
   6. Define the style for the topic header, including color, padding, text alignment, border style, height, width, and float.
   7. Define the style for the main holder, including position, top, left, z-index, and float.

b. ***The Registration page***
   The registration page is created as follows and the page sample is given in Figure 4.5.
   1. Import necessary libraries and stylesheets.
   2. Set background image for the body.
   3. Define styles for the registration form and its elements.
   4. Create a form with input fields for username, password, email, phone number, and birth date.
   5. Add a submit button for the form.
   6. Import JavaScript libraries for form functionality.

c. ***The Login page is created as follows and the page is given in Figure 4.6.***
   1. Import necessary libraries and stylesheets.
   2. Set background image for the body.
   3. Define styles for the registration form and its elements.

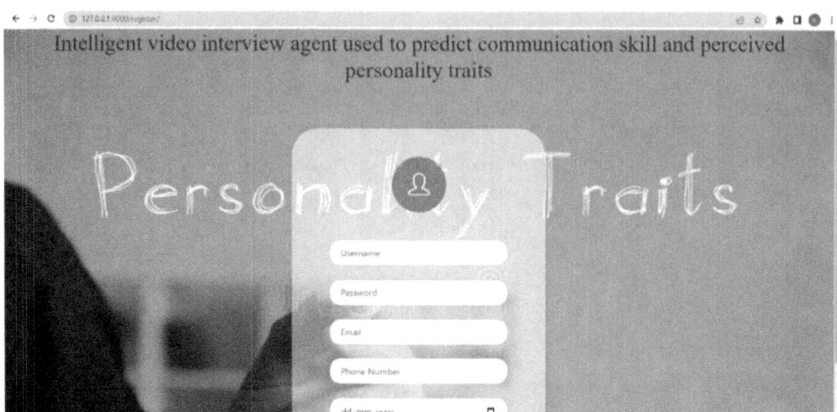

Figure 4.5 Registration page.

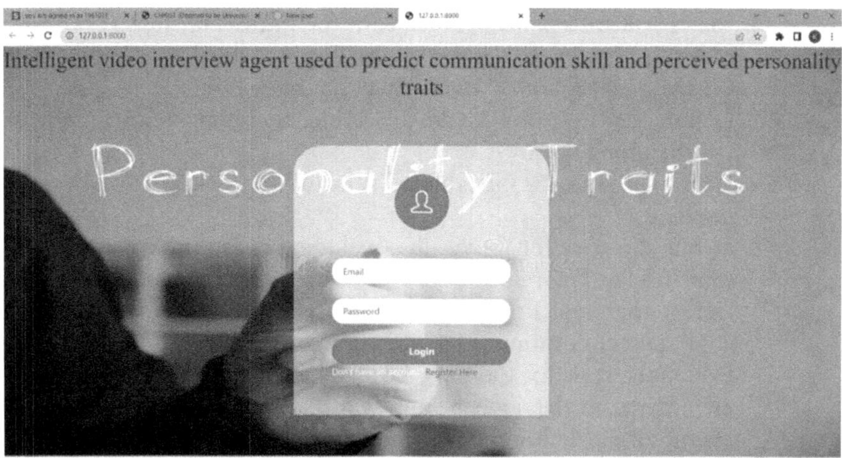

Figure 4.6 Login page.

4. Create a form with input fields for email and password.
5. Add a submit button for the form.
6. Include a link to the registration page.
7. Import JavaScript libraries for form functionality.

d. ***Environment Setup***: The following pseudocode give the Django setup for video capture and summarization.

```
BASE_DIR = ‘give dir’
SECRET_KEY = ‘give key’
DEBUG = True
ALLOWED_HOSTS = []
INSTALLED_APPS [‘list of needed supporting apps’]
```

```
TEMPLATES = [ {
   'BACKEND': ['directories and messages needed']
WSGI_APPLICATION = 'video_personality_trait.wsgi.application'
DATABASES = { 'default': {
   'ENGINE': 'django.db.backends.mysql', 'NAME':
       'video_summarization',
   'USER': 'root', 'PASSWORD': '', 'HOST': 'localhost',  'PORT': '3306',
       }}
AUTH_PASSWORD_VALIDATORS = [  ]
LANGUAGE_CODE = 'en-us'
TIME_ZONE = 'UTC'
USE_I18N = True
USE_L10N = True
USE_TZ = True
STATIC_URL = '/static/'
STATICFILES_DIRS = [os.path.join(BASE_DIR, 'asset/static')]
MEDIA_URL = '/media/'
MEDIA_ROOT = os.path.join(BASE_DIR, 'asset/media')
```

e. ***Processing of Extraction Images***

The following functions were used to perform various image processing tasks such as translation, rotation, resizing, edge detection, and contour extraction.

1. Translate: Translates (shifts) an image by a specified number of pixels along the x and y axes.
2. Rotate: Rotates an image by a specified angle around a specified center point.
3. Rotate_bound: Rotates an image by a specified angle around its center, ensuring the entire rotated image fits within the original image boundaries.
4. Resize: Resizes an image to a specified width and height.
5. Skeletonize: Applies a morphological skeletonization operation to an image.
6. Opencv2matplotlib: Converts an image from opencv's BGR color format to Matplotlib's RGB color format.
7. Url_to_image: Downloads an image from a URL and converts it to a numpy array.
8. Auto_canny: Applies automatic Canny edge detection to an image.
9. Grab_contours: Extracts the contours from an image.
10. Is_cv2: Checks if the current opencv version is 2.x.
11. Is_cv3: Checks if the current opencv version is 3.x.
12. Is_cv4: Checks if the current opencv version is 4.x.
13. Get_opencv_major_version: Returns the major version number of the installed opencv library.

14. Check_opencv_version: Deprecated. Checks if the installed opencv version matches a specified major version number.
15. Build_montages: Converts a list of single images into a list of "montage" images with specified rows and columns.

f. ***Summarization*:** The process of summarization of the video is given as follows.
   1. Select Video: Open a file dialog to select a video file.
   2. Extract Audio: Extract the audio from the selected video file and save it as a WAV file.
   3. Detect Emotions: Use a pre-trained model to detect emotions in the video frames.
   4. Transcribe Audio: Transcribe the audio from the video into text using Google's Speech Recognition API.
   5. Summarize Text: Summarize the transcribed text using an API like Aylien.
   6. Display Results: Display the summarized text along with any other relevant information (e.g., emotion detection results) to the user.

The emotions of the students and the model parameters have been realized in tensor board – an opensource platform. The results are also obtained from the same. AN overall interaction of the components are as shown in Figure 4.7.

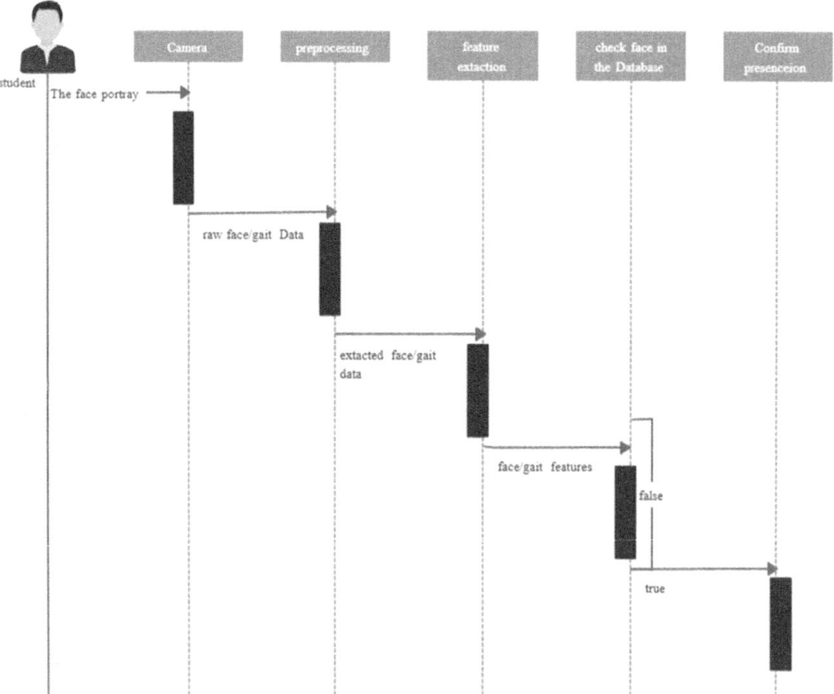

Figure 4.7 Overall interaction of the system.

4.4 RESULTS & ANALYSIS

The results and analysis for an intelligent video interview agent can be done by evaluating its performance on a separate dataset of video data. The following steps can be taken for analyzing the results of the system:

Accuracy Metrics: Calculate the accuracy of the intelligent video interview agent using various metrics such as precision, recall, F1 score, and accuracy.
Confusion Matrix: Construct a confusion matrix to visualize the performance of the system in terms of true positives, false positives, true negatives, and false negatives.
Bias Analysis: Perform bias analysis to ensure that the system is fair and unbiased. Check if the system is performing equally well on candidates from different backgrounds, genders, and ethnicities.
Error Analysis: Conduct an error analysis to identify the errors made by the system and their root causes. This can help in improving the system's performance and accuracy.
User Feedback: Gather feedback from hiring managers and recruiters on the effectiveness and usability of the system.

Figure 4.8 gives the analysis of the person taking up the interview.

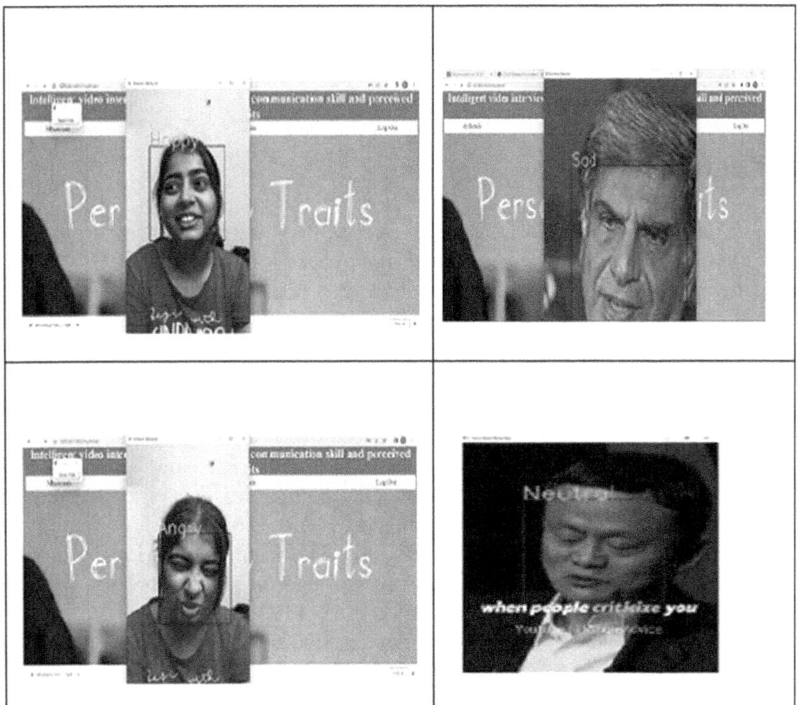

Figure 4.8 Output analysis with different emotions.

4.5 CONCLUSIONS

In conclusion, intelligent video interview agents have the potential to revolutionize the job interview process by providing an automated, efficient, and objective solution. With the advancements in computer vision and machine learning techniques, intelligent video interview agents can analyze candidate responses, facial expressions, and behaviors to predict their suitability for the job. The development of intelligent video interview agents involves various stages such as data collection, preprocessing, feature extraction, model training, and evaluation. The use of open-source libraries like OpenCV and Tensor Flow can significantly reduce the development time and cost. While intelligent video interview agents have many advantages, there are also some limitations to be considered. Moreover, it is important to ensure that candidates are informed about the use of intelligent video interview agents and given the option to opt-out if they feel uncomfortable. Intelligent video interview agents have the potential to streamline the hiring process and provide valuable insights into candidate suitability.

ACKNOWLEDGEMENTS

The authors are grateful for the facilities provided by CHRIST (Deemed to be University), Bengaluru and Bharath Institute of Higher Education and Research (Deemed to be University), Chennai for the facilities offered to carry out this work.

REFERENCES

[1] Will, P., Krpan, D., and Lordan, G. (2023). People versus machines: introducing the HIRE framework. *Artificial Intelligence Review*, 56(2), 1071–1100.
[2] Mhadgut, S., Koppikar, N., Chouhan, N., Nalawade, R., and Dhamdhere, D. (2022). An automated smart recruitment webapp using machine learning. In *Proceedings of the International Conference on Innovative Trends in Computer Engineering (ITCE)*, IIIT Kottayam (pp. 263–268). IEEE.
[3] Zhang, L., and Yencha, C. (2022). Examining perceptions towards hiring algorithms. *Technology in Society*, 70, 101809.
[4] Kim, J. Y., and Heo, W. G. (2021). Artificial intelligence video interviewing for employment: Perspectives from applicants, companies, developer and academicians. *Information Technology and People*, 34(5), 1632–1652.
[5] Kammerer, B. (2021). Hired by a robot: The legal implications of artificial intelligence video interviews and advocating for greater protection of job applicants. *Iowa Law Review*, 106(3), 1133–1173.
[6] Ajunwa, I. (2021). Automated video interviewing as the new phrenology. *Berkeley Technology Law Journal*, 36(1), 293–326.
[7] Suen, H. Y., Chen, M. Y. C., and Lu, S. H. (2019). Does the use of synchrony and artificial intelligence in video interviews affect interview ratings and applicant attitudes. *Computers in Human Behavior*, 98, 202–213.

[8] Burke, S. L., Bresnahan, T., Li, T., Epnere, K., Rizzo, A., and D'Cruz, D. (2018). Using virtual interactive training agents (ViTA) with adults with autism and other developmental disabilities. *Journal of Autism and Developmental Disorders*, 48(5), 1470–1483.

[9] Zavattaro, S. M., French, P. E., and Mohanty, S. D. (2015). A sentiment analysis of US local government tweets: The connection between tone and citizen involvement. *Government Information Quarterly*, 32(4), 441–452.

Chapter 5

Visualizing food quality and safety

A dynamic dashboard approach with near-infrared imaging and machine learning

Brighty Ebenezer L, Sasithradevi A, Chanthini Baskar, and Kanimozhi S

Vellore Institute of Technology, Chennai, India

5.1 INTRODUCTION

In today's world, consumers demand not only delicious and nutritious food but also complete assurance of its quality and safety. This necessitates robust and efficient methods for food analysis throughout the supply chain, from farm to fork. Traditional methods, while valuable, often have limitations, being time-consuming, destructive, and labor-intensive. Near-infrared (NIR) imaging has emerged as a game-changer in this domain, offering a non-invasive, rapid, and cost-effective solution for visualizing and analyzing food quality. This chapter delves into the transformative power of NIR imaging, exploring its potential to revolutionize food quality control practices. Also, we explore how NIR imaging, unlike its counterpart NIR spectroscopy which directly quantifies components, leverages the intriguing interplay between light and matter to generate visual representations that are highly sensitive to chemical variations within food samples. This innovative technique empowers the food industry to make informed and real-time decisions regarding food quality at various critical stages of the supply chain. Furthermore, we discuss the core principles underlying NIR imaging, demystifying how it sheds light on crucial compositional parameters like moisture content, protein and fat levels, and the presence of carbohydrates. However, the potential of NIR imaging extends far beyond simply assessing inherent quality. We discover how it excels at detecting and identifying contaminants and adulterants, safeguarding consumer health. Additionally, NIR imaging plays a vital role in verifying the geographical origin of food products, fostering transparency and trust within the food system. Also, we delve into the exciting realm of real-time monitoring capabilities offered by NIR imaging, enabling proactive interventions along the supply chain to ensure product integrity and safety. To further enhance the power of NIR imaging, we explore its synergy with machine learning algorithms, tackling the complexities of data analysis and leading to more accurate and robust quality assessments.

DOI: 10.1201/9781003542735-5

While acknowledging the challenges associated with initial investment costs and the need for specialized expertize, we emphasize the continuous advancements being made in NIR imaging technology. Looking towards the future, we explore exciting opportunities for integrating NIR imaging with other cutting-edge technologies, paving the way for a dynamic dashboard approach. This approach promises to transform food quality control by offering real-time, interactive visualizations to make data-driven decisions and ensure a safer, more efficient, and transparent food system for all.

5.2 RELATED WORKS

In recent years, there has been a growing demand for accurate and rapid analytical methods to ensure the quality and safety of food products. Traditional methods are often time-consuming and expensive, prompting the need for alternative approaches that offer quick results, simplicity, and cost-effectiveness. This literature survey explores various analytical techniques, including near-infrared spectroscopy (NIR), hyperspectral imaging (HSI), and fluorescent probes, along with their applications in assessing the nutritional content, adulteration, cultivation practices, and safety of food products. A study focused on Cape gooseberries highlighted the potential of NIR-HSI coupled with chemometrics for predicting key nutritional parameters such as titratable acidity (TA), vitamin C content, firmness, and soluble solids content (SSC). The study showcased the efficacy of Partial Least Square Regression (PLSR) and Support Vector Machine Regression (SVMR) in precisely forecasting these variables, pinpointing the lower portion of berries as the prime area for image capturing. This method exhibits potential in forecasting the physicochemical composition of Cape gooseberries, providing valuable insights into their nutritional content [1]. Another review article discussed the importance of rapid and precise analytical methods in detecting adulteration in various food matrices. NIR spectroscopy and HIS (HyperSpectral Imaging) emerged as economically preferred technologies due to their simplicity, rapid results, and non-destructive nature. The review focused on applications of these techniques in detecting adulteration in fruit juices, edible oils, spices, dairy products, honey, and infant formula, highlighting their potential to enhance food quality control measures in the industry [2]. Another investigation delved into the potential of non-invasive optical methodologies, such as HSI and Fourier Transform (FT)-NIR spectroscopy, for categorizing sustainably produced tomatoes based on growing practices and water use efficiency (WUE) [3]. The study utilized partial least squares discriminant analysis (PLS-DA) to distinguish between various cultivation systems and levels of Water Use Efficiency (WUE) and Partial Factor Productivity of Nutrients (PFP). The results showcased the efficacy of these methods in distinguishing tomato fruits cultivated under different water and fertilizer levels, underscoring their potential to facilitate the implementation of low-input growing practices. To achieve real-time detection of defective apples within fruit sorting machines, NIR cameras were employed in conjunction with the YOLO V4 deep learning algorithm. By employing channel and layer pruning techniques, the pruned YOLO V4 network demonstrates reduced model size and inference time, while achieving improved defect detection accuracy, making it suitable for commercial fruit packing lines [4]. Another research endeavor concentrated on crafting a mitochondrial-targeted near-infrared fluorescent probe utilizing cyanine derivatives (PC-H2S) to detect hydrogen sulfide (H2S) in both food and biological environments. The probe exhibited a strong fluorescent response upon reacting with H2S, enabling the detection of H2S generated during food spoilage. Moreover, PC-H2S showcased remarkable mitochondrial targeting capabilities and proved to be effective in imaging living cells and zebrafish, showcasing its potential for physiological and pathological research in food safety [5]. To conduct non-destructive analysis of powdered chicken egg samples, the authors introduced an approach utilizing digital images, NIR spectra, and multivariate calibration methods for the determination of protein content, moisture content, and phosphorus contents. The study demonstrated the feasibility of predicting these parameters accurately using digital images and NIR spectra, offering a non-laborious and cost-effective alternative to traditional chemical analysis methods. The techniques align with the principles of Green Chemistry, emphasizing sustainability and reduced environmental impact [6]. A groundbreaking near-infrared (NIR) emission fluorescent probe (Dpyt) designed for ultrafast detection of bisulfate and organic amines in food samples was introduced by the authors. Dpyt exhibited distinct color changes upon reacting with these analytes, enabling visual detection by the naked eye. The probe showed encouraging potential for use in imaging pollutants in living cells and detecting food contaminants such as sugar and red wine bisulphate. Moreover, indicator labels constructed with Dpyt showed promise in real-time monitoring of seafood freshness, offering a practical tool for food quality assessment [7]. Advancements in analytical techniques, including NIR spectroscopy, hyperspectral imaging, fluorescent probes, and nanoprobes, hold great promise for ensuring food quality and safety. These techniques offer rapid, non-destructive, and cost-effective solutions for assessing nutritional content, detecting adulteration, monitoring cultivation practices, and identifying contaminants in food products. Further research and validation are essential to harness the full potential of these technologies and support their widespread application in the food industry. Table 5.1 displays a summary of the related studies visualized using the power BI.

Table 5.1 An overview of NIR imaging and its applications

Application	*Methodology*	*NIR camera*	*References*
Application of visible near-infrared spectrum for the detection of adulterated beef.	LS-SVM with IWO, CARS, and GA for wavelength selection in the 496–1000 nm spectral region.	Vis-NIR	[8]
Bacterial foodborne pathogen classification using agar plates.	Channel and layer pruning techniques, non-maximum suppression based on L1 norm, optimal wavelength selection techniques like CARS and GA.	NIR	[9]
Bisulfate and organic amines detection in food and living cells	NIR emission fluorescent probe (Dpyt). Probe exhibits color change and fluorescence response upon detection.	NIR, Fluorescence Spectroscopy	[7]
Cape gooseberry quality assessment	Utilization of chemometrics (PLSR and SVMR) in conjunction with near-infrared hyperspectral imaging (NIR-HSI) to forecast physicochemical attributes (vitamin C concentration, hardness, SSC, TA).	NIR-HSI	[1]
Detection of complete bruise regions on apples for quality assessment.	PCA for data compression, RF algorithm for classification, selection of characteristic wavebands, and ROI-based extraction models.	HSI	[10]
Determining the moisture and protein/ phosphorus content of powdered chicken egg samples	Digital images and NIR. Multivariate calibration methods employed for analysis.	NIR Spectroscopy, Digital Imaging	[11]
Food adulteration detection	Utilization of hyperspectral imaging (HSI) and NIR spectroscopy. Chemometrics methods employed for analysis.	NIR Spectroscopy, Hyperspectral Imaging	[2]
Hydrogen sulfide detection in food and biological systems	PC-H2S is a near-infrared fluorescent probe that targets mitochondria. Zebrafish and live cells were used for endogenous and exogenous H2S imaging.	NIR, Fluorescence Spectroscopy	[5]

(Continued)

Table 5.1 (Continued)

Application	*Methodology*	*NIR camera*	*References*
Nitrosamine-induced acute liver injury detection	Activatable probe (BHC-Lut). Probe is capable of NIR II fluorescence and optoacoustic imaging when activated by the hydrogen peroxide.	NIR, Fluorescence Spectroscopy	[12]
Real-time detection of defective apples in fruit sorting	NIR cameras and YOLO V4 deep learning algorithm	NIR	[4]

5.3 NIR IMAGING: A NON-INVASIVE AND INSIGHTFUL TECHNIQUE

NIR imaging utilizes the electromagnetic spectrum's near-infrared range (780-2500 nm) to analyze food samples. Unlike traditional methods, which can be destructive and time-consuming, NIR imaging is non-invasive and rapid. It works by capturing the interactions between light and food molecules at specific wavelengths, generating a unique "spectral fingerprint" for each sample. This fingerprint reveals valuable information about the food's chemical composition, including:

Moisture content: Crucial for determining freshness and shelf-life, especially in fruits and vegetables.
Protein and fat content: Essential markers in meat, dairy, and other products.
Carbohydrate levels: Influence taste and texture, particularly in fruits and vegetables.

NIR imaging goes beyond just measuring these core components. It excels at:

Detecting contaminants and adulterants: Foreign objects, microbial pathogens, and chemical adulterants possess distinct spectral signatures, allowing NIR imaging to identify them and safeguard consumer health.
Verifying geographical origin: By analyzing unique chemical fingerprints, NIR imaging can help verify the geographical origin of high-value foods, protecting consumers from fraud.
Real-time monitoring: Integrating NIR imaging into production lines enables real-time quality inspections, facilitating corrective actions, minimizing waste, and ensuring product consistency. The utilization of NIR imaging within the food industry is underscored in the dynamic dashboard environment illustrated in Figure 5.1. This dashboard, crafted using smart art features, showcases the multifaceted applications of NIR imaging, ranging from quality control to compositional analysis.

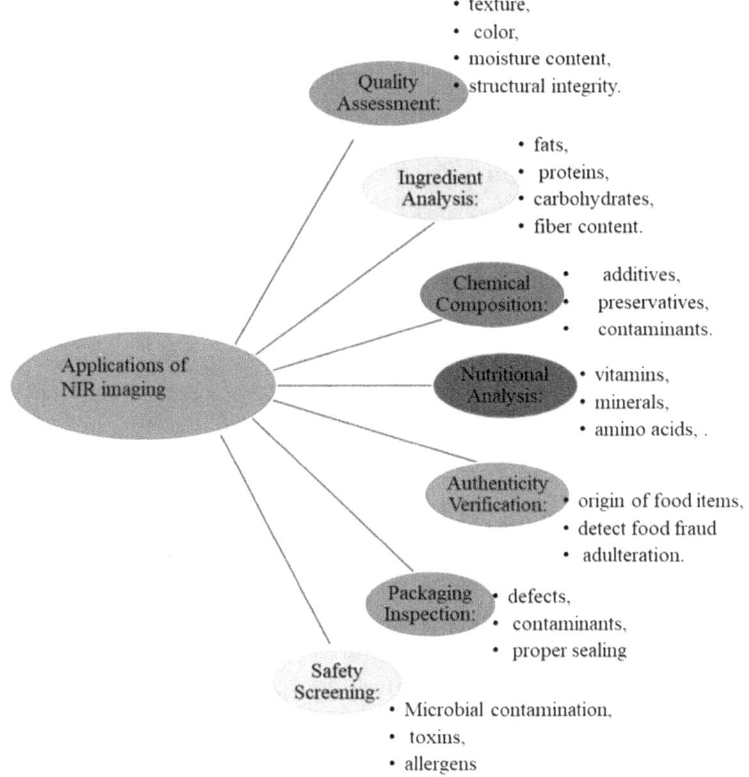

Figure 5.1 Applications of NIR imaging.

The technologies of NIR (Near-Infrared) spectroscopy and NIR imaging are closely connected that harness the power of near-infrared light to analyze materials. However, they differ fundamentally in how they collect and present information. NIR spectroscopy focuses on measuring the absorption of near-infrared light by a sample as a function of wavelength. This results in a spectrum – a graph showing variations in absorption across different wavelengths of the NIR region. This spectrum acts as a unique chemical fingerprint, revealing the composition of the sample (e.g., levels of moisture, fat, and protein). In contrast, NIR imaging captures spatial information by forming images using reflected or transmitted near-infrared light. Each pixel in the image has its own spectrum, allowing for both chemical analysis and visualization of how those chemical components are distributed over the sample's surface. Essentially, if NIR spectroscopy tells us what is in a sample, NIR imaging tells you both what is there and where it is located. The comparative display of a sample NIR image alongside its RGB counterpart is showcased within the interactive dashboard environment, as depicted in Figure 5.2.

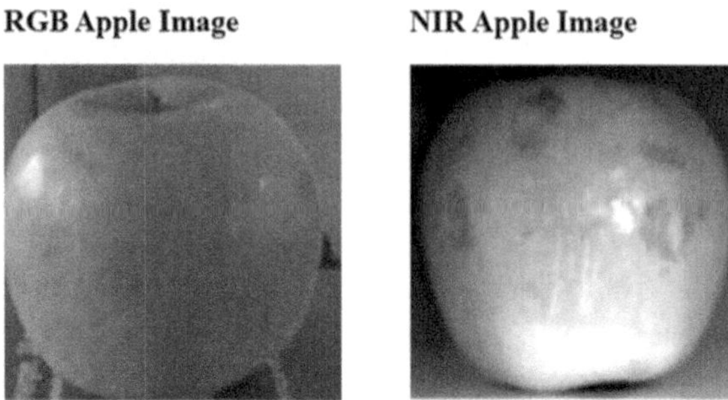

Figure 5.2 Sample of RGB image with NIR image.

5.3.1 Wavelengths of NIR imaging

The Near-Infrared spectrum lies just beyond what the human eye can see and is generally divided into four ranges: NIR, SWIR, MWIR, and LWIR. Near-infrared (NIR) light refers to electromagnetic radiation with wavelengths ranging from 0.75 to 1.4 microns (750 to 1400 nm). It is extensively utilized in machine vision, security systems, and certain medical applications due to its ability to penetrate various materials and provide detailed imaging capabilities. Short-Wavelength Infrared (SWIR) light spans from 1.4 to 3 microns (1400 to 3000 nm) and finds applications in moisture detection, telecommunications, and specialized material analysis, leveraging its ability to interact with molecular vibrations. Mid-Wavelength Infrared (MWIR) light, with wavelengths between 3 and 8 microns (3000 to 8000 nm), offers unique advantages in thermal imaging applications. Long-Wavelength Infrared (LWIR) light, ranging from 8 to 15 microns (8000 to 15000 nm), is primarily associated with thermal imaging due to its ability to detect thermal radiation emitted by objects, making it invaluable for night vision and environmental monitoring [13]. Each segment of the infrared spectrum serves distinct purposes across various industries, ranging from imaging and communication to scientific research and surveillance, and are showcased in Figure 5.3.

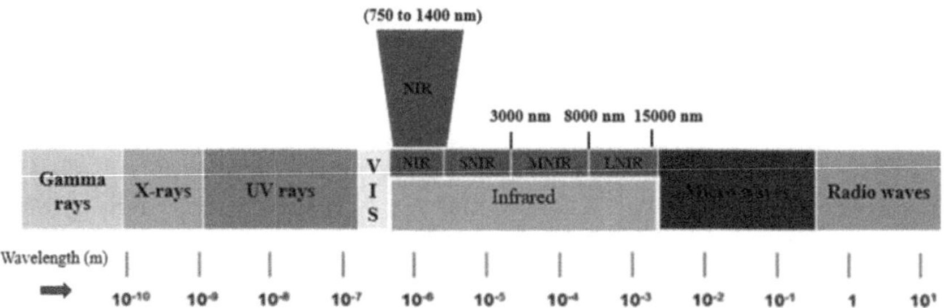

Figure 5.3 Wavelengths of NIR imaging.

5.3.2 Techniques in NIR imaging

Near-Infrared (NIR) imaging encompasses various techniques that utilize NIR light for imaging and analysis. The types of NIR imaging include:

Reflective NIR imaging

Reflective NIR imaging functions similarly to how we perceive objects with visible light. In this method, NIR light is emitted onto an object, and a specialized NIR-sensitive camera captures the reflected NIR light. This approach is commonly used in applications such as surveillance, material sorting, and quality control. Reflective NIR imaging enables the detection of surface features and characteristics that may not be visible to the naked eye, providing valuable insights in industrial and security settings.

Hyperspectral NIR imaging

Hyperspectral NIR imaging involves capturing images across a broad range of the NIR spectrum, and sometimes beyond. Unlike traditional imaging methods that produce visual images, hyperspectral imaging provides a detailed "spectral signature" of objects. This allows for far more precise material identification and analysis of chemical composition. Hyperspectral NIR imaging finds applications in fields such as agriculture, environmental monitoring, and biomedical research, where accurate material characterization is essential for understanding complex systems.

NIR spectroscopy

NIR spectroscopy focuses on analyzing the specific wavelengths of NIR light absorbed and reflected by an object. This technique involves studying how different materials interact with NIR radiation to determine their chemical composition. NIR spectroscopy is widely used in laboratories, industrial settings, and field applications to analyze biological samples, assess plant health, and identify substances based on their spectral properties. It plays a crucial role in research, quality control, and process optimization across various industries.

5.3.3 NIR imaging system

The hardware components of a Near-Infrared (NIR) system vary depending on its specific application and complexity but typically include a Light Source, Beam Shaper (Optional), Sample Presentation Unit, NIR Camera, Data Acquisition System, and computer for processing the captured data, and generating results. The depiction of a standard NIR imaging system is exemplified in Figure 5.4, ingeniously illustrated within a dynamic dashboard environment utilizing smart art features.

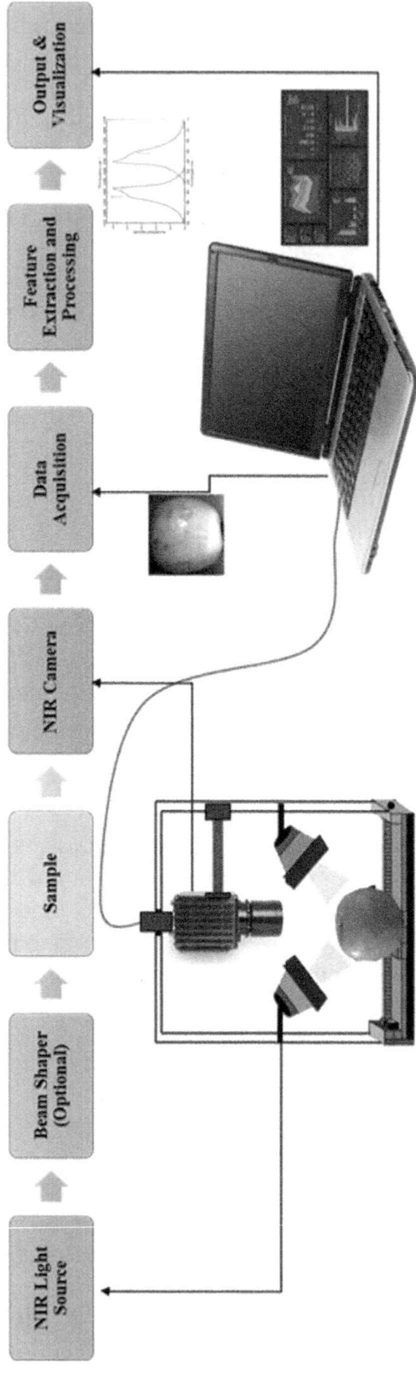

Figure 5.4 NIR imaging system.

NIR light source

The NIR light source is a crucial component that emits near-infrared light to illuminate the food sample being analyzed. Depending on the specific application, different types of light sources are utilized. Tungsten halogen lamps are a common and affordable option, although they may generate unwanted heat. LED lamps offer better energy efficiency, precise control over the emitted light spectrum, and generate less heat compared to tungsten halogen lamps. Laser sources provide highly focused and collimated light, making them ideal for applications requiring precise illumination.

Camera

The camera in an NIR imaging system captures the near-infrared light that is reflected or transmitted by the food sample. Specialized NIR-sensitive cameras are used as they are sensitive to the longer wavelengths of the near-infrared spectrum, unlike standard cameras. There are two main types of NIR cameras used: InGaAs (Indium Gallium Arsenide) cameras, which offer high sensitivity in the near-infrared range but are typically more expensive, and SWIR (Short-wave Infrared) cameras, which are less expensive but have lower sensitivity compared to InGaAs cameras.

Spectrometer (Hyperspectral systems only)

In hyperspectral imaging systems, a spectrometer is a critical component that separates the captured light into its individual wavelengths, similar to how a prism separates visible light into a rainbow. This creates a unique "spectrum" for each pixel in the image, revealing the spectral fingerprint of the material at that point. By analyzing these spectra, the system can identify the chemical composition of different components within the food sample, providing detailed analysis capabilities.

Computer and software

The computer and software serve as the brain of the NIR imaging system, responsible for processing images, analyzing spectral data, and interpreting the results. The software performs various tasks, including image processing (such as noise reduction and enhancement), spectral analysis (extracting information from individual spectra or comparing spectra to reference databases), and utilizing machine learning algorithms to classify food products, detect contaminants, or predict ripeness based on learned models.

Calibration of the entire system is essential to ensure accurate and reliable measurements. This involves periodic calibration using reference materials with known properties to adjust the system's response. The software often includes functionalities for data visualization, allowing users to visualize results in various ways, such as displaying raw images, spectral plots, or color-coded maps representing different chemical components within the food sample. By understanding the function of each component and their interaction, NIR imaging systems offer a powerful tool to assess food quality and safety effectively.

5.4 A SYMBIOTIC RELATIONSHIP: NIR IMAGING AND MACHINE LEARNING

Machine learning unlocks the full potential of NIR data by enabling advanced image analysis, object classification, and predictive modeling. NIR imaging, in turn, provides a unique, non-invasive data source with hidden information about molecular composition, perfect for training and improving machine learning models. Together, they hold immense potential in diverse fields like precision agriculture, food quality, medical diagnostics, and material science, offering valuable insights and driving innovation. Machine learning can further enhance the vast datasets generated by NIR imaging systems. By training algorithms on known samples, machine learning can significantly improve the accuracy and robustness of quality predictions and classifications based on NIR imaging data. This empowers researchers to:

- Automate routine quality control tasks, freeing up valuable resources for other critical activities.
- Gain deeper insights from complex data patterns, potentially identifying emerging issues or optimizing processes for enhanced quality.

NIR imaging, combined with machine learning techniques, has revolutionized various fields by enabling the extraction of valuable insights from complex data. Deep learning models, including Generative Adversarial Networks (GANs) and Convolutional Neural Networks (CNNs), excel at tasks like object detection, classification, and segmentation within NIR images. These models are particularly useful for applications like identifying diseased plant leaves or segmenting tumors in medical imaging. Non-deep learning models like Support Vector Machines (SVMs) and Random Forests are efficient for tasks such as material classification and regression analysis, providing robust performance and interpretability. Additionally, techniques like Principal Component Analysis (PCA) and K-Means clustering offer dimensionality reduction and clustering capabilities, respectively, further enhancing the analysis of NIR data. Several studies are there utilizing advanced techniques, such as machine learning techniques paired with NIR and hyperspectral imaging, for various applications. One such study focused on the classification of bacterial pathogens and demonstrated the value of meta-heuristic optimization algorithms and optimal wavelength selection strategies in raising classification accuracy, especially when paired with Support Vector Machine (SVM) and Linear Discriminant Analysis (LDA) models. [9]. Another research addressed beef adulteration detection, highlighting the success of multivariate statistical analysis methods coupled with wavelength selection techniques like competitive adaptive reweighted sampling (CARS), invasive weed optimization (IWO), and genetic algorithm (GA) in simplifying models and achieving high prediction accuracy [8]. Moreover, another research delves into soil spatial sampling evaluation for digital soil mapping (DSM) models, emphasizing the impact of sampling characteristics on prediction performances and uncertainty estimation. The study explores various spatial sampling sizes, distributions, and evaluation methods to optimize sampling strategies and improve DSM model reliability [14]. The task of apple bruise extraction using hyperspectral imaging and Principal Component Analysis (PCA) were employed. Random Forest (RF) algorithm was utilized for classification. The study identified characteristic wavebands related to bruise regions, demonstrating high accuracy in bruise extraction and potential for enhancing apple grading and sorting efficiency [10]. Similarly, another study focused on detecting fungal contamination in maize kernels using hyperspectral imaging (HSI). Using Artificial Neural Networks (ANN), 1D-Convolutional Neural Networks (1D-CNN), and Partial Least Square-Discriminant Analysis (PLS-DA) models, the research evaluates prediction capabilities based on the orientation of maize germ relative to the HSI lens. Results indicate improved prediction accuracy, particularly with germ-up orientation, offering a promising method for fungal contamination detection in maize kernels [15]. In the realm of water pollution detection, an improved convolutional neural network (CNN) architecture for near-infrared (NIR) spectroscopy-based assessment was introduced. Incorporating a shallow design with a decision tree algorithm, this CNN model enhances prediction accuracy for detecting water pollution levels rapidly. Its adaptability to dynamic data and large datasets makes it a valuable tool for assessing agricultural water resources [16]. Another study explored NIR spectroscopy-based fruit quality assessment, employing domain-invariant partial least squares (di-PLS) regression to maintain model performance under varying conditions. By allowing unsupervised adaptation to new conditions without reference measurements, di-PLS regression enhances prediction accuracy and compensates for instrumental and environmental changes. This approach showed promise for ensuring consistent fruit quality assessment across different scenarios [17]. Furthermore, the rapid prediction of spoilage bacteria in chicken flesh using high-resolution NIR spectroscopy utilized partial least squares (PLS) algorithm and pre-processing methods including multiplicative scatter correction, standard normal variate, and Savitzky-Golay convolution smoothing. The study achieved high prediction accuracy for total counts of spoilage bacteria. This non-destructive method holds potential for enhancing food safety in poultry products [6]. Platforms for estimating plant nitrogen status in potato crops, using parametric and random forest regression algorithms was done by [18]. Results demonstrated the superiority of UAV data, with random forest regression showing robustness and high accuracy in predicting plant nitrogen status, offering valuable insights for precision nutrient management in agriculture. The concise overview of Machine Learning algorithms for NIR imaging is presented in Table 5.2, elegantly displayed via Power BI. The most appropriate model will be selected based on the particular task at hand, the properties of the data, and the available computational resources. By leveraging these diverse machine learning techniques, researchers and professionals can unlock the full potential of NIR imaging data across a wide range of applications.

5.5 INTERACTIVE DASHBOARDS: TRANSFORMING DATA INTO ACTIONABLE INSIGHTS

While NIR imaging generates valuable data, its true potential is unlocked when combined with interactive dashboards. Imagine a dynamic platform displaying real-time NIR imaging data from various stages of the food supply chain. This dashboard would allow to:

- Visualize real-time quality parameters like moisture, protein, and fat content using color-coded heat maps or 3D models.
- Track trends over time and identify potential quality issues before they escalate.
- Compare data from different batches or production lines, facilitating process optimization and informed decision-making.
- Drill into specific data points to gain granular details and investigate potential anomalies.

5.5.1 Tools for creating a dashboard

For visualizing food quality and safety utilizing NIR imaging, a comprehensive dashboard can be crafted either through Python's powerful visualization libraries or within MATLAB's GUI environment, providing insightful representations of critical parameters for quality assurance and risk mitigation in food products.

5.5.1.1 Creating dashboards with python

Dash (by Plotly): Dash is a robust framework for building web-based dashboards in Python. It seamlessly integrates with Plotly for creating interactive visualizations and offers straightforward deployment options. With Dash, we can define the layout of your dashboard using Python code and incorporate various interactive components such as graphs, dropdowns, and sliders.

Table 5.2 Machine learning algorithms for NIR imaging in food quality and safety

Category	*Algorithm*	*Description*	*Applications*	*References*
Classification	Support Vector Machines (SVMs)	Efficiently classifies food based on NIR spectral data.	• Classification of bacterial foodborne pathogens on agar plates. • Detection of beef adulteration	[8, 9]
	Random Forests	Robust to overfitting and interpretable, good for feature selection.	• Evaluation and optimization of soil spatial sampling for DSM models. • Detection of complete bruise regions on apples.	[10, 14]
	Convolutional Neural Networks (CNNs)	Particularly adept at image recognition and classification.	• Detection of fungal contamination in maize kernels. • Rapid detection of water pollution for agricultural irrigation.	[15, 16]
Regression	Partial Least Squares Regression (PLSR)	Relates NIR spectral data to specific quality parameters.	• Fruit quality assessment. • Detection of spoilage bacteria in chicken flesh	[6, 17]
	Random Forests (Regression)	Can predict continuous variables based on NIR data.	• Assessment of potato nitrogen status	[18]
Unsupervised learning	**K-Means Clustering**	Groups similar NIR images or data points based on their features.	• Characterization of pulse flours for food industry. • Precise oil content estimation in olives	[19, 20]

In this context, the dashboard created using Dash likely serves as a tool for displaying and interacting with data in a user-friendly manner. Figure 5.5 represents the screenshot showcasing how this dashboard looks, including its layout, visualizations, and any interactive elements. Following the execution of feature extraction, which involves deriving relevant information or features from the data, the output is displayed. Figure 5.6 provides a visual representation of the output, illustrating the output prediction and relation between the extracted features. This includes graphs, charts, or other visualizations that highlight the extracted features and the output prediction.

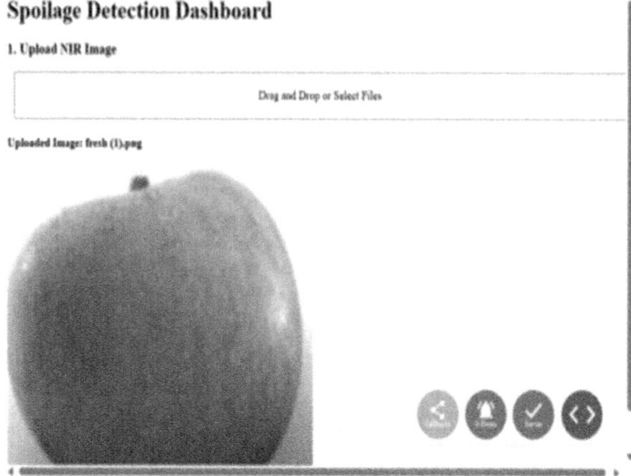

Figure 5.5 Our new dashboard for apple spoilage detection.

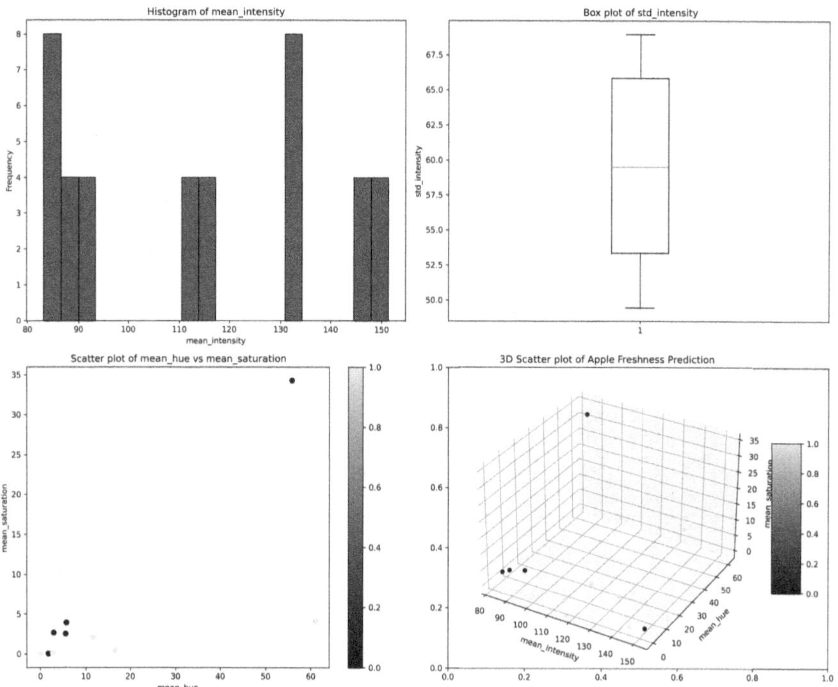

Figure 5.6 Visual representation of output prediction and the relation between the extracted features.

Streamlit: Streamlit is a lightweight framework designed for rapid development of data apps and dashboards. It simplifies the process of building interactive web applications in Python, allowing users to create dashboards without needing a separate server. Streamlit's simple syntax and intuitive interface make it ideal for prototyping and sharing data-driven insights.

Panel/Hvplot: Panel and Hvplot are libraries that enable users to create interactive visualizations and dashboards directly within Jupyter Notebooks. Panel combines the strengths of Plotly and Bokeh to provide a versatile platform for building customizable dashboards. Hvplot, on the other hand, offers a high-level interface for quickly creating interactive plots using Panda's data frames.

5.5.1.2 Creating dashboards with MATLAB

MATLAB App Designer: MATLAB App Designer is a built-in tool for creating graphical user interfaces (GUIs) with interactive elements. While primarily not designed for pure dashboards, it allows users to incorporate various visualization tools like charts, gauges, and tables into their GUI applications. With MATLAB App Designer, you can design and deploy simple dashboards directly within the MATLAB environment.

GUIDE (Graphical User Interface Design Environment): GUIDE is a more advanced toolbox in MATLAB for building custom GUI applications, including dashboards. While it offers greater flexibility and control over the design and functionality of the dashboard, it requires a deeper understanding of MATLAB programming and UI design principles. Users can leverage GUIDE to create sophisticated dashboards with interactive elements and dynamic data visualization capabilities.

Creating dashboards with Python and MATLAB involves importing data, designing the layout, adding visualizations, and optionally implementing interactivity features. Python libraries like Dash, Streamlit, and Panel/Hvplot offer intuitive ways to create web-based dashboards, while MATLAB provides built-in tools like App Designer and GUIDE for building GUI-based dashboards directly within the MATLAB environment.

5.5.2 Revolutionizing the food industry: The road ahead

While NIR imaging technology holds immense potential, there are still challenges to address. Initial investment costs and the need for specialized expertize in data analysis can be hurdles. However, continuous advancements are making NIR imaging more accessible and cost-effective.

Looking ahead, the future lies in integrating NIR imaging with other technologies like hyperspectral imaging, 3D imaging, and sensor fusion. This combined approach will provide even greater depth and precision in food quality analysis. Furthermore, the synergy between NIR imaging, interactive dashboards, and machine learning promises to revolutionize food quality control by offering:

- Real-time, data-driven decision-making for stakeholders across the food supply chain.
- Enhanced transparency and trust within the food system.
- A safer and more efficient food system for consumers worldwide.
- By embracing the potential of NIR imaging and its integration with interactive dashboards and machine learning, the food industry can take a significant leap forward, ensuring consistent quality, safety, and transparency for consumers, from farm to fork.

5.6 CONCLUSION

In conclusion, NIR imaging stands as a powerful and transformative technology for ensuring food quality and safety throughout the supply chain. Its non-destructive, rapid, and cost-effective nature empowers stakeholders across the food industry to make informed decisions at every stage. By offering insights into key compositional parameters, detecting contaminants, and enabling real-time monitoring, NIR imaging safeguards consumer health and product integrity. Additionally, its integration with machine learning algorithms further enhances the accuracy and robustness of quality assessments. While challenges like initial investment costs and data analysis expertize exist, continuous advancements are streamlining the technology. As we explore the integration of NIR imaging with other technologies within dynamic dashboards, we pave the way for a safer, more efficient, and transparent food system for the future.

REFERENCES

[1] R. Blas Saavedra, J. P. Cruz-Tirado, H. M. Figueroa-Avalos, D. F. Barbin, J. M. Amigo, and R. Siche, "Prediction of physicochemical properties of cape gooseberry (Physalis peruviana L.) using near infrared hyperspectral imaging (NIR-HSI)," *J. Food Eng.*, vol. 371, no. November 2023, 2024, https://doi.org/10.1016/j.jfoodeng.2024.111991

[2] H. Nobari Moghaddam, Z. Tamiji, M. Akbari Lakeh, M. R. Khoshayand, and M. Haji Mahmoodi, "Multivariate analysis of food fraud: A review of NIR based instruments in tandem with chemometrics," *J. Food Compos. Anal.*, vol. 107, no. April 2021, p. 104343, 2022, https://doi.org/10.1016/j.jfca.2021.104343

[3] H. Fazayeli et al., "Potential application of hyperspectral imaging and FT-NIR spectroscopy for discrimination of soilless tomato according to growing techniques, water use efficiency and fertilizer productivity," *Sci. Hortic. (Amsterdam).*, vol. 328, no. August 2023, p. 112928, 2024, 10.1016/j.scienta.2024.112928

[4] S. Fan et al., "Real-time defects detection for apple sorting using NIR cameras with pruning-based YOLOV4 network," *Comput. Electron. Agric.*, vol. 193, p. 106715, 2022, https://doi.org/10.1016/j.compag.2022.106715
[5] P. Sun, Y. Cui, J. Yang, T. Wu, J. Zhang, and Y. Zhou, "A cyanine derivative-based NIR fluorescent probe for hydrogen sulfide bioimaging and food spoilage monitoring," *Dyes Pigments*, vol. 219, no. July, p. 111644, 2023, https://doi.org/10.1016/j.dyepig.2023.111644
[6] S. Jiang et al., "Quick assessment of chicken spoilage based on hyperspectral nir spectra combined with partial least squares regression.," vol. 14, no. 1 PP-Beijing DT-Int. J. Agric. Biol. Eng., pp. 243–250.
[7] K. Zhong et al., "A colorimetric and NIR fluorescent probe for ultrafast detecting bisulfite and organic amines and its applications in food, imaging, and monitoring fish freshness," *Food Chem.*, vol. 438, no. August 2023, p. 137987, 2024, https://doi.org/10.1016/j.foodchem.2023.137987
[8] H.-T. Zhao, Y.-Z. Feng, W. Chen, and G.-F. Jia, "Application of invasive weed optimization and least square support vector machine for prediction of beef adulteration with spoiled beef based on visible near-infrared (Vis-NIR) hyperspectral imaging," *Meat Sci.*, vol. 151, pp. 75–81, 2019, https://doi.org/10.1016/j.meatsci.2019.01.010
[9] E. Bonah, X. Huang, R. Yi, J. H. Aheto, and S. Yu, "Vis-NIR hyperspectral imaging for the classification of bacterial foodborne pathogens based on pixel-wise analysis and a novel CARS-PSO-SVM model," *Infrared Phys. Technol.*, vol. 105, p. 103220, 2020, https://doi.org/10.1016/j.infrared.2020.103220
[10] W. Che et al., "Pixel based bruise region extraction of apple using Vis-NIR hyperspectral imaging," *Comput. Electron. Agric.*, vol. 146, pp. 12–21, 2018, https://doi.org/10.1016/j.compag.2018.01.013
[11] L. B. Ramo, N. O. Nóbrega, D. D. S. Fernandes, W. S. Lyra, P. H. G. D. Diniz, and A. C. U. Araújo, "Determination of moisture and total protein and phosphorus contents in powdered chicken egg samples using digital images, NIR spectra, data fusion, and multivariate calibration," *J. Food Compos. Anal.*, vol. 127, no. August 2023, p. 105940, 2023, https://doi.org/10.1016/j.jfca.2023.105940
[12] C. Zeng et al., "An activatable probe for detection and therapy of food-additive-related hepatic injury via NIR-II fluorescence/optoacoustic imaging and biomarker-triggered drug release," *Anal. Chim. Acta*, vol. 1208, p. 339831, 2022, https://doi.org/10.1016/j.aca.2022.339831
[13] A. Mouahid, "Infrared thermography used for composite materials," *MATEC Web Conf.*, vol. 191, p. 11, Jan. 2018, https://doi.org/10.1051/matecconf/201819100011
[14] P. Lagacherie, D. Arrouays, H. Bourennane, C. Gomez, and L. Nkuba-Kasanda, "Analysing the impact of soil spatial sampling on the performances of digital soil mapping models and their evaluation: A numerical experiment on quantile random forest using clay contents obtained from Vis-NIR-SWIR hyperspectral imagery," *Geoderma*, vol. 375, p. 114503, 2020, https://doi.org/10.1016/j.geoderma.2020.114503
[15] S. M. Mansuri, S. K. Chakraborty, N. K. Mahanti, and R. Pandiselvam, "Effect of germ orientation during Vis-NIR hyperspectral imaging for the detection of fungal contamination in maize kernel using PLS-DA, ANN and 1D-CNN modelling," *Food Control*, vol. 139, p. 109077, 2022, https://doi.org/10.1016/j.foodcont.2022.109077

[16] H. Chen et al., "A deep learning CNN architecture applied in smart near-infrared analysis of water pollution for agricultural irrigation resources," *Agric. Water Manag.*, vol. 240, p. 106303, 2020, https://doi.org/10.1016/j.agwat.2020.106303
[17] P. Mishra and R. Nikzad-Langerodi, "Partial least square regression versus domain invariant partial least square regression with application to near-infrared spectroscopy of fresh fruit," *Infrared Phys. Technol.*, vol. 111, p. 103547, 2020, https://doi.org/10.1016/j.infrared.2020.103547
[18] J. Peng, K. Manevski, K. Kørup, R. Larsen, and M. N. Andersen, "Random forest regression results in accurate assessment of potato nitrogen status based on multispectral data from different platforms and the critical concentration approach," *F. Crop. Res.*, vol. 268, p. 108158, 2021, https://doi.org/10.1016/j.fcr.2021.108158
[19] C. Sivakumar, M. M. A. Chaudhry, and J. Paliwal, "Classification of pulse flours using near-infrared hyperspectral imaging," *LWT*, vol. 154, p. 112799, 2022, https://doi.org/10.1016/j.lwt.2021.112799
[20] C. Fredes et al., "A model based on clusters of similar color and NIR to estimate oil content of single olives," *Foods*, vol. 10, no. 3. 2021. https://doi.org/10.3390/foods10030609

Chapter 6

Interactive dashboard and 3D visualization using t-SNE dimensionality reduction technique

Praveen Gujjar J and Lubna Ambreen
JAIN (Deemed-to-be University), Bengaluru, India

Vani Hiremani
Symbiosis International (Deemed), University, Pune, India

Raghavendra M Devadas
Manipal Institute of Technology Bengaluru, Bengaluru, Karnataka, India

INTRODUCTION

A dimensionality reduction technique called t-SNE t-Distributed Stochastic Neighbor Embedding (t-SNE) is used to aid in the visualization of high dimensional data. It's not usually the main technique for model training. Rather, it's frequently the initial stage of exploratory clustering, visualization, and data analysis. By simplifying high-dimensional data, t-SNE offers a natural way to comprehend it. This understanding can inform the choice and use of further techniques for a more in-depth and targeted analysis. One of the key components of unsupervised learning in data science and machine learning is dimensionality reduction. This step is essentially necessary when the data has extremely large dimensions and we need to project the data into a lower-dimensional space in order to tell its story. For dimensionality reduction, there are several methods available, including PCA, SVD, truncated SVD, LDA, and others. Another method for reducing dimensionality is t-SNA. For data exploration and high-dimensional data visualization, t-SNE is an unsupervised non-linear dimensionality reduction technique. Due to the algorithm's non-linear dimensionality reduction, data that cannot be divided by a straight line can be divided. The dimensional reduction methods t-SNE and PCA both have distinct mechanisms and are most effective when applied to particular kinds of data. The linear method known as Principal Component Analysis, or PCA, functions best with data that has a linear structure. It uses large pairwise distance preservation, variance minimization, and projection onto lower dimensions to find the underlying

DOI: 10.1201/9781003542735-6

principal components in the data. View our Principal Component Analysis (PCA) tutorial with examples in R to learn how the algorithms function internally. However, pairwise similarities between data points are preserved in a lower-dimensional space using the nonlinear approach known as t-SNE. While PCA concentrates on maintaining large pairwise distances to maximize variance, t-SNE is more concerned with maintaining small pairwise distances.

3D VISUALIZATION

While t-SNE is typically visualized in two dimensions, it is possible to extend the technique to 3D visualization to capture additional information and perspectives. The process involves embedding the high-dimensional data into a three-dimensional space. Three-dimensional (3D) visualization refers to the representation of data in a three-dimensional space, allowing for the depiction of volume and depth. This type of visualization is often used to gain a more comprehensive understanding of complex structures, relationships, or patterns in data. In 3D visualization, depth is represented along the third dimension, allowing viewers to perceive the relative distances between objects or data points. This depth perception can enhance the understanding of spatial relationships. 3D visualization is particularly useful when dealing with complex structures or datasets. It allows for the representation of intricate patterns, clusters, or shapes that may not be as apparent in two-dimensional visualizations. Many 3D visualization tools offer interactive features, allowing users to rotate, zoom, and pan the view. This interactivity facilitates a more dynamic exploration of the data, enabling users to inspect details from different angles. Three dimensions provide the opportunity to represent and visualize multiple variables simultaneously. For example, different colors, sizes, or shapes can be used to convey additional information about data points. 3D visualizations can create a sense of realism and immersion, especially in virtual environments or simulations. This is beneficial in fields such as scientific visualization, medicine, and engineering 3D visualization is employed in a wide range of fields, including scientific research, engineering, geospatial analysis, medical imaging, molecular biology, and computer-aided design (CAD). It helps professionals in these fields to analyze and interpret complex datasets. When presenting data or findings to a diverse audience, 3D visualization can enhance communication by providing a more intuitive representation. It allows viewers to grasp spatial relationships and patterns more easily than traditional 2D plots. Beyond its analytical uses, 3D visualization is also utilized for artistic and creative expression. Artists, designers, and multimedia creators use 3D graphics to produce visually compelling and immersive content.

LITERATURE REVIEW

Two popular techniques for visualizing large, multidimensional datasets are the t-SNE and UMAP (Uniform Manifold Approximation and Projection). UMAP performs better in terms of computing efficiency and maintaining more meaningful linkages between data points than t-SNE because of its emphasis on maintaining both local and global frameworks. The trade-offs and design decisions involved in these approaches have been examined recently, highlighting the significance of sophisticated knowledge in creating algorithms that effectively balance the preservation of both local and global structure in low-dimensional embedded data [1]. The crucial responsibility of reducing higher-dimensional data, which is significant across numerous sectors including eCommerce, financial services, healthcare, and telecommunications. Using data from telecommunications gateways, the study compares the effectiveness of two well-known non-linear dimension reduction techniques, UMAP and t-SNE, with a linear method, PCA. The research delves deeper into 3D visualization, highlighting its application in the analysis of handwritten digit images [2]. A deep-learning learning (DL) model based on mammography was applied to predict breast cancer in a population of people of different races and risks. Competitive regions of the receiver operator characteristic curve (AUC) values show how well Mirai performs, particularly in identifying precancerous alterations. The study emphasizes the DL model's potential for early detection of breast cancer symptoms in a high-risk setting by focusing on its performance in a population of African American patients who carry the BRCA mutation [3]. UMAP outperforms existing methods on a simulated dataset for reducing dimensionality in the segmentation of core-loss electrons energy loss spectroscopic (EELS) spectrum pictures. The study also highlights the utility of HDBSCAN and UMAP in deciphering a real experimental core-shell nanoparticle dataset, highlighting the program's capacity to reveal a variety of aspects of complex datasets. The triple combination of not positive matrix factorization, UMAP, and HDBSCAN is investigated, suggesting the complementary application of various methods to achieve a comprehensive understanding of the dataset under study [4]. City-scale energy data is analyzed using dimensionality reduction techniques, such as t-SNE and UMAP, to identify significant features of high demand and generation locations based on building characteristics. The evaluation reveals that while t-SNE is susceptible to perplexity factors, UMAP is stable in low-dimensional shapes. Using a real dataset from a Dutch city, the proposed method shows how effective UMAP is at identifying high solar generation and consumption zones, giving neighborhood-level grid managers and energy managers valuable information [5]. The impact of distributed energy resources, or DERs, on distribution feeders is examined by distribution network designers. In order to identify sample medium-voltage (MV) feeders for DER penetrating research, it presents a nonlinear clustering approach. Given the complexity and intermittent nature of DERs, it emphasizes the significance of efficient computing methods. The paper highlights how popular linear clustering techniques like k-mean and hierarchy clustering were in earlier studies, but also how vulnerable they were to outliers and how difficult it was to handle clusters with different densities and sizes [6]. Using neural network architectures to date family album photos in the IMAGO collection automatically, gleaning socio-historical information from photos taken in Italy in the 20th century. A number of Convolutional Neural Network models are analyzed in this study, and the evaluation is extended to cross-dataset trials in order to show temporal changes and potential cross-cultural ramifications. In the context of dating photos, Uniform Contour Estimation and Projection (UMAP) enhances the qualitative comprehension of cross-dataset interpretations, demonstrating the potential of deep learning for cross-cultural research [7]. the impact of different dimension reduction (DR) techniques on visual cluster analysis, emphasizing localization and linearity. Tests are conducted on twelve sample DR approaches, such as UMAP and t-SNE, on tasks like distance and density comparison, membership identification, and cluster recognition.

METHODOLOGY

In both higher and lower dimensional space, the t-SNE algorithm determines the similarity measure between pairs of instances. It then attempts to maximize two similarity metrics. All of that is accomplished in three steps.

1. In both higher and lower dimensions, t-SNE models the selection of a point as a neighbor of another point. Using a Gaussian kernel, it first computes the pairwise similarity between each data point in the high-dimensional space. Compared to points that are close together, points that are farther apart are less likely to be chosen.
2. Next, while maintaining pairwise similarities, the algorithm attempts to map higher dimensional data points onto lower dimensional space.
3. The process involves reducing the difference in probability distributions between the initial high-dimensional and lower-dimensional distributions. The algorithm minimizes the divergence by using gradient descent. An optimal state for the lower-dimensional embedding is reached.

Figure 6.1 shows the t-SNE working principles.

The working principles are outlined below.

a. Define the Similarity/Dissimilarity between Data Points: For each pair of high-dimensional data points, calculate their similarity or dissimilarity. Commonly used measures include Euclidean distance or conditional probabilities based on the similarity of pairwise data points.

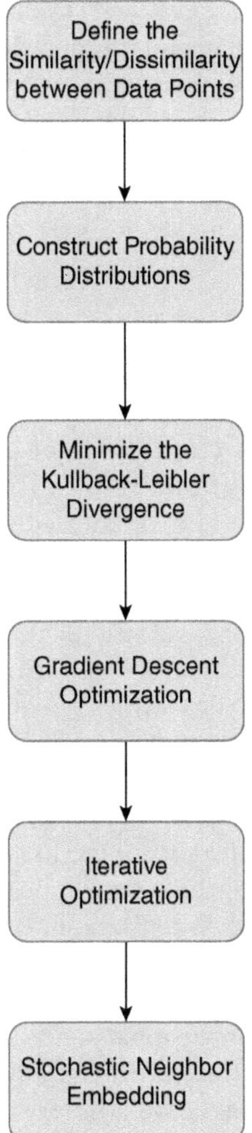

Figure 6.1 t-SNE working principles.

b. Construct Probability Distributions: Convert the pairwise similarities/dissimilarities into conditional probabilities that represent similarities in the lower-dimensional space. This is done using a Gaussian distribution for the similarity in the lower-dimensional space.
c. Minimize the Kullback-Leibler Divergence: The algorithm aims to minimize the Kullback-Leibler divergence between the conditional probability distributions in the high-dimensional and low-dimensional spaces. This helps in preserving the pairwise similarities as much as possible.
d. Gradient Descent Optimization: Utilize gradient descent to optimize the positions of the data points in the lower-dimensional space, reducing the divergence between the high-dimensional and low-dimensional probability distributions.

e. Iterative Optimization: Iteratively update the positions of the data points in the lower-dimensional space, adjusting for better alignment with the original high-dimensional relationships.
f. Stochastic Neighbor Embedding: The "stochastic" aspect in t-SNE refers to the use of a Student's t-distribution to model pairwise similarities in the lower-dimensional space, which helps to handle the crowding problem and makes t-SNE particularly effective for visualizing clusters in data.

RESULTS AND DISCUSSION

t-SNE is a dimensionality reduction technique commonly used for visualizing high-dimensional data in lower dimensions. The Iris dataset is a well-known dataset in machine learning and is often used for testing and demonstrating various algorithms. The dataset consists of measurements of four features (sepal length, sepal width, petal length, and petal width) for three species of iris flowers (setosa, versicolor, and virginica). t-SNE can be applied to this dataset for dimensionality reduction and visualization. t-SNE provides an effective way to visualize the similarities between data points. In the case of the Iris dataset, you would anticipate that flowers of the same species would be close together in the t-SNE plot, reflecting their similar measurements. While the Iris dataset is relatively clean and well-behaved, t-SNE can still help identify potential outliers or instances that do not conform to the general patterns observed in the data. Outliers may appear as points that are relatively far from their expected cluster. t-SNE might offer insights into the relationships between petal and sepal measurements. It could reveal if certain combinations of petal and sepal characteristics contribute more to the separation of iris species in the lower-dimensional space. It's important to consider the impact of the perplexity parameter when applying t-SNE to the Iris dataset. Different perplexity values might yield different visualizations, and selecting an appropriate value is crucial for obtaining meaningful insights. t-SNE may exhibit the crowding problem, especially if there are instances where flowers of different species have similar measurements. In such cases, points in the t-SNE plot might be closer together, and interpreting their relationships should be done cautiously. t-SNE is particularly effective at preserving local structures, but it might not always accurately represent the global structure of the data. The visualization should be interpreted with an understanding that t-SNE focuses on capturing local relationships. Figures 6.2 and 6.3 show the visualization Iris dataset in 2D and 3D format, respectively.

Given a t-SNE visualization of the Iris dataset, interactive exploration tools can be valuable. Tools that allow you to hover over points to see their original feature values or highlight specific clusters can enhance the interpretability of the results.

The 2D Distribution of Iris Dataset are outlines in Figure 6.5. It has been clearly shown from the graph that t-SNE can make the proper classification

Figure 6.2 2D visualization of Iris dataset for t-SNE classification.

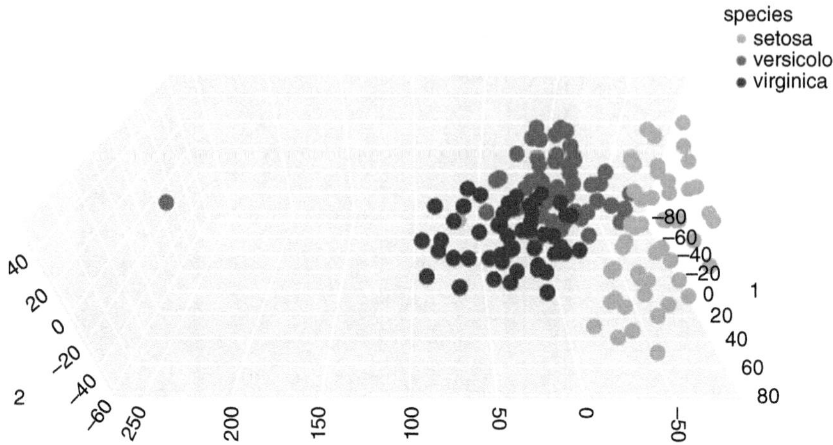

Figure 6.3 3D visualization of Iris dataset.

and same can be bring into the dashboard by connecting it through tableau or Power BI. The other view of 3D visualization of Iris Dataset are shown in the Figure 6.4.

The violin plot for the Iris Dataset are shown in the Figure 6.6. The violin plot highlights the distribution of the data with respect to sepal length, sepal width, petal length and petal width.

Histogram of Iris dataset are shown in the Figure 6.7. It has been observed the only the sepal width is normally distributed. Sepal length and petal length are not normally distributed

Scatter plot for the Iris Dataset are shown in the Figure 6.8. The scatter plot with respect to sepal length and sepal width, Sepal width and Petal length, Petal length and Petal width are shown in the Figure 6.8.

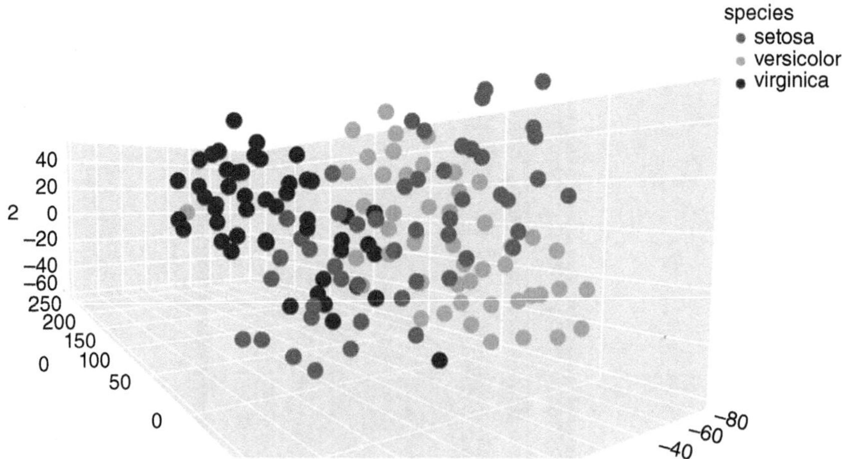

Figure 6.4 Other view of 3D visualization of Iris dataset.

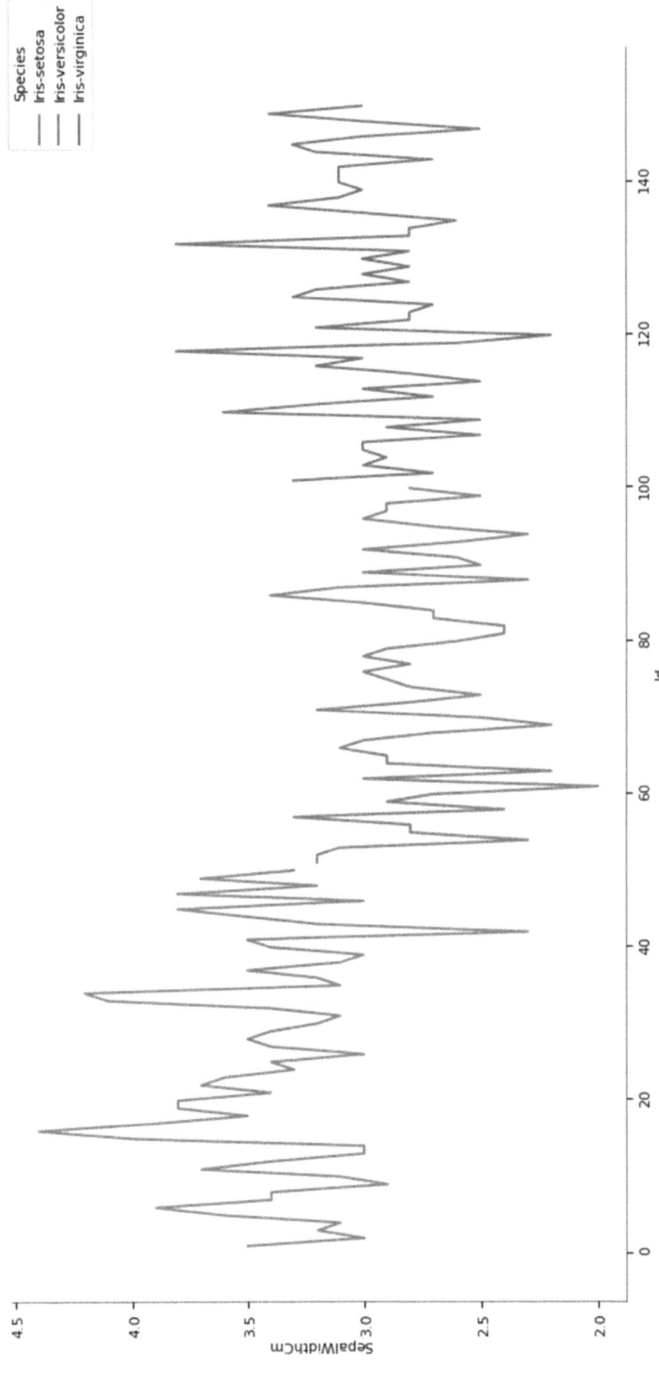

Figure 6.5 2D distributions of Iris dataset.

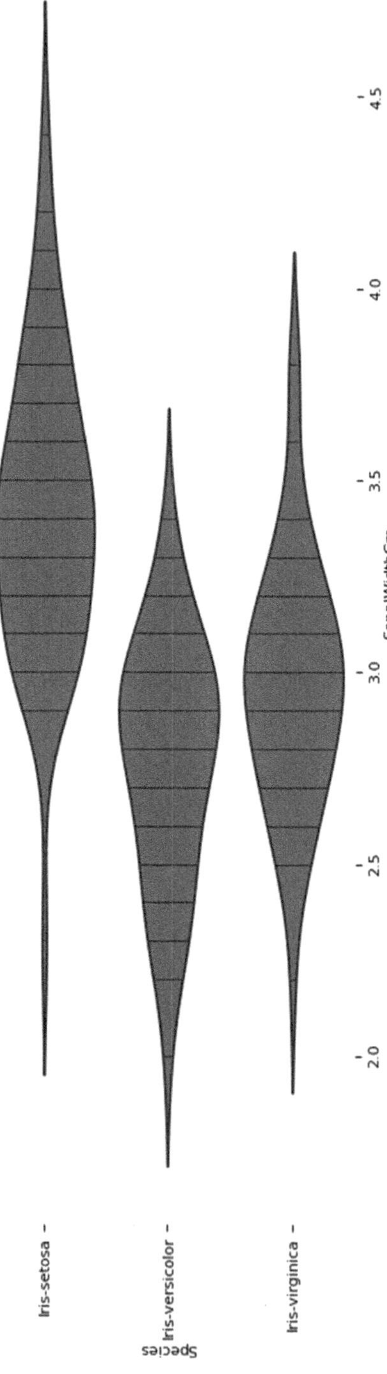

Figure 6.6 Violin plot for the Iris dataset.

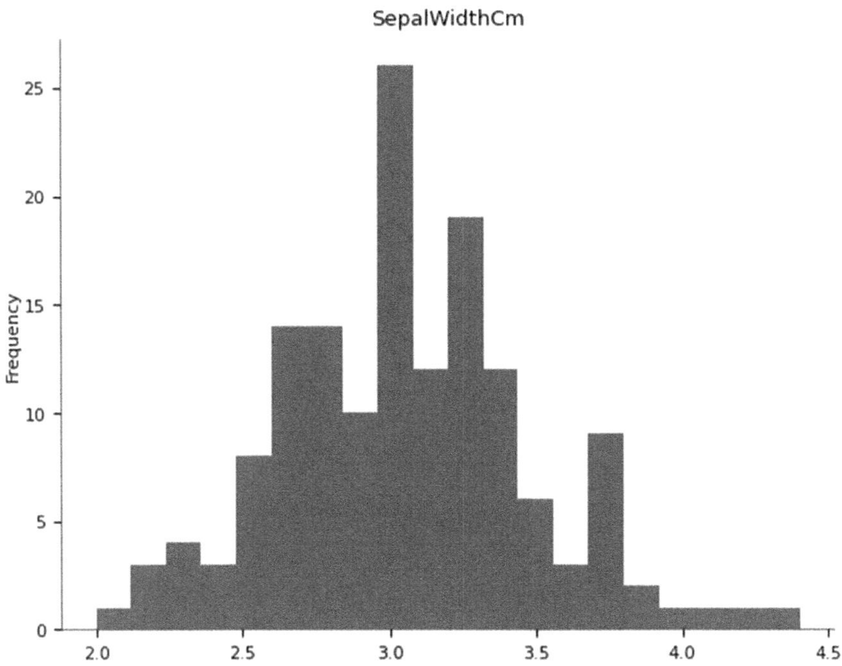

Figure 6.7 Histogram of the Iris dataset.

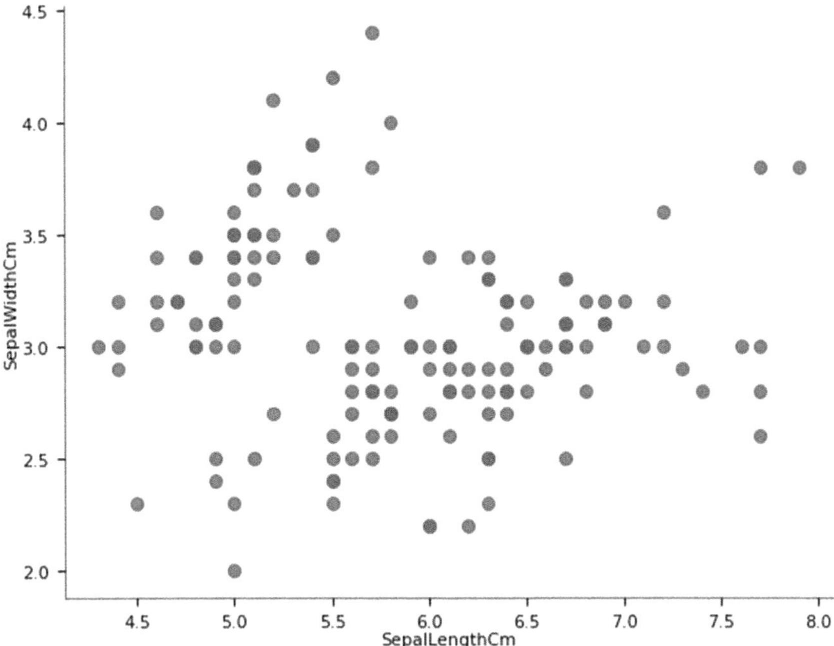

Figure 6.8 Scatter plot of the Iris dataset.

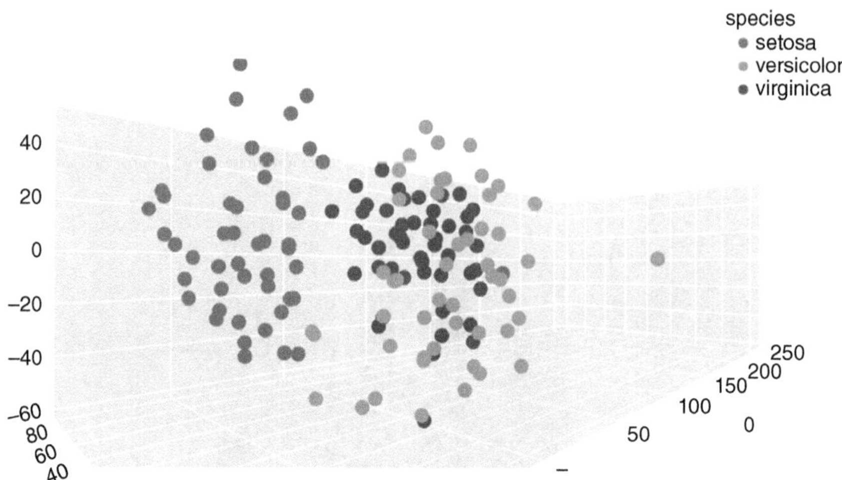

Figure 6.9 3D visualization of classification.

The 3D view of classification of Iris dataset with t-SNE are shown in the Figure 6.9. The t-SNE classify the data by reducing the dimension. Relatively t-SNE perform well when compared with principal component analysis.

DIFFERENCE BETWEEN PCA AND t-SNE

Principal Component Analysis (PCA) and t-Distributed Stochastic Neighbor Embedding (t-SNE) are both dimensionality reduction techniques used in machine learning and data analysis.

Sl. No	*Parameter*	*PCA*	*t-SNE*
1	Objective	It aims to capture the maximum variance in the data by identifying the principal components, which are linear combinations of the original features.	It focuses on preserving the local structure of the data, emphasizing the relationships between nearby points in high-dimensional space.
2	Linearity	PCA is a linear technique, meaning it works well when the relationships between features are linear. It may not capture complex non-linear relationships effectively.	t-SNE is non-linear and excels at capturing intricate relationships, making it suitable for visualizing clusters and groups in complex data.

(*Continued*)

Sl. No	Parameter	PCA	t-SNE
3	Global vs. local structure	PCA is good for capturing the global structure of the data. It provides a global view of the relationships between data points.	t-SNE is designed to emphasize local structure, making it useful for visualizing clusters and patterns in the data, especially in cases where global relationships are less critical.
4	Interpretability	Principal components have clear interpretations as linear combinations of original features, making it easy to understand the contribution of each feature to the components.	The output of t-SNE is not easily interpretable in terms of the original features. It is primarily used for visualization rather than feature interpretation.
5	Computation time	PCA is computationally efficient and can handle large datasets. It has a time complexity of O(N^2), where N is the number of features.	t-SNE can be computationally expensive, especially for large datasets, due to its time complexity of approximately O(N^2). However, approximations like the Barnes-Hut algorithm can improve scalability.
6	Preservation of variance	PCA preserves the maximum variance in the data and is often used when retaining as much information as possible is important.	t-SNE does not preserve variance; instead, it prioritizes preserving local similarities. This makes it useful for visualizing clusters in high-dimensional space.
7	Hyperparameters	PCA has no hyperparameters, making it simpler to use. The principal components are directly obtained from the covariance matrix.	t-SNE has hyperparameters, such as perplexity, that need to be carefully tuned. The choice of perplexity can impact the visualization results.
8	Use cases	It is commonly used for tasks like feature engineering, noise reduction, and retaining most of the information in a lower-dimensional space.	It is often used for visual exploration and cluster identification in high-dimensional datasets, particularly in scenarios where non-linear relationships are crucial.

CHALLENGES OF t-SNE

While t-SNE is a powerful technique for visualizing high-dimensional data, it comes with certain challenges and considerations:

Stochastic Nature: t-SNE is inherently stochastic, meaning that different runs with the same data may produce different results. This randomness can make it challenging to precisely reproduce a visualization, and users should be cautious about overinterpreting specific configurations.

Crowding Problem: In the lower-dimensional space, t-SNE tends to suffer from the crowding problem. This occurs when points are too close together, leading to distorted representations. It's crucial to be aware that distances between points in the t-SNE visualization may not accurately reflect the original high-dimensional distances.

Hyperparameter Sensitivity: t-SNE has hyperparameters, such as the perplexity, which controls the balance between preserving global and local relationships. Selecting an appropriate perplexity value is crucial, and different datasets may require different settings. Choosing inappropriate values can result in suboptimal visualizations.

Computationally Intensive: t-SNE can be computationally intensive, particularly for large datasets. The algorithm has a time complexity of approximately O(N^2) due to the pairwise comparisons, making it less scalable for extremely large datasets compared to other dimensionality reduction techniques.

Interpretation Challenges: While t-SNE is excellent for visual exploration, interpreting the exact meaning of distances in the lower-dimensional space can be challenging. Points that are close in the t-SNE visualization may represent similar relationships, but the distances do not necessarily correspond to absolute quantitative measures.

Memory Requirements: The memory requirements of t-SNE can be substantial, especially for large datasets. Users need to ensure that their computational resources are sufficient to handle the memory demands of the algorithm.

Global Structure Preservation: t-SNE is primarily designed for preserving local structures in the data, which means that it may not always capture the global structure accurately. For datasets where global relationships are crucial, other dimensionality reduction methods like PCA (Principal Component Analysis) may be more suitable.

Optimal Cluster Representation: While t-SNE is effective for visualizing clusters in data, the choice of hyperparameters and the stochastic nature of the algorithm can impact the optimal representation of clusters. Users should be cautious about drawing definitive conclusions based solely on t-SNE visualizations.

CONCLUSION

Employing the dimensionality reduction technique using t-SNE in conjunction with 3D visualization proves to be a powerful approach for gaining a nuanced understanding of complex datasets. By harnessing the capabilities of t-SNE to capture local structures and relationships, and complementing it with the depth and interactive exploration afforded by 3D visualization, this methodology unveils intricate patterns and spatial arrangements within high-dimensional data. The integration of t-SNE and 3D visualization is particularly valuable when dealing with datasets characterized by intricate structures or subtle clusters that may elude traditional two-dimensional representations. The depth perception provided by the third dimension adds a layer of richness to the visual exploration, enabling a more holistic interpretation of the underlying data. The interactive nature of 3D visualization allows users to dynamically navigate and inspect the data from various angles, fostering a more immersive and insightful analysis. This combination proves beneficial across diverse domains, ranging from scientific research and engineering to artistic and creative endeavors. While this approach offers compelling insights, it is essential to judiciously select parameters, validate results, and interpret findings with a contextual understanding of the data. As with any visualization technique, the choice between dimensionality reduction and 3D visualization should align with the specific goals of the analysis and the inherent characteristics of the dataset.

REFERENCES

1. Y. Wang, H. Huang, C. Rudin, and Y. Shaposhnik, "Understanding how dimension reduction tools work: An empirical approach to deciphering T-SNE, UMAP, TriMAP, and PACMAP for data visualization," *arXiv (Cornell University)*, Dec. 2020, https://doi.org/10.48550/arxiv.2012.04456
2. K. Pal and M. Sharma, "Performance evaluation of non-linear techniques UMAP and t-SNE for data in higher dimensional topological space," *2020 Fourth International Conference on I-SMAC (IoT in Social, Mobile, Analytics and Cloud) (I-SMAC)*, Oct. 2020, https://doi.org/10.1109/i-smac49090.2020.9243502
3. O. J. Omoleye et al., "External evaluation of a mammography-based deep learning model for predicting breast cancer in an ethnically diverse population," *Radiology:ArtificialIntelligence*Vol.5,No.6,Nov.2023,https://doi.org/10.1148/ryai.220299
4. J. B. Portals, F. Peiró, and S. Estradé, "Strategies for EELS data analysis. Introducing UMAP and HDBSCAN for dimensionality reduction and clustering," *Microscopy and Microanalysis*, vol. 28, no. 1, pp. 109–122, Nov. 2021, https://doi.org/10.1017/s1431927621013696

5. W. A. Khan, S. Walker, and W. W. Zeiler, "A bottom-up framework for analysing city-scale energy data using high dimension reduction techniques," *Sustainable Cities and Society*, vol. 89, p. 104323, Feb. 2023, https://doi.org/10.1016/j.scs.2022.104323
6. O. Ramos-Leaños, J. Jneid, and B. Fazio, "Non-linear clustering of distribution feeders," *Energies*, vol. 15, no. 21, p. 7883, Oct. 2022, https://doi.org/10.3390/en15217883
7. L. Stacchio, A. Angeli, G. Lisanti, and G. Marfia, "Analyzing cultural relationships visual cues through deep learning models in a cross-dataset setting," *Neural Computing and Applications*, Sep. 2023, https://doi.org/10.1007/s00521-023-08966-3

Chapter 7

Dynamic dashboard creation for sales trends and optimize pricing strategies

Sasithradevi A, Kanimozhi S, and Srinath Srinivasan
Vellore Institute of Technology, Chennai, India

7.1 INTRODUCTION

In today's competitive business landscape, organizations rely on data-driven insights to make informed decisions and gain a competitive edge. Dynamic dashboards have emerged as powerful tools for visualizing and analyzing sales trends, enabling businesses to track performance, identify opportunities, and optimize pricing strategies in real-time [1]. This chapter provides an overview of the significance of dynamic dashboards in sales analytics and outlines the objectives and scope of the study.

7.2 LITERATURE REVIEW

7.2.1 Introduction to dynamic dashboards

Dynamic dashboards have gained significant traction in recent years as organizations seek to harness the power of data visualization and analytics for decision-making. This section provides an overview of the literature surrounding dynamic dashboards, focusing on their definition, features, and benefits. Dynamic dashboards are interactive data visualization tools that allow users to explore and analyze data in real-time, enabling agile decision-making and insights discovery [2]. Key features of dynamic dashboards include interactivity, customization, and real-time data updates, which empower users to drill down into details, explore relationships, and identify trends effortlessly. The literature highlights the benefits of dynamic dashboards in improving data accessibility, enhancing decision-making, and driving organizational performance.

7.2.2 Power BI as a dynamic dashboard tool

Power BI, developed by Microsoft, is one of the leading tools for creating dynamic dashboards [3]. This section examines the literature related to Power BI, focusing on its features, capabilities, and applications in dynamic

DOI: 10.1201/9781003542735-7

dashboard development. Power BI offers a rich set of visualization options, data connectors, and analytical capabilities, making it a versatile tool for creating interactive and visually compelling dashboards. Studies have demonstrated the effectiveness of Power BI in various domains, including sales analytics, marketing, finance, and operations. The literature emphasizes Power BI's ease of use, scalability, and integration with other Microsoft products, making it a preferred choice for organizations seeking to implement dynamic dashboard solutions.

7.2.3 Design principles for dynamic dashboards

Effective dashboard design is crucial for maximizing usability, readability, and user engagement. This section reviews literature on design principles and best practices for creating dynamic dashboards using Power BI. Key considerations include layout design, color schemes, typography, and data visualization techniques [4]. Studies have highlighted the importance of simplicity, clarity, and consistency in dashboard design, as well as the need to tailor visualizations to the specific needs and preferences of users. The literature also explores advanced design concepts, such as storytelling, dashboard interactivity, and mobile optimization, which enhance the overall user experience and effectiveness of dynamic dashboards.

7.2.4 Evaluation metrics for dynamic dashboards

Measuring the effectiveness of dynamic dashboards is essential for assessing their impact on decision-making and organizational performance. This section examines literature on evaluation metrics and methodologies for assessing the usability, usefulness, and user satisfaction of dynamic dashboards developed using Power BI [5]. Common evaluation metrics include task completion time, error rates, user satisfaction surveys, and usability testing. Studies have emphasized the importance of involving end-users in the evaluation process to gather feedback and identify areas for improvement. Additionally, the literature discusses the role of analytics tools and tracking mechanisms in monitoring dashboard usage, user interactions, and performance metrics over time [6].

7.2.5 Challenges and future directions

Despite the numerous benefits of dynamic dashboards, several challenges and opportunities for future research exist. This section reviews literature on challenges such as data quality, security, scalability, and user adoption, which can impact the effectiveness and success of dynamic dashboard implementations [7]. Future research directions include exploring advanced analytics techniques, integrating predictive analytics models, enhancing dashboard interactivity, and addressing emerging trends such as augmented analytics and natural language processing. By addressing these challenges and embracing emerging technologies, organizations can unlock the full potential of dynamic dashboards in driving data-driven decision-making and organizational performance.

This literature review provides a comprehensive overview of the existing research and practices related to dynamic dashboard development using Power BI [8]. By synthesizing insights from the literature, this chapter lays the foundation for the subsequent chapters, which will focus on the design, implementation, and evaluation of dynamic dashboards for sales trends and pricing strategy optimization.

7.3 METHODOLOGY

Overall of workflow of proposed study deals with creation of dashboard using dynamic property to analyze the sales trends is shown in Figure 7.1. Initially the data acquisition is carried out by collecting the raw sales data from source system such as CRM, ERP and other transaction database [9]. Followed by data cleaning along with removal of duplicates, eliminate missing data and ensure data consistency. These cleaned data is then integrated together to ensure the compatibility and consistence in terms of unified dataset. Next is data modeling which uses the power BI tools to define the relationships, measures and calculated columns. Once all these pre-processing works are completed the designing of dashboard work is initiated with drag and drop interface, incorporating charts, graphs and tables [10].

After design, the interactiveness features such as slicers, filters and drill-down capabilities were incorporated in the created dashboard. To obtain the real-time updates data refresh option is enabled to ensure that the dashboard reflects the latest sales data. For measuring the performance of the proposed dashboard is evaluated by conducting user testing and feedback sessions [11]. Finally the dynamic dashboard is deployed to power BI service or on-premises server for access by stakeholders and decision-makers. Regular monitoring and maintenance is done to ensure optimal functionality.

7.3.1 Data collection

To perform analysis over trending sales, a dataset is collected which consist of 15 features that are related to purchase order of various branches [12]. Also each branch in relate with various different product line is gathered to aggregated and visualize the price variation over each product in each store which is shown in Figure 7.2.

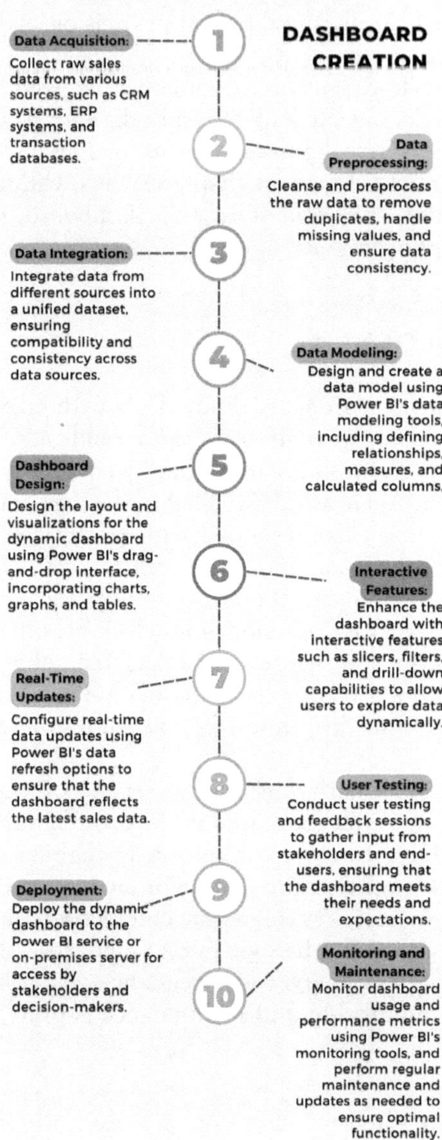

Figure 7.1 Workflow diagram of proposed dynamic dashboard.

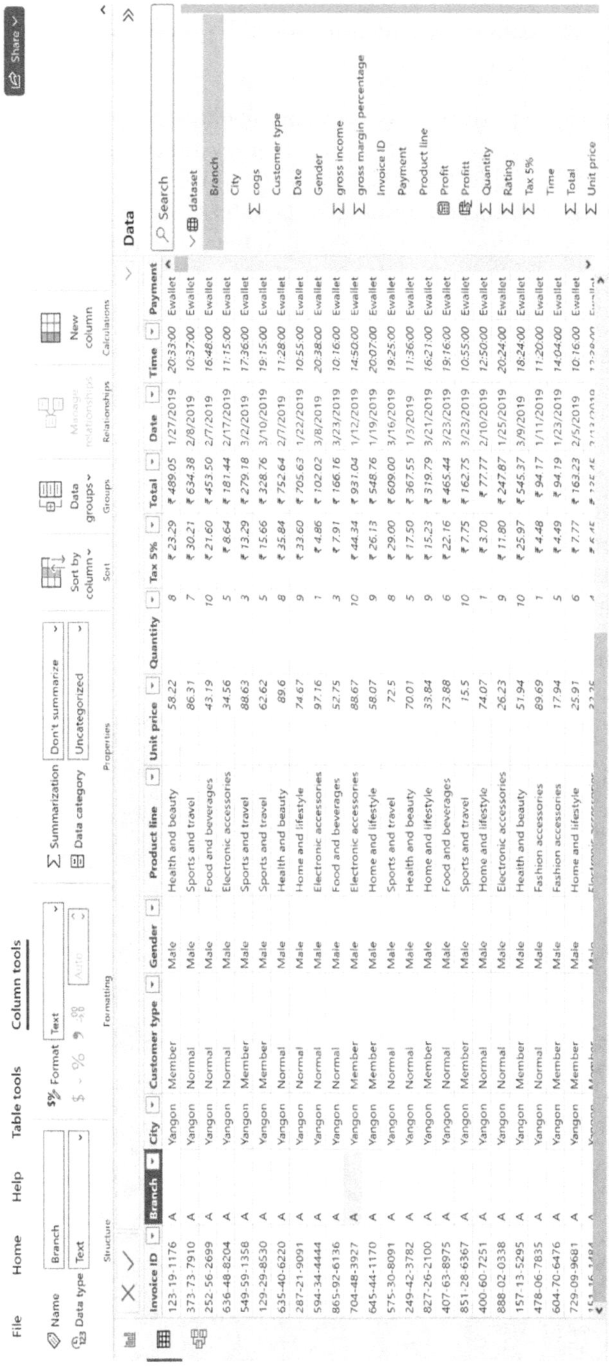
File Home Help Table tools Column tools
Share
Name Branch
Data type Text
Format Text
Summarization Don't summarize
Data category Uncategorized
Structure
Formatting
Properties
Sort by column
Sort
Data groups
Groups
Manage relationships
Relationships
New column
Calculations
Invoice ID | Branch | City | Customer type | Gender | Product line | Unit price | Quantity | Tax 5% | Total | Date | Time | Payment
123-19-1176 A Yangon Member Male Health and beauty 58.22 8 ₹23.29 ₹489.05 1/27/2019 20:33:00 Ewallet
373-73-7910 A Yangon Normal Male Sports and travel 86.31 7 ₹30.21 ₹634.38 2/8/2019 10:37:00 Ewallet
252-56-2699 A Yangon Normal Male Food and beverages 43.19 10 ₹21.60 ₹453.50 2/7/2019 16:48:00 Ewallet
636-48-8204 A Yangon Normal Male Electronic accessories 34.56 5 ₹8.64 ₹181.44 2/17/2019 11:15:00 Ewallet
549-59-1358 A Yangon Member Male Sports and travel 88.63 3 ₹13.29 ₹279.18 3/2/2019 17:36:00 Ewallet
129-29-8530 A Yangon Member Male Sports and travel 62.62 5 ₹15.66 ₹328.76 3/10/2019 19:15:00 Ewallet
635-40-6220 A Yangon Normal Male Health and beauty 89.6 8 ₹35.84 ₹752.64 2/7/2019 11:28:00 Ewallet
287-21-9091 A Yangon Normal Male Home and lifestyle 74.67 9 ₹33.60 ₹705.63 1/22/2019 10:55:00 Ewallet
594-34-4444 A Yangon Normal Male Electronic accessories 97.16 1 ₹4.86 ₹102.02 3/8/2019 20:38:00 Ewallet
865-92-6136 A Yangon Normal Male Food and beverages 52.75 3 ₹7.91 ₹166.16 3/23/2019 10:16:00 Ewallet
704-48-3927 A Yangon Member Male Electronic accessories 88.67 10 ₹44.34 ₹931.04 1/12/2019 14:50:00 Ewallet
645-44-1170 A Yangon Member Male Home and lifestyle 58.07 9 ₹26.13 ₹548.76 1/19/2019 20:07:00 Ewallet
575-30-8091 A Yangon Normal Male Sports and travel 72.5 8 ₹29.00 ₹609.00 3/16/2019 19:25:00 Ewallet
249-42-3782 A Yangon Normal Male Health and beauty 70.01 5 ₹17.50 ₹367.55 1/3/2019 11:36:00 Ewallet
827-26-2100 A Yangon Member Male Home and lifestyle 33.84 9 ₹15.23 ₹319.79 3/21/2019 16:21:00 Ewallet
407-63-8975 A Yangon Normal Male Food and beverages 73.88 6 ₹22.16 ₹465.44 3/23/2019 19:16:00 Ewallet
851-28-6367 A Yangon Member Male Sports and travel 15.5 10 ₹7.75 ₹162.75 3/23/2019 10:55:00 Ewallet
400-60-7251 A Yangon Normal Male Home and lifestyle 74.07 1 ₹3.70 ₹77.77 2/10/2019 12:50:00 Ewallet
888-02-0338 A Yangon Normal Male Electronic accessories 26.23 9 ₹11.80 ₹247.87 1/25/2019 20:24:00 Ewallet
157-13-5295 A Yangon Member Male Health and beauty 51.94 10 ₹25.97 ₹545.37 3/9/2019 18:24:00 Ewallet
478-06-7835 A Yangon Normal Male Fashion accessories 89.69 1 ₹4.48 ₹94.17 1/11/2019 11:20:00 Ewallet
604-70-6476 A Yangon Member Male Fashion accessories 17.94 5 ₹4.49 ₹94.19 1/23/2019 14:04:00 Ewallet
729-09-9681 A Yangon Member Male Home and lifestyle 25.91 6 ₹7.77 ₹163.23 2/5/2019 10:16:00 Ewallet
Table: dataset (1,000 rows) Column: Branch (3 distinct values)
Data
Search
dataset
Branch
City
cogs
Customer type
Date
Gender
gross income
gross margin percentage
Invoice ID
Payment
Product line
Profit
Profitt
Quantity
Rating
Tax 5%
Time
Total
Unit price

Figure 7.2 Sample dataset of a superstore.

7.3.2 Algorithm for filling NaN and outlier values

To perform the missing value identification a procedure wash proposed which consist of following steps:

- Use power query editor after loading the dataset
- Identify NaN values by using power query functions such as "Table. SelectRows", "Table. SelectColumns" or "Table.AddColumn" to identify rows or columns containing NaN values.
- Replace NaN values using "Table.FillDown" to replace NaN values with appropriate values such mean or median of the columns.

Also to handle the outliers, a set of precise steps were deployed such as,

- Determine the criteria for identifying outliers using Z-Score, standard deviation.
- Use the statistical functions such as "List.Min", "List.Max", "List. Average" to determine the outliers.
- Use conditional statements such as "Table.SelectRows" to identify outlier and replace with median, mean and fixed values
- Implement smoothing techniques such as moving average to reduce the impact of outliers on visualizations.

A set of sample output about implemented fillers is shown in Figure 7.3.

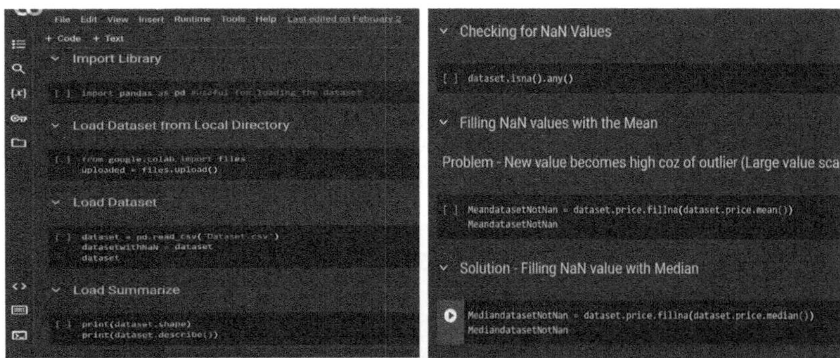

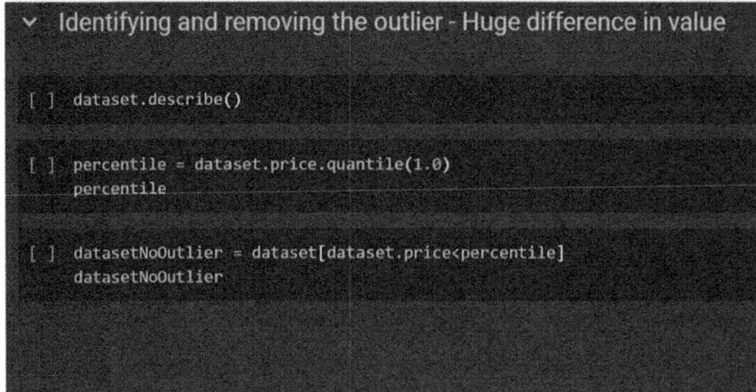

Figure 7.3 Algorithm for checking and filling Nan and outlier values in Google Colab.

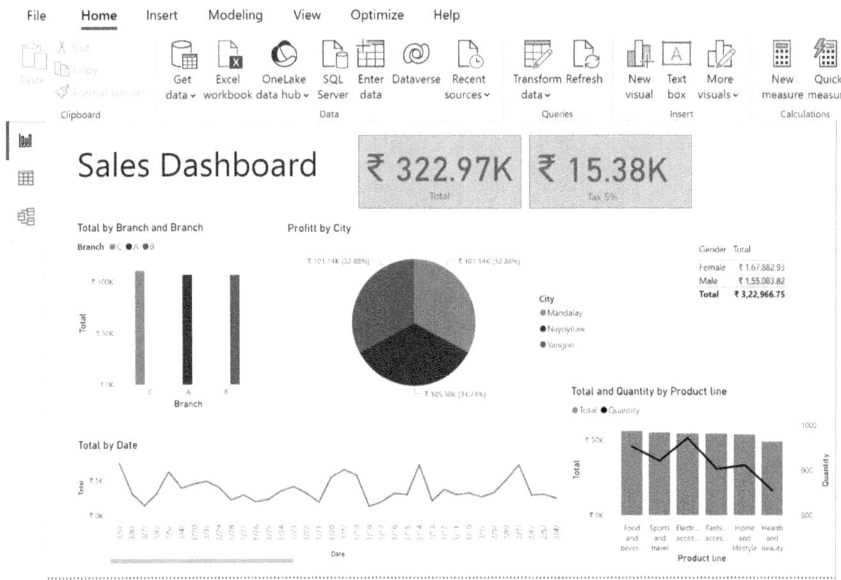

Figure 7.4 Power BI sales trends dashboard.

7.3.3 Dashboard design

Figure 7.4 describes about the overall sales dashboard for the supermarket dataset. In this various visualization is deployed such as line graph, pie chart, bar in relate with line graph and also aggregated feature values of specific columns along with overall sales. Hence visualizing this dashboard one can understand the sales strategy along with trending product with respect to gender and division.

7.4 RESULTS

The dynamic dashboard creation process using Power BI for sales trends and pricing strategy optimization culminated in the development of an interactive and insightful dashboard solution. This section presents the key outcomes and findings of the project, highlighting the effectiveness, usability, and impact of the dynamic dashboard on decision-making and business performance.

7.4.1 Dashboard design and features

The dynamic dashboard was meticulously designed to provide stakeholders with intuitive access to critical sales data and pricing insights. Leveraging Power BI's rich visualization capabilities, the dashboard featured interactive charts, graphs, and KPIs that enabled users to explore sales trends, analyze pricing dynamics, and identify actionable insights at a glance [13]. Key features included:

- Sales Performance Overview: Interactive visualizations depicting sales trends over time, regional sales distribution, and product performance metrics.
- Real-Time Data Updates: Seamless integration with data sources and real-time data refresh capabilities ensured that the dashboard reflected the latest sales data and pricing information, enabling timely decision-making.
- Drill-Down and Filtering Options: Flexible filtering and drill-down options empowered users to analyze sales data by various dimensions, such as product category, customer segment, and sales channel, facilitating deeper insights and targeted analysis.

7.4.2 User feedback and adoption

User feedback and adoption played a crucial role in assessing the effectiveness and usability of the dynamic dashboard solution [14]. Stakeholders and end-users praised the dashboard's intuitive interface, interactive features, and actionable insights, highlighting its value in guiding strategic decisions and optimizing pricing strategies. Key feedback included:

- Improved Decision-Making Users reported that the dynamic dashboard provided valuable insights into sales trends and pricing dynamics, enabling more informed decision-making and strategy formulation.
- Enhanced Collaboration: The interactive nature of the dashboard fostered collaboration among cross-functional teams, facilitating data-driven discussions and alignment on sales and pricing strategies.
- Increased Efficiency: Users noted that the dashboard's real-time data updates and customizable features streamlined their workflow and enhanced efficiency in analyzing sales data and monitoring pricing performance.

7.4.3 Business impact

The dynamic dashboard solution had a tangible impact on business performance, driving improvements in sales effectiveness, pricing strategy optimization, and overall competitiveness [15]. Key business outcomes included:

- Revenue Growth: By leveraging insights from the dynamic dashboard, organizations were able to identify sales opportunities, optimize pricing strategies, and drive revenue growth across key market segments and product categories.

- Competitive Advantage: The ability to monitor competitor pricing trends and adjust pricing strategies in real-time provided organizations with a competitive advantage, enabling them to respond swiftly to market dynamics and customer preferences.
- Customer Satisfaction: By aligning pricing strategies with customer needs and market demand, organizations were able to enhance customer satisfaction and loyalty, driving long-term value and profitability.

7.5 CONCLUSION

Dynamic dashboard creation using Power BI for sales trends and pricing strategy optimization represents a pivotal advancement in the realm of business intelligence. This chapter reflects on the journey undertaken, highlighting key insights, contributions, and implications derived from the exploration of dynamic dashboard development.

7.5.1 Recapitulation of findings

Throughout this study, we delved into the intricacies of dynamic dashboard creation, leveraging Power BI's robust features and capabilities. We reviewed literature, examined related works, and synthesized best practices to design and implement effective dynamic dashboards tailored for sales analytics and pricing strategy optimization.

7.5.2 Key insights

Our investigation unveiled the significance of dynamic dashboards in facilitating agile decision-making, empowering stakeholders to glean actionable insights from complex sales data. By harnessing Power BI's intuitive interface, interactive visualizations, and real-time data connectivity, organizations can gain a competitive edge in today's dynamic marketplace [16].

7.5.3 Contributions to practice and research

This research contributes to both practice and research by offering practical guidance and theoretical insights into dynamic dashboard development with Power BI. Practitioners can leverage the findings to design and deploy impactful dashboard solutions, while researchers can build upon the knowledge gained to advance the field of business intelligence and analytics.

7.5.4 Implications for business

The implications of dynamic dashboards extend beyond mere visualization; they serve as strategic tools for driving business performance and competitiveness. By leveraging real-time insights into sales trends and pricing dynamics, organizations can optimize strategies, enhance customer satisfaction, and achieve sustainable growth in an ever-evolving market landscape.

7.5.5 Future directions

Looking ahead, the journey of dynamic dashboard development with Power BI continues, with opportunities for further innovation and exploration. Future research directions may include the integration of advanced analytics techniques, such as machine learning and predictive modeling, to augment dashboard capabilities and unlock deeper insights.

REFERENCES

1. Microsoft. (n.d). Power BI: Business Analytics Solutions. Retrieved from https://powerbi.microsoft.com/
2. Few, S. (2006). *Information Dashboard Design: Displaying Data for At-a-Glance Monitoring*. O'Reilly Media.
3. Stephen Few. (2013). *Information Dashboard Design: The Effective Visual Communication of Data*. O'Reilly Media.
4. Kaur, G., & Rani, P. (2016). Comparative Study of Data Visualization Tools. *International Journal of Advanced Research in Computer Science and Software Engineering*, 6(9), 7–11.
5. Vinoth, M., Arunkumar, R., & Sriram, P. (2017). A Survey on Business Intelligence using Power BI. *International Journal of Advanced Research in Computer and Communication Engineering*, 6(7), 64–69.
6. Turban, E., Sharda, R., & Delen, D. (2014). *Decision Support and Business Intelligence Systems*. Pearson Education.
7. Cichocki, A., & Unbehauen, R. (2009). *Neural Networks for Optimization and Signal Processing*. John Wiley & Sons.
8. Armstrong, J. S. (2001). *Principles of Forecasting: A Handbook for Researchers and Practitioners*. Springer Science & Business Media.
9. Haghighi, H., & Boucher, T. (2019). *Pricing Optimization with Big Data: Techniques and Use Cases*. Springer.
10. Montano, V. E., & Mercado, J. L. L. (2023). Dynamic Relationships: E-Commerce Sales and Key Exogenous Variables in the Philippines. *European Journal of Management and Marketing Studies*, 8(3), pp 5–10.
11. Ratner, S. E. (2023). *Progress You Can See: Measuring for Social Change*. Practical Action Publishing.
12. Gonçalves, C. T., Gonçalves, M. J. A., & Campante, M. I. (2023). Developing Integrated Performance Dashboards Visualisations Using Power BI as a Platform. *Information*, 14(11), 614.
13. Bobylev, T. (2023). *Dashboard for Data-Driven Decision Support in Small and Medium Enterprises: A Web-Based Approach. Bachelor programme in Digital Service Innovation*. Luleå University of Technology.

14. Kanagaraj, K., & Venkatesh, R. (2023, November). Enhancing Business Performance: A Comprehensive Study of Sales and Distribution Analytics in Speciality Retail Sectors. In *2023 International Conference on Research Methodologies in Knowledge Management, Artificial Intelligence and Telecommunication Engineering (RMKMATE)* (pp. 1–5). IEEE.
15. Sachs, J. D., Kroll, C., Lafortune, G., Fuller, G., & Woelm, F. (2022). *Sustainable Development Report 2022*. Cambridge University Press.
16. Murugesan, R., Shanmugaraja, V., & Vadivel, A. (2022). Forecasting Bitcoin Price Using Interval Graph and ANN Model: A Novel Approach. *SN Computer Science*, 3(5), 411.

Chapter 8

Scaling up the business with the aid of power query tool

Narayanan Ganesh, Manickavasagam Suruthi, and Ganesh Hariharan
Vellore Institute of Technology, Chennai, India

8.1 INTRODUCTION

There are several organizations coming up with many new dashboard presentation tools. The dashboards can be created using any of the Business Intelligence tools such as Power query, Power BI, QiiK Sense, TIBCO Spotfire, Oracle Cloud Analytics and so on. The idea of creating an interactive dashboard is to have the data in a visual format thereby making the values easily understandable. The data can be extracted from any of the backend sources such as CSV file format, Java Script Notation (JSON) format or Microsoft Excel format. There is a proud saying as, "A picture is worth thousand words". Similarly, graphs and other pictorial representations are the vital means of representing the data and this pictorial representation will act as an aid for the effective decision making in businesses. Power Query transcends mere automation. It acts as a powerful refinery, cleaning and filtering the data to ensure accuracy and consistency. Power Query fosters a culture of data-driven decision-making, empowering individuals across departments to contribute valuable insights.

8.2 LITERATURE STUDY

Not much literature is available exclusively in the use of power query. But there are several literatures available on Business Intelligence tools in general. Kazemi et al. [1] mentions about how the Business Intelligence tool acts as an aid in information processing and making precise decision making at the organization level. Jones [2] states the vital role of business intelligence tool in optimizing the financial data and he also claims that the business intelligence tools have become the foundation stone in making precise decisions. Kankaew et al. [3] in his book describes about the use of business intelligence tool in the entrepreneurial orientation. Mohammed et al. [4] discusses the vitality of using the business intelligence tools in the banking industries so that the banker can by himself analyze the business performance. Adevumi et al. [5] demarcates the adaptability of the business

 DOI: 10.1201/9781003542735-8

intelligence tools in the United States of America and Africa and discusses about the importance of business intelligence tools in shaping the finance sectors across the regions. Mezhoud [6] in his book expresses the impact of business intelligence tool in the small and medium enterprise industrial zones of Setif in Algeria wherein he put forth the implications of cost management tools when used with the business intelligence tools. Jimenez and Medina [7] in one of their articles has expressed the leveraging impact of the business intelligence tools in terms of formulating the strategies and evolution of building new ideas which will in turn result in effective decision making. Ramirez and de Oliviera [8] in their article make a review of the success models through the incorporation of business intelligence tools. Moreno and Hernández [9] specifies the diligence in selecting the appropriate technologies and tools which would encompass the organizational objectives and organizational goals. Campante et al. [10] in his presentation in a conference discusses the versatility in combining operational data with that of the analytical tools. Gurcan et al. [11] in his paper narrates about the strategies, best practices and the latest trends carried out using Business Intelligence (BI). Al-Okaily et al. [12] in his writing explains about the effectiveness of data analysis through business analytic tools. Ragazou et al. [13] enunciates the fact of how the small, medium enterprises get benefitted in terms of effectively streamlining the good business decisions and thereby able to gain more profits in their businesses. Maaitah [14] discusses the importance of gathering the data that could have an impact in the business operations. Xu et al. [15] claims that the business intelligence tool as a powerful tool in the Information Technology sector and the way it impacts the startup firms in their businesses. Lee et al. [16] claims that good design displays will yield better results in analyzing the industries that would lead to the support of innovation policies.

8.3 RESEARCH APPROACH

Business intelligence tools play a vital role in representing the results graphically. There are several business intelligence tools such as qlik sense, Tibco Spotfire, Oracle Cloud Analytics, Tableau, Power Query and Power BI. The Power Query editor is an inbuilt editor available through Microsoft Excel. But many are not aware of its existence, and they focus on different other tools. But in this article, the awareness of the power query tool and its effectiveness in handling large volumes of business data, thereby creating the interactive dashboards will be discussed in detail. The illustration of power query editor and the importance in handling business will be discussed by taking up a case of the multinational bakery industry named "Guru Bakers and Fries".

8.4 RESULTS AND DISCUSSIONS

The spreadsheets containing transaction data of the yesteryears of 2018 and 2023 of Guru Bakers are stored in a folder. Figure 8.1 represents the same.

A new excel workbook is opened and the transaction data is populated. To have a better understanding two years of transaction data viz., 2022 and 2023 are populated. In Microsoft Office 365, the Data tab is selected in which Get Data is selected in which whether a person would like to take the data from File or Folder is determined. From the scenario taken, the Folder option is selected as shown in Figure 8.2.

Figure 8.1 Transaction history.

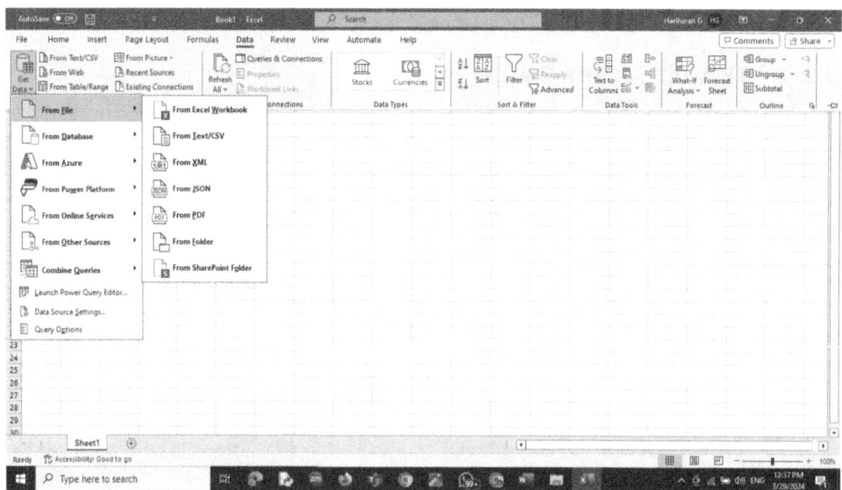

Figure 8.2 Populating data.

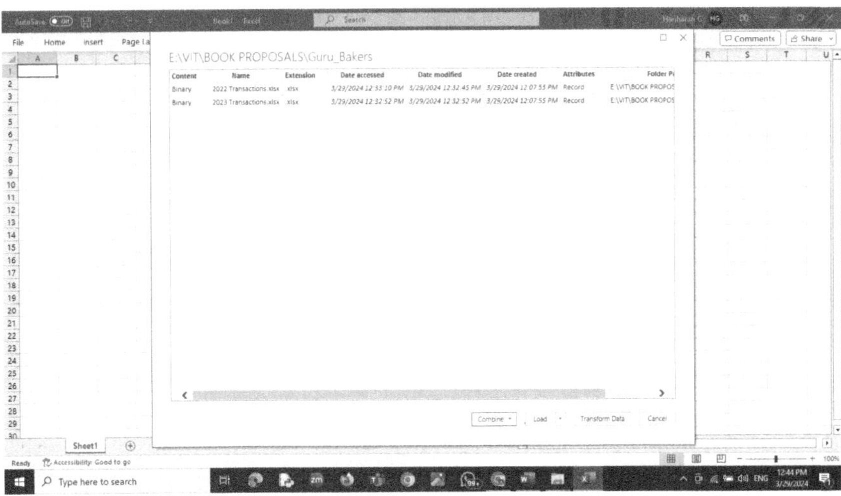

Figure 8.3 Transaction files.

When the Guru Bakers' folder containing the transaction data is selected, the following screen will be obtained which will have the list of files inside the folder as shown in Figure 8.3.

Before working with any data, the data must be cleaned and transformed according to the user's needs. The effective Extract – Transform – Load [ETL] tool will be effectively handled by Microsoft Excel by itself. There will be several options, out of which the Combine and Transform data option, if given, will open the Power query editor, provided the correct file which has the relevant set of data is considered. The power query editor will open after the query is evaluated and processed. Figures 8.4–8.8 determines

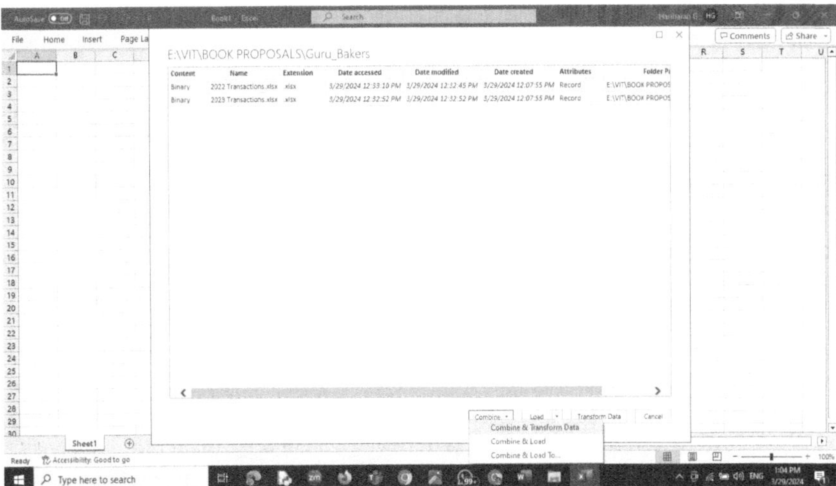

Figure 8.4 Combining and transforming the data.

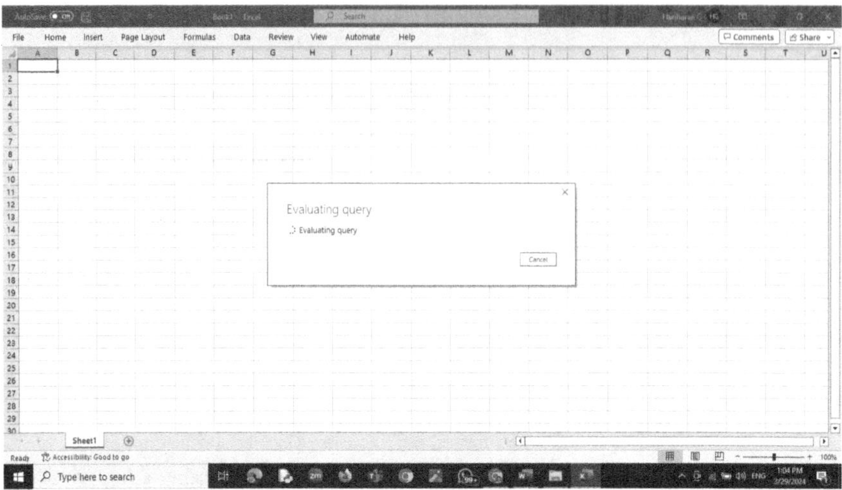

Figure 8.5 Evaluating query.

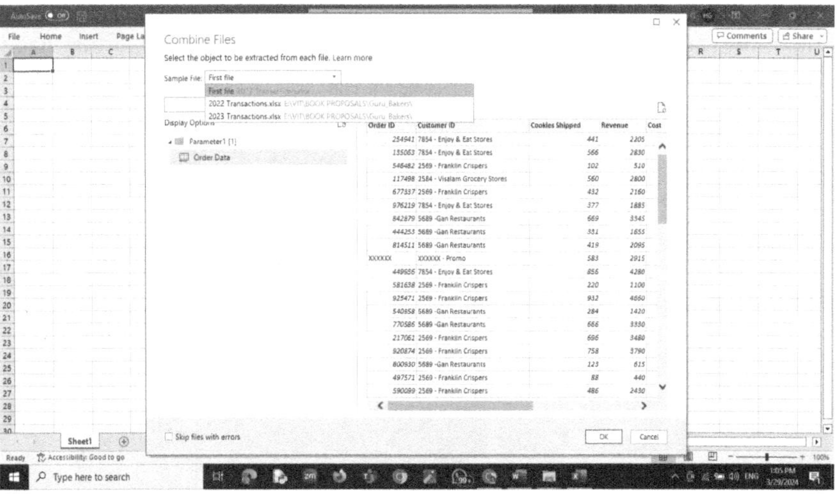

Figure 8.6 Selecting the relevant file.

the flow of the query wherein the power query editor will open, and further transaction will be carried out in the power query editor.

Though Power Query editor is an addon feature available in Excel to create interactive dashboards, this power query editor window will be a unique window which has its own tabs and tools. Figure 8.9 shows the first look of a power query editor. From this Figure 8.9, it can be viewed that power query editor is different from the Microsoft Excel spreadsheet.

The Source file column can be removed, unnecessary order data can be removed, the column can be split with the help of split column tool available.

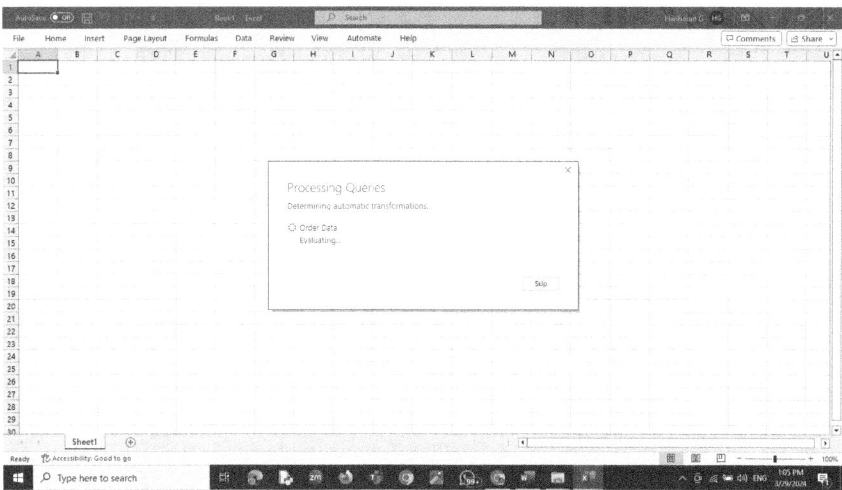

Figure 8.7 Query processing.

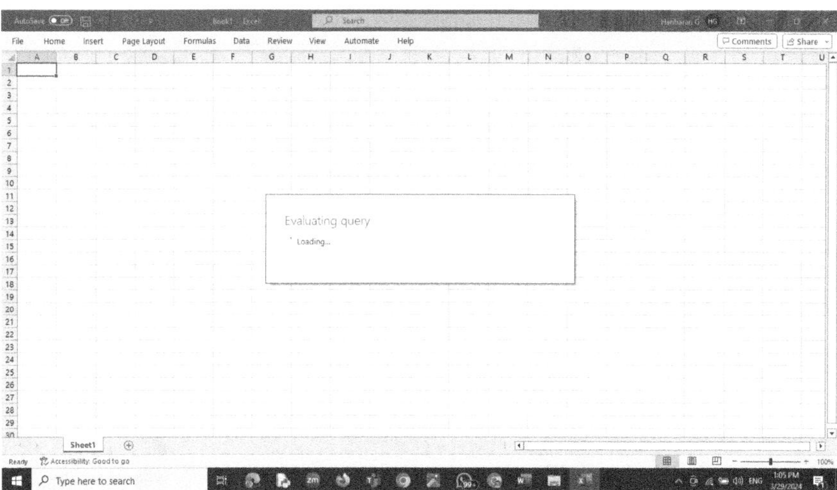

Figure 8.8 Query evaluation.

Figures 8.10–8.15 will show the data cleaning process that could take place in the power query editor.

Data manipulation is carried out from Figures 8.16–8.19. The profit in the business of Guru bakers and fries is calculated as follows:

$$\text{Profit} = \text{Revenue} - \text{Cost} \tag{1}$$

$$\text{Shipment Days} = \text{Ship date} - \text{Order date} \tag{2}$$

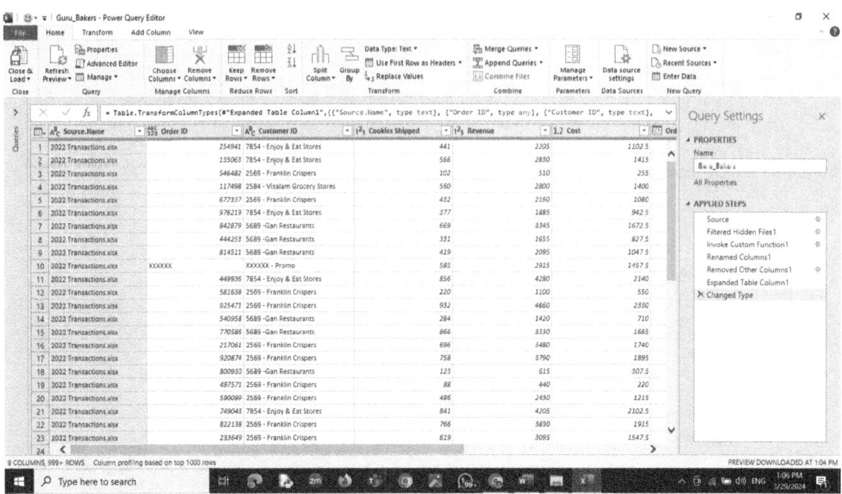

Figure 8.9 Power query editor.

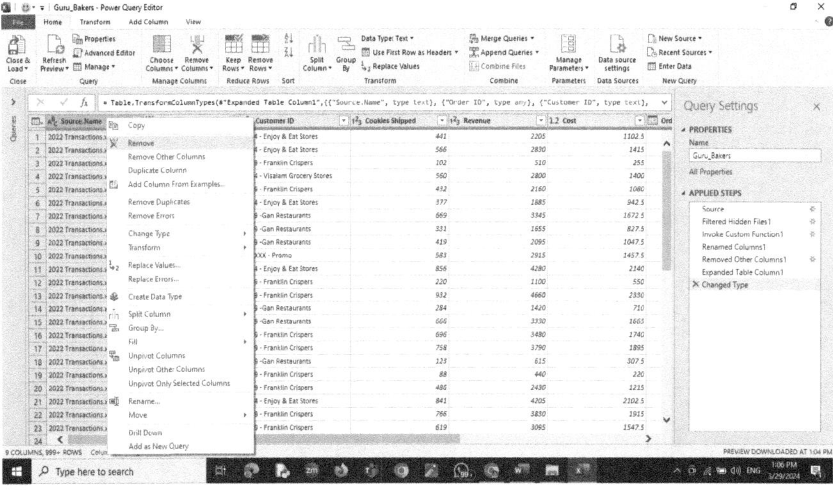

Figure 8.10 Data cleaning 1.

Unlike Excel, in power query, there is an option called "Add Column". When clicked, the Standard toolbar appears. On clicking, the subtraction between Revenue and Cost could be done. Similarly, in power query, for subtracting the days can be done through subtract days option. This is clearly shown in the Figures 8.15–8.19.

All the changes made will be made available at the right-side tab. If any changes need to be undone within the power query editor, then the same could be deleted at a single click on the applied steps.

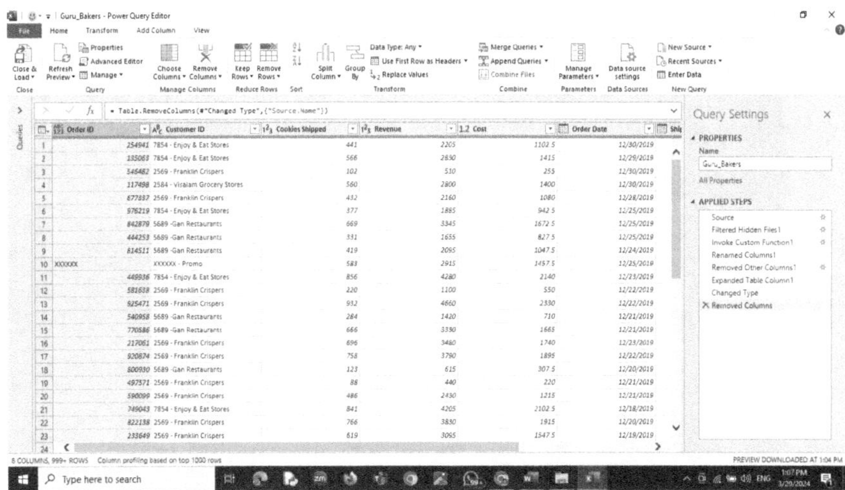

Figure 8.11 Data cleaning 2.

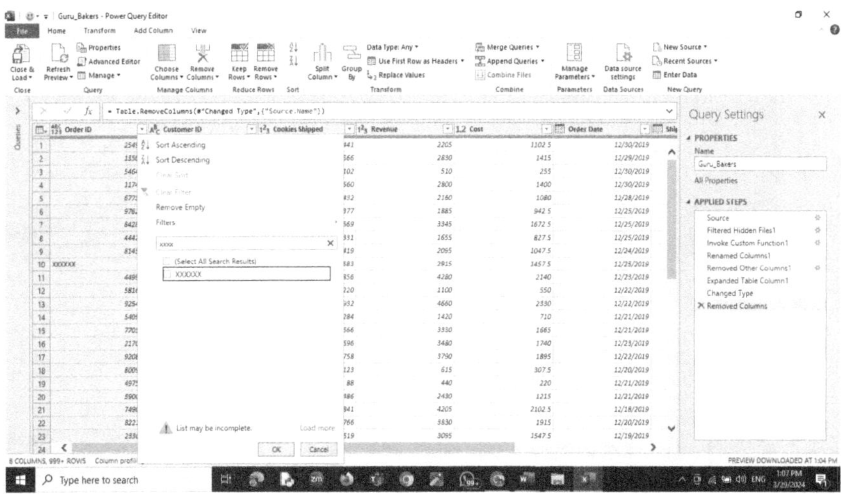

Figure 8.12 Data cleaning 2(1).

The power query editor can be closed by pressing the close and load button available at the corner of the editor as shown in Figure 8.20.

After closing the power query editor, all the data that were cleaned and transformed will get loaded in the Excel (Figure 8.21).

In Excel, a pivot table has to be created. The purpose of pivoting is to group similar items so that the results will be better viewed. Pivot table can

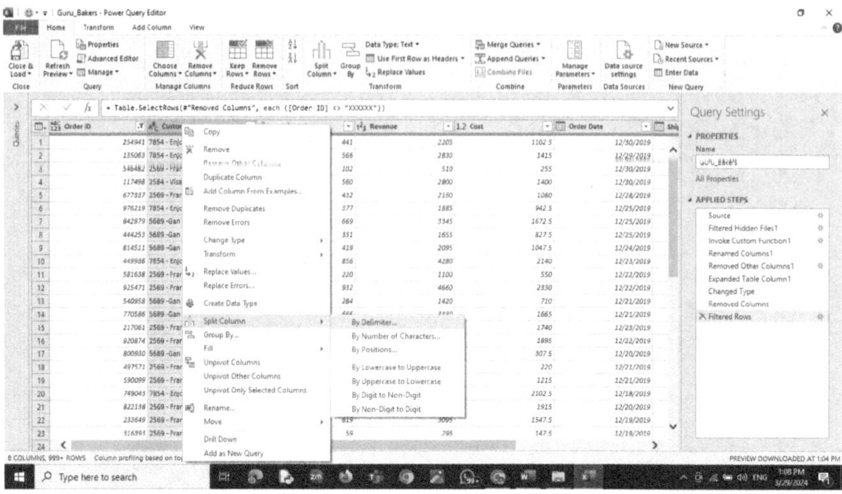

Figure 8.13 Data cleaning 3.

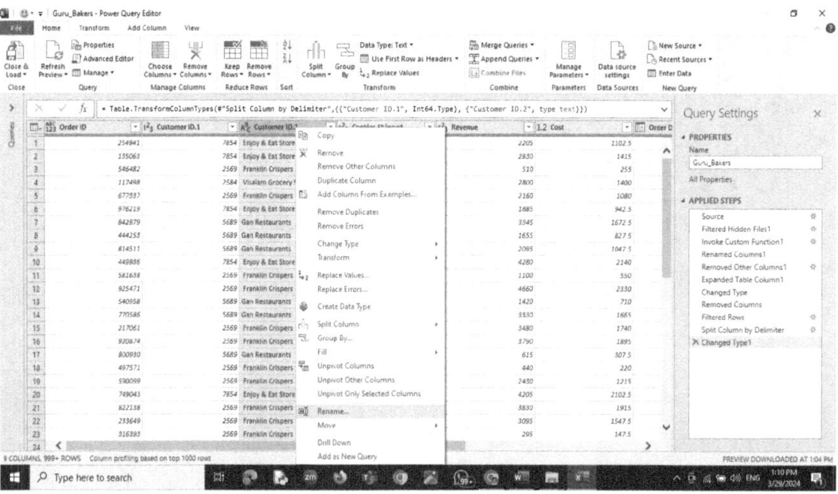

Figure 8.14 Data cleaning 3(1).

be found under the Insert menu or as a tool it will be available as shown in Figure 8.22.

Usually to have a better clarity, pivoting should be done in a new sheet of the excel workbook (Figure 8.23).

The right side of the excel sheet will have the pivot window wherein the transformed data can be grouped according to the user's needs. It can be

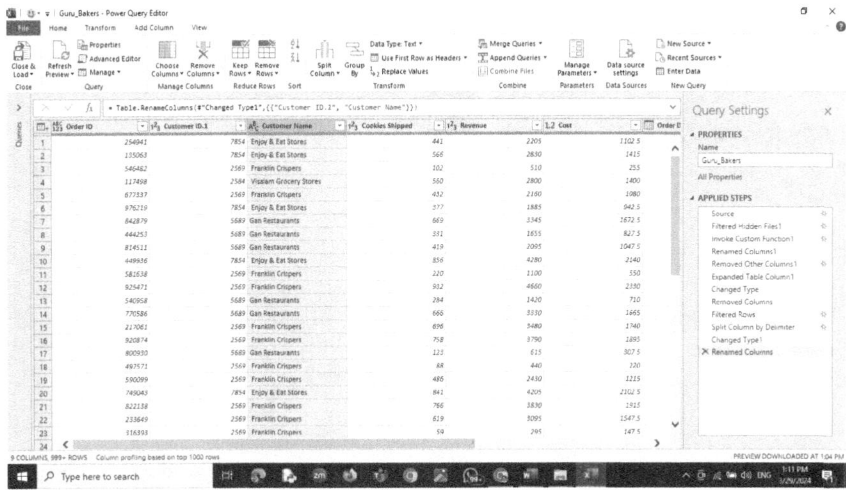

Figure 8.15 Data cleaning 3(2).

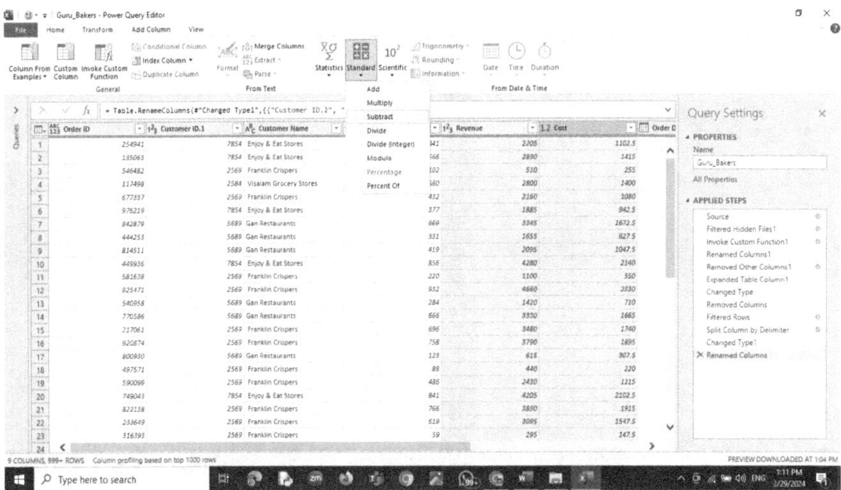

Figure 8.16 Manipulation of data 1.

placed under any of the four boxes such as Filters, Columns, Rows and Values in accordance with the user dashboard requirements. Figure 8.24 represents the same.

Figure 8.25 shows the sample pivoted table wherein the values are clubbed and classified based on month. The specialty of creating power query is that,

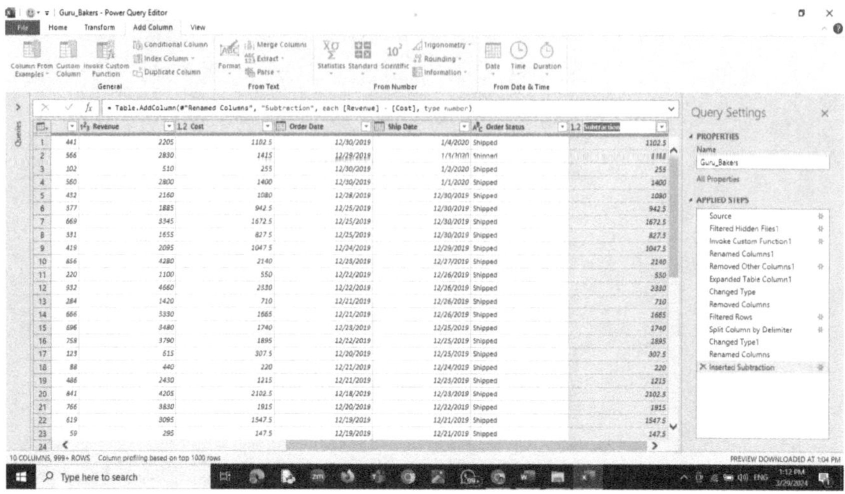

Figure 8.17 Manipulation of data 2.

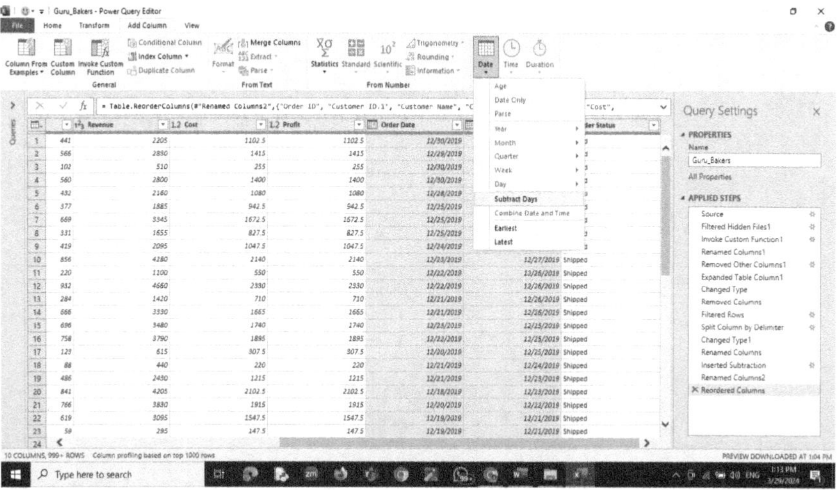

Figure 8.18 Manipulation of data 3.

if created once, there is no need to create again and again. The new data needs to be populated in the Guru Bakers' folder and need to press the Refresh All button available in Excel (Figure 8.26).

Figure 8.27 will show the type of compatible charts that can be chosen in creation of a pivot chart. The clustered columnar graph will be one of the traditional options in creating a pivot chart.

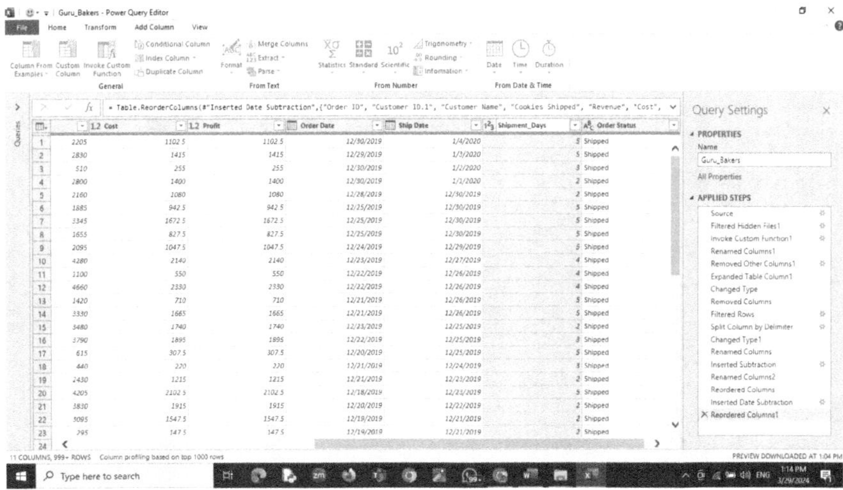

Figure 8.19 Manipulation of data 4.

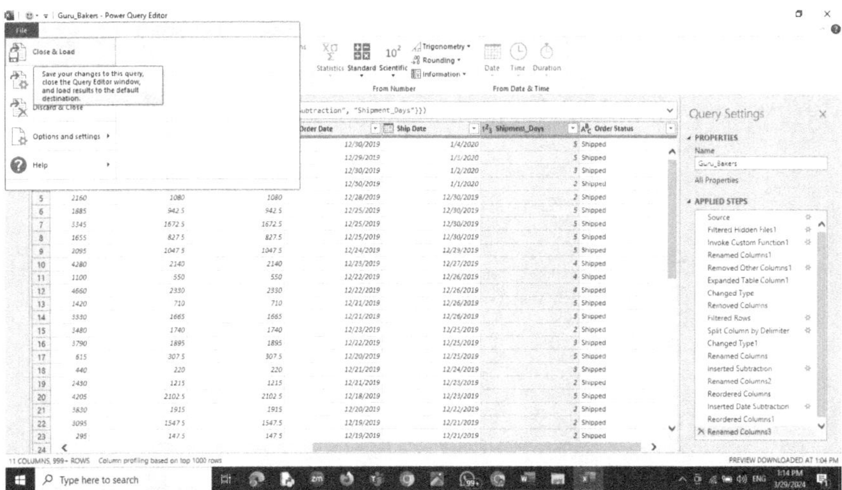

Figure 8.20 Close and load.

Slicing will help the user to have a different look of the values in the identified chart. In the given scenario, the slicing is done based on Customer name, Profit, Order Date and the Ship Date as shown in Figure 8.28. The Slicers can be picked from the Insert menu.

Similarly, timeline need to be chosen which will give the user about the sales happened in a particular month or in a particular week or in a

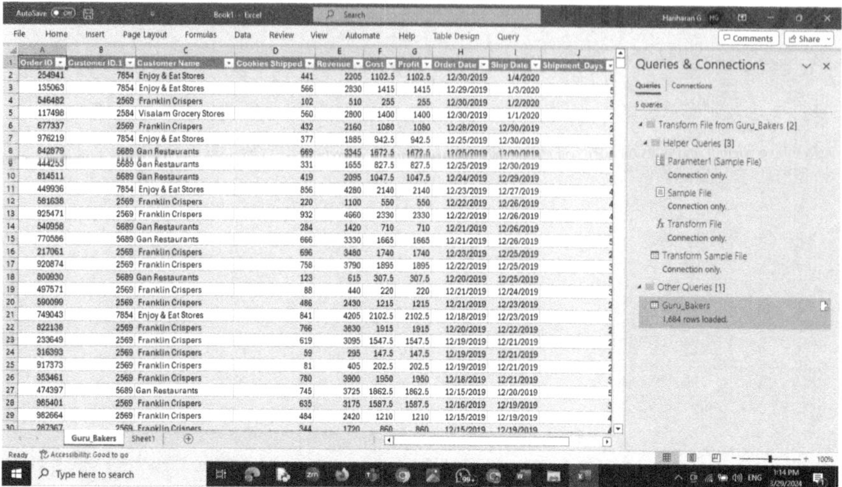

Figure 8.21 ETL data loaded back to Excel.

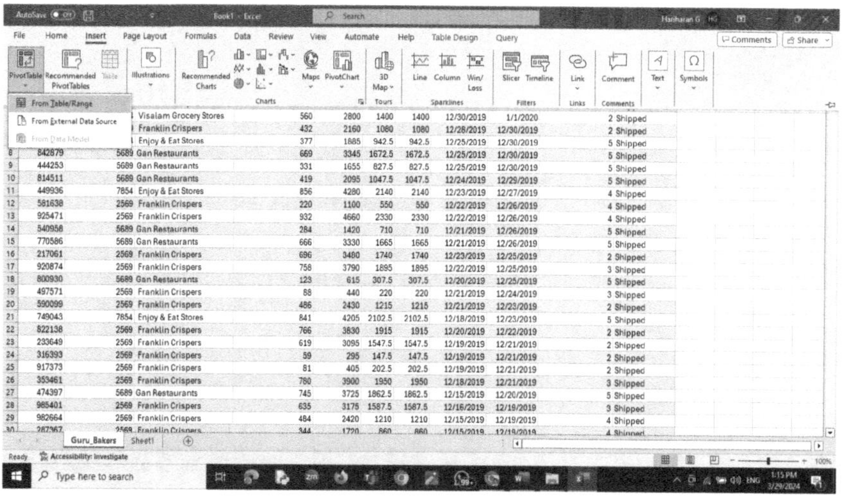

Figure 8.22 Pivot table location.

particular day. The timeline can be picked from the insert menu. But, the timeline will not work if the data points are not selected and wit will result in a connection error as stated in Figure 8.29.

Figure 8.30 represents the Timeline. The timeline can be chosen based on user's needs and on project requirements.

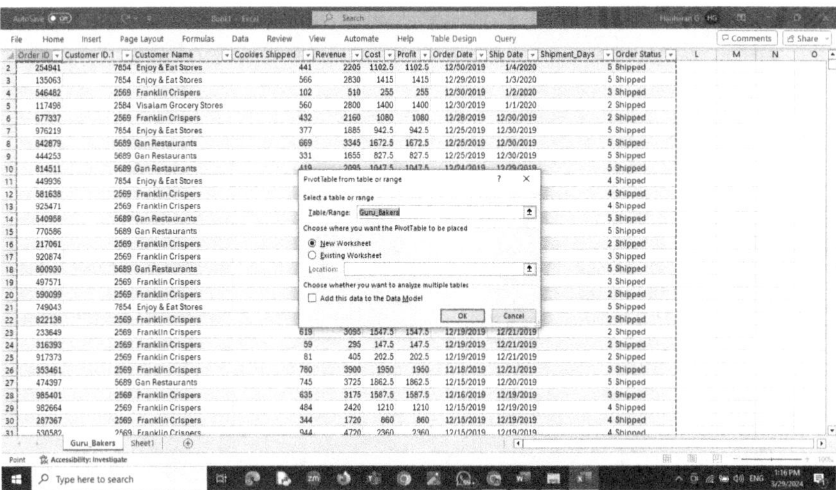

Figure 8.23 Pivoting in a new sheet.

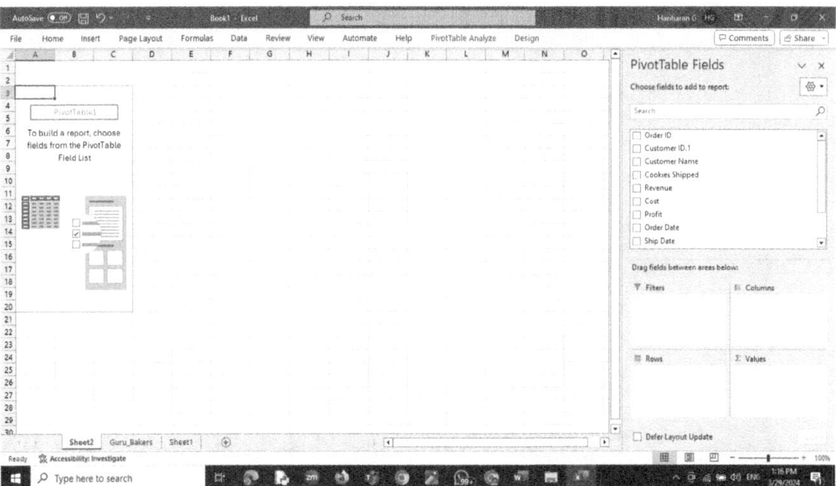

Figure 8.24 Pivoting based on user's requirements.

All the data models including the pivoted chart, slicers and timelines can be placed in a new sheet for better clarity. In the "view" menu, the checkmark of the gridlines and headings, if taken, will give a good look and feel of the interactive dashboard. Figure 8.31 shows the interactive dashboard with slicers and timelines.

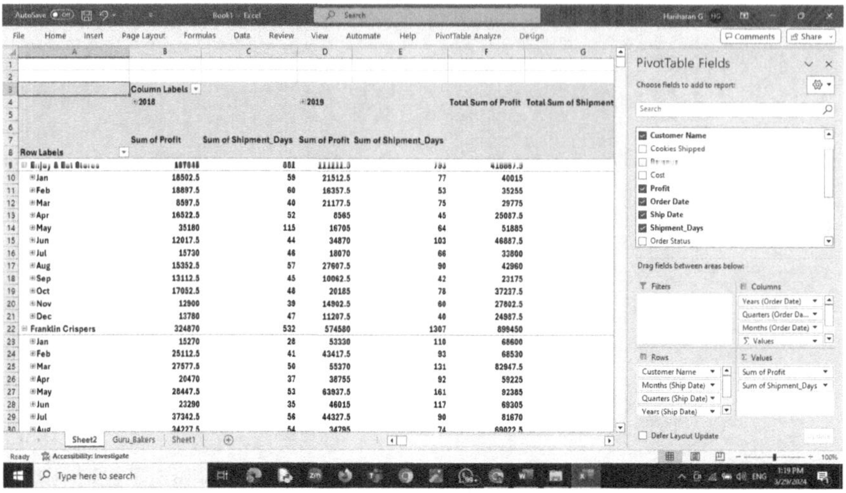

Figure 8.25 Pivoting of values.

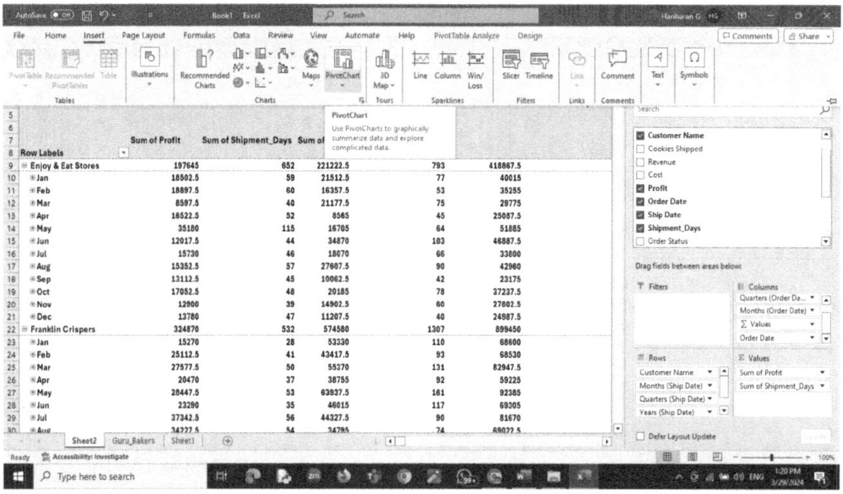

Figure 8.26 Pivot chart.

The graphical pivot chart will change based on the user's selection on the slicer and timeline, thus making the dashboard to be purely interactive. Figure 8.32 shows the changes in graphs based on the user's selection on the slicers and timelines.

The charts may change based on the geographical locations. The given scenario has the details which shows that the business is done across the

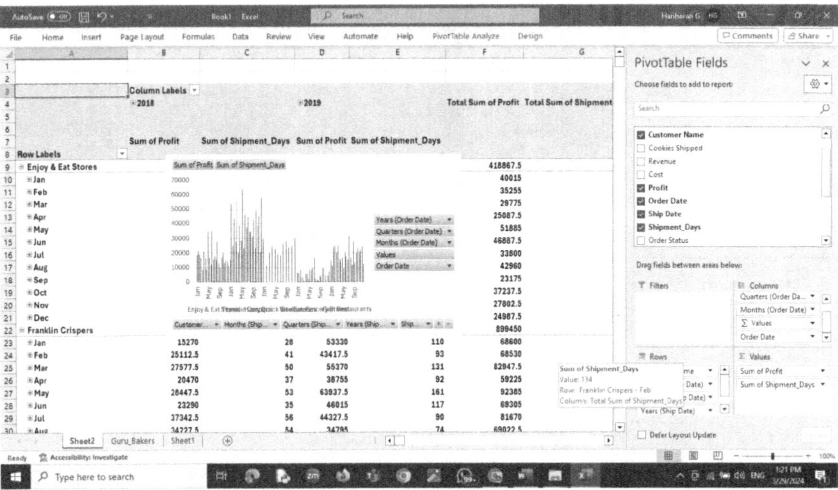

Figure 8.27 Selecting the exact pivot chart.

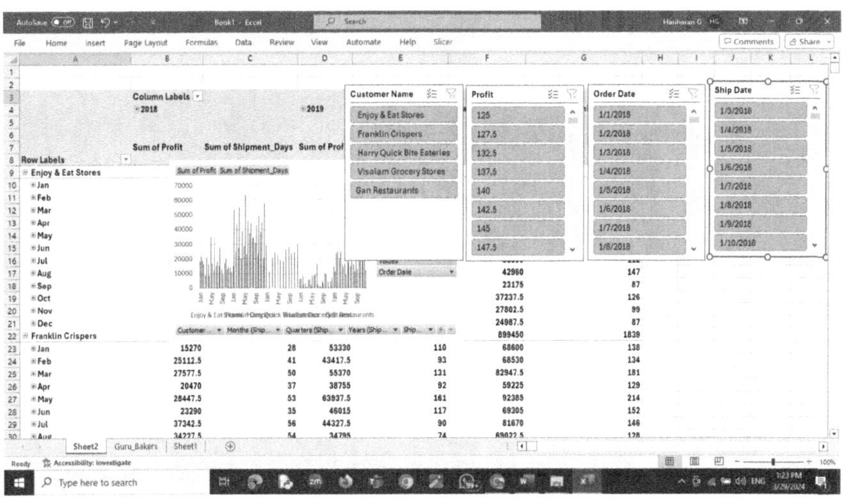

Figure 8.28 Slicing according to user needs.

globe. The 3-Dimensional Map tour is possible in Microsoft Office 365. The data in Figure 8.33 shows the Pie Chart of the sales happened in the locations wherever Guru Bakers and Fries have their businesses (Figure 8.34).

Figures 8.35 and 8.36 will show Cookies selling values specific to a particular Country like India and Canada. Similarly, other Countries can be chosen.

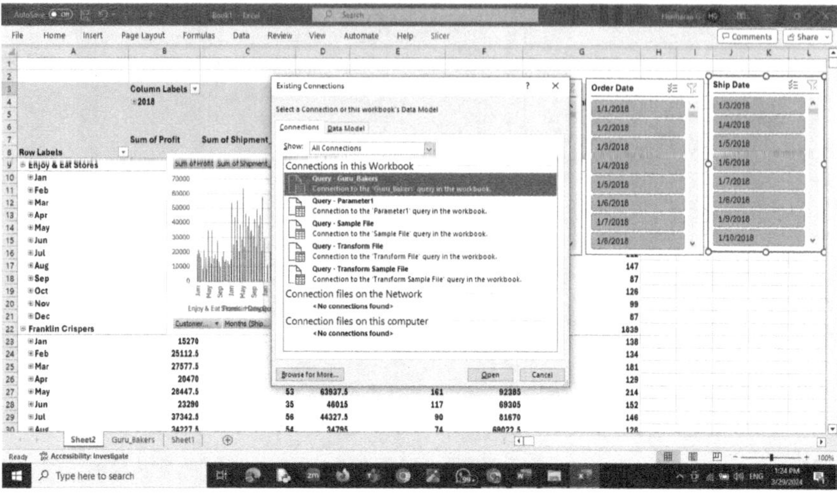

Figure 8.29 Connection error of not choosing the data model.

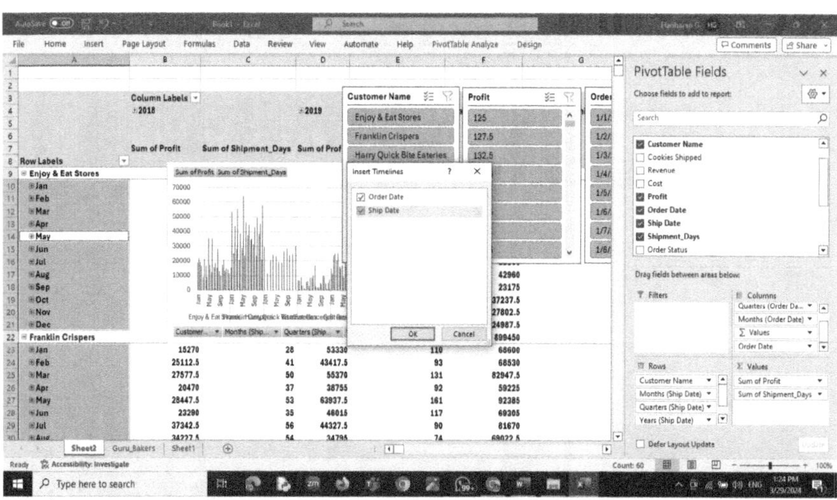

Figure 8.30 Choosing timeline.

Figure 8.37 shows the decorated glossy interactive dashboard so that the customer when uses, might get delighted by witnessing the changing colors of the dashboard. This figure also shows the Performance Dashboard that calculates the profit by market with respect to the type of cookies sold. The X-axis determines the Country sales, and the Y-axis determines the various varieties of cookies. The dashboard will also show the number of units sold each month and the monthly profit with respect to the available projects.

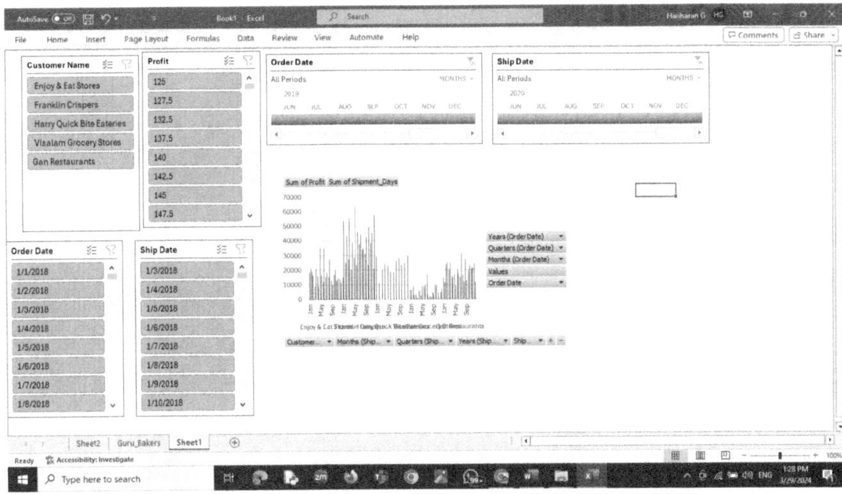

Figure 8.31 Interactive dashboard with slicers and timelines.

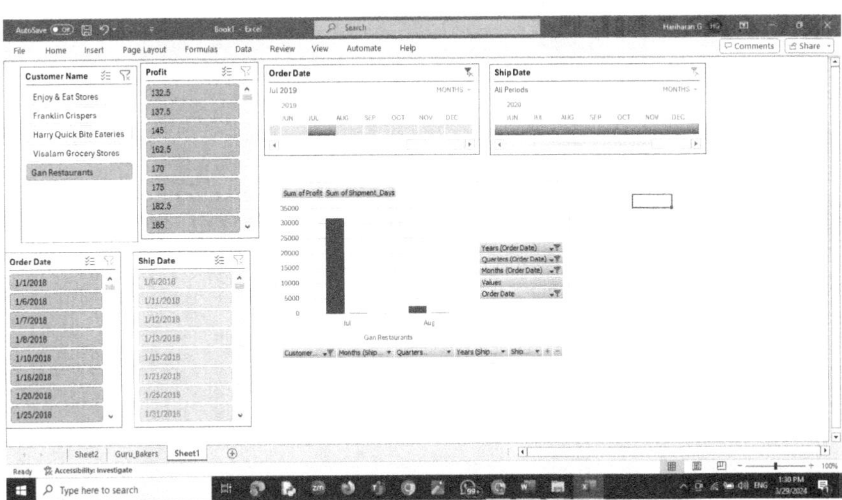

Figure 8.32 Changing graph trend based on selection.

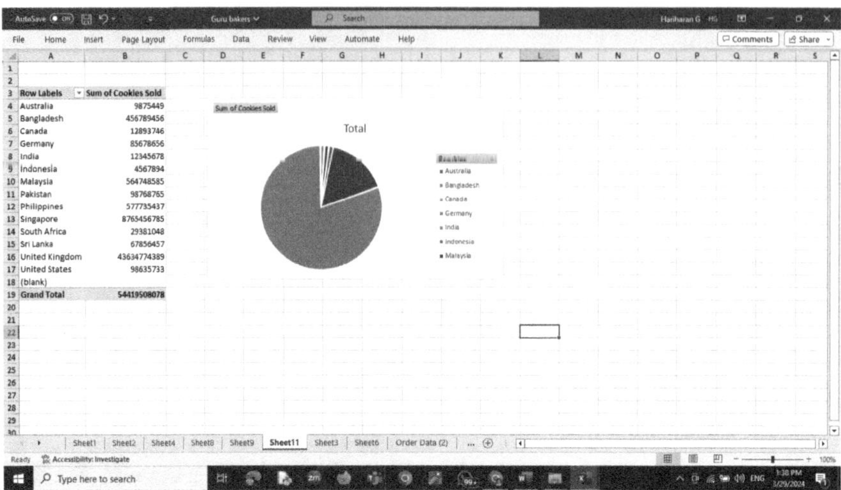

Figure 8.33 Shows the highlighted map data with the all Countries chosen in the Slicer

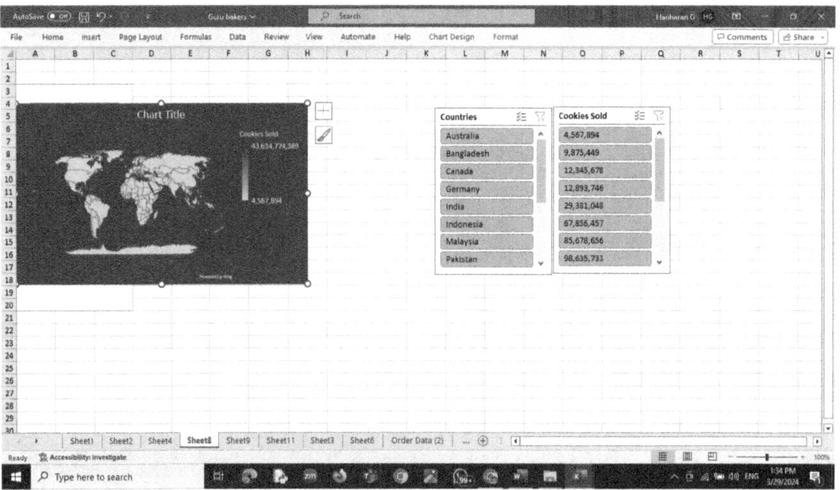

Figure 8.34 All countries map data.

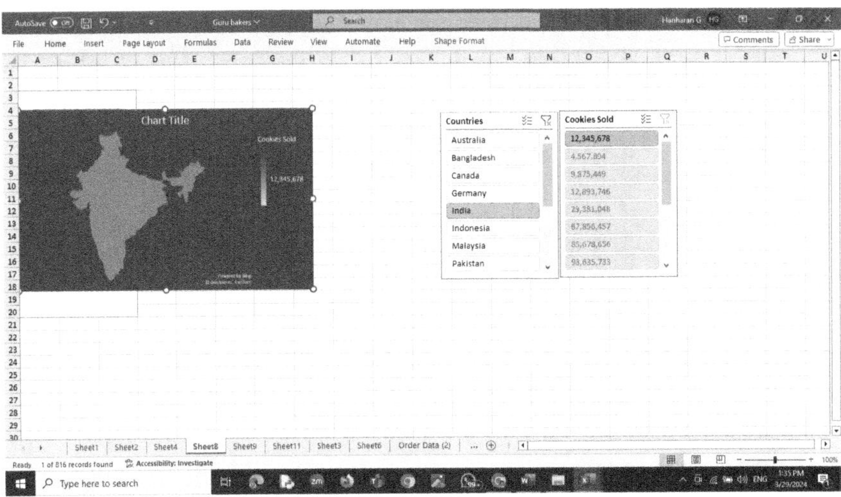

Figure 8.35 Map data with India.

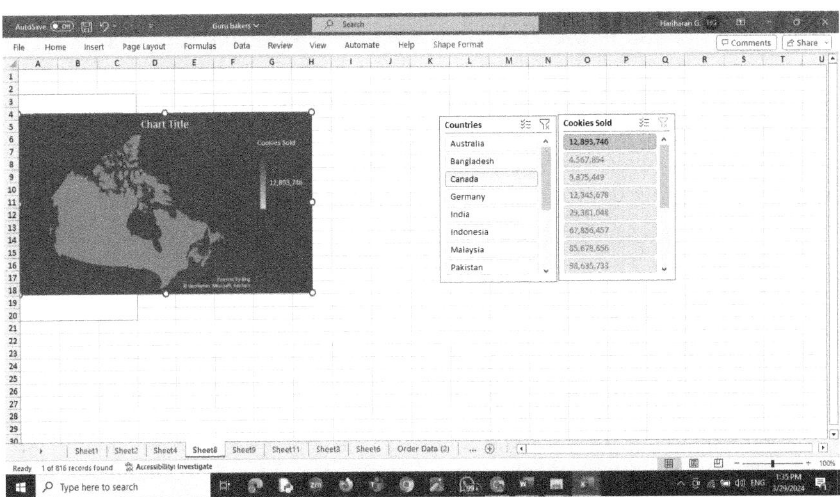

Figure 8.36 Map data with Canada.

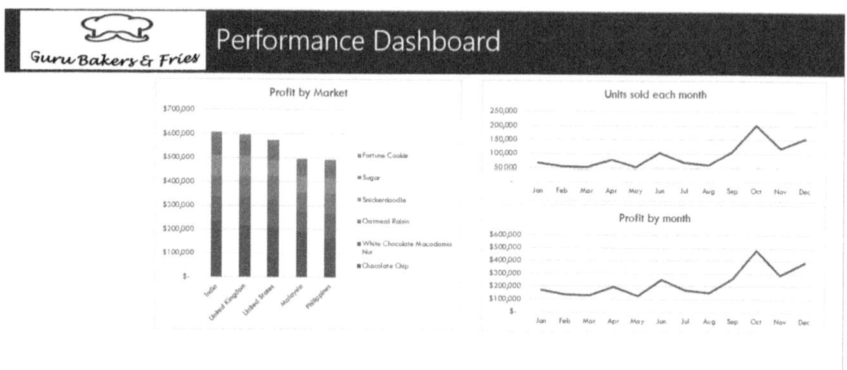

Figure 8.37 Glossy interactive performance dashboard.

8.5 CONCLUSION

Power Query is a vital tool that acts as an aid in showing the business performances in terms of revenue, turnover and profit margins which could be showcased using interactive dashboards. The advantage of having the interactive dashboard is that the graph will change in accordance with the change of selection in the slicer and timeline. Visualizing the data based on the user needs is the specialty in working with such business intelligence tools such as power query. Such data visualization will yield better formulations in making strategic decisions. As a future work, the same dashboard created using power query can be checked with the other business intelligence tools and a comparative analysis of all the tools can also be made.

REFERENCES

1. Kazemi, A., Kazemi, Z., Heshmat, H., Nazarian-Jashnabadi, J., & Tomášková, H. Ranking factors affecting sustainable competitive advantage from the business intelligence perspective: Using content analysis and F-TOPSIS. *Journal of Soft Computing and Decision Analytics*, 2(1), pp: 39–53, 2024.
2. Jones, V. A. Business intelligence solutions for enhanced accounting decision-making in digital transformation. *Engineering Science Letter*, 3(01), pp: 11–15, 2024.
3. Kankaew, K., Nakpathom, P., Chnitphattana, A., Pitchayadejanant, K., & Kunnapapdeelert, S. *Applying Business Intelligence and Innovation to Entrepreneurship*. IGI Global, 2024.
4. Mohammed, A. B., Al-Okaily, M., Qasim, D., & Al-Majali, M. K. Towards an understanding of business intelligence and analytics usage: Evidence from the banking industry. *International Journal of Information Management Data Insights*, 4(1), p: 100215, 2024.

5. Adewumi, A., Oshioste, E. E., Asuzu, O. F., Ndubuisi, N. L., Awonnuga, K. F., & Daraojimba, O. H. Business intelligence tools in finance: A review of trends in the USA and Africa. *World Journal of Advanced Research and Reviews*, *21*(3), pp: 608–616, 2024.
6. Mezhoud, H. Requirements of adopting SMEs for business intelligence systems: A field study in the industrial zone of Setif in Algeria. In *Applying Business Intelligence and Innovation to Entrepreneurship*. IGI Global, pp: 125–154, 2024.
7. Jiménez-Partearroyo, M., & Medina-López, A. Leveraging business intelligence systems for enhanced corporate competitiveness: Strategy and evolution. *System*, *12*(3), p: 94, 2024.
8. Ramirez-Aristizabal, C., & de Oliveira Moraes, R. Işık's and Popovič's business intelligence success models: A review, consolidation, and expansion. *Journal of Decision Systems*, *33*(1), pp: 130–163, 2024.
9. Moreno, M., & Hernández, W. G. The business intelligence blueprint: Integrating analytics, big data, and project management for organizational triumph. *Journal of Environmental Science and Technology*, *3*(1), pp: 161–175, 2024.
10. Campante, M. I., Gonçalves, C. T., & Gonçalves, M. J. A. Business Intelligence Tools to Improve Business Strategy. In *International Conference on Tourism, Technology and Systems*. Singapore: Springer Nature Singapore, pp: 251–268, 2023.
11. Gurcan, F., Ayaz, A., Menekse Dalveren, G. G., & Derawi, M. Business intelligence strategies, best practices, and latest trends: Analysis of scientometric data from 2003 to 2023 using machine learning. *Sustainability*, *15*(13), p: 9458, 2023.
12. Al-Okaily, Aws, Ai Ping Teoh, and Manaf Al-Okaily. "Evaluation of data analytics-oriented business intelligence technology effectiveness: An enterprise-level analysis." *Business Process Management Journal* 29(3), pp: 777–800, 2023.
13. Ragazou, K., Passas, I., Garefalakis, A., & Zopounidis, C. Business intelligence model empowering SMEs to make better decisions and enhance their competitive advantage. *Discover Analytics*, *1*(1), pp: 1–2, 2023.
14. Maaitah, T. The role of business intelligence tools in the decision-making process and performance. *Journal of Intelligence Studies in Business*, *13*(1), pp: 43–52, 2023.
15. Xu, Y., Li, X., bin Mustakim, F., Alotaibi, F. M., & Abdullah, N. N. Investigating the business intelligence capabilities' and network learning effect on the data mining for start-up's function. *Information Processing & Management*, *59*(5), p: 103055, 2022.
16. Lee, S., Lim, D., Moon, Y., Lee, H., & Lee, S. Designing a business intelligence system to support industry analysis and innovation policy. *Science and Public Policy*, *49*(3), pp: 414–426, 2022.

Chapter 9

Interactive visualization techniques for thermal imaging analysis in ophthalmology

Comparative insights and future directions

Persiya J, Sasithradevi A, and Shoba S
Vellore Institute of Technology, Chennai, India

Prakash P
Madras Institute of Technology, Anna University, Chennai, India

9.1 INTRODUCTION

Sight is arguably our most crucial sense, shaping our perception of the world and enabling us to navigate and interact with our surroundings. For centuries, ophthalmologists have relied on the power of light-based imaging techniques to diagnose eye diseases [1]. However, these methods may miss subtle changes or require visible symptoms to be present. Thermal imaging emerges as a powerful complementary tool, offering the potential to detect pre-symptomatic signs of various eye conditions, paving the way for earlier intervention and improved patient outcomes. Specialized thermal cameras are employed to record the surface temperature of the eye, producing radiometric images known as thermograms [2]. These thermograms undergo additional analysis to derive valuable insights. This chapter delves into the potential of thermal imaging, comparing it to established techniques and exploring how advancements in technology and AI are revolutionizing its role in ophthalmology. Beyond the limitations of visual light, thermal imaging offers a unique perspective, capturing the intricate dance of heat signatures within the eye [3]. This thermal fingerprint holds the key to detecting hidden patterns associated with various ocular pathologies, often before they become visible to conventional methods. Imagine identifying the telltale signs of dry eye disease or uveitis before symptoms even arise, empowering early intervention and improved patient outcomes. This chapter serves as a comparative guide, meticulously dissecting the strengths and limitations of each imaging technique. We will demystify the principles of thermal imaging, revealing its distinct advantages in ophthalmology. We will shed light on the unique insights it offers, complementing established methods for a more comprehensive understanding of ocular health. But our

DOI: 10.1201/9781003542735-9

journey doesn't stop there. We will delve into the exciting advancements in thermal imaging technology, exploring cutting-edge image processing algorithms and the transformative power of artificial intelligence. We will unveil the next generation of interactive visualization techniques, where thermal data comes alive, empowering clinicians with unparalleled diagnostic precision and personalized patient care. Looking ahead, we will chart the course for the future of thermal imaging in ophthalmology, highlighting emerging applications and transformative potential. By weaving together this comparative analysis, we aim to illuminate the transformative potential of thermal imaging, paving the way for a brighter future in eye care.

9.1.1 Evolution of ophthalmological imaging techniques

Over the years, advancements in imaging techniques have been closely linked to our understanding of ocular anatomy. Initially, examinations were conducted purely through direct observation, but the invention of the ophthalmoscope in 1850 was a revolutionary step that allowed medical professionals to examine the retina's intricate details for the first time. This breakthrough catalyzed a cascade of innovation, leading to the emergence of several pivotal techniques: Fundus photography, which captures static retinal images, initially employing traditional film cameras before transitioning to digital platforms; Fluorescein angiography (FA) helps visualize vascular obstructions by injecting a fluorescent dye and then imaging its circulation. Ultrasound offers cross-sectional perspectives of ocular structures, especially useful in cases involving opaqueness. Optical coherence tomography (OCT) uses light waves to create detailed, high-resolution cross-sectional images of ocular layers, offering unprecedented insight into ocular anatomy. Optical coherence tomography angiography (OCTA) utilizes OCT principles to depict blood flow within ocular vasculature, shedding light on circulatory irregularities. Despite their remarkable contributions to our comprehension of ocular health, each technique possesses inherent limitations.

The timeline depicted in Figure 9.1 illustrates the remarkable evolution of ophthalmic technology, covering a span of more than four centuries. This timeline diagram is drawn using draw.io dashboard. A Draw.io dashboard can provide a visually engaging and informative overview of key milestones in the field. By leveraging Draw.io's intuitive interface and versatile diagramming capabilities, users can create a timeline diagram that highlights significant advancements over time. With customizable styling options and collaborative features, Draw.io empowers users to design a timeline dashboard that effectively conveys complex information in a clear and compelling manner. It begins in the 16th century with the conceptualization of the camera obscura, laying the groundwork for future

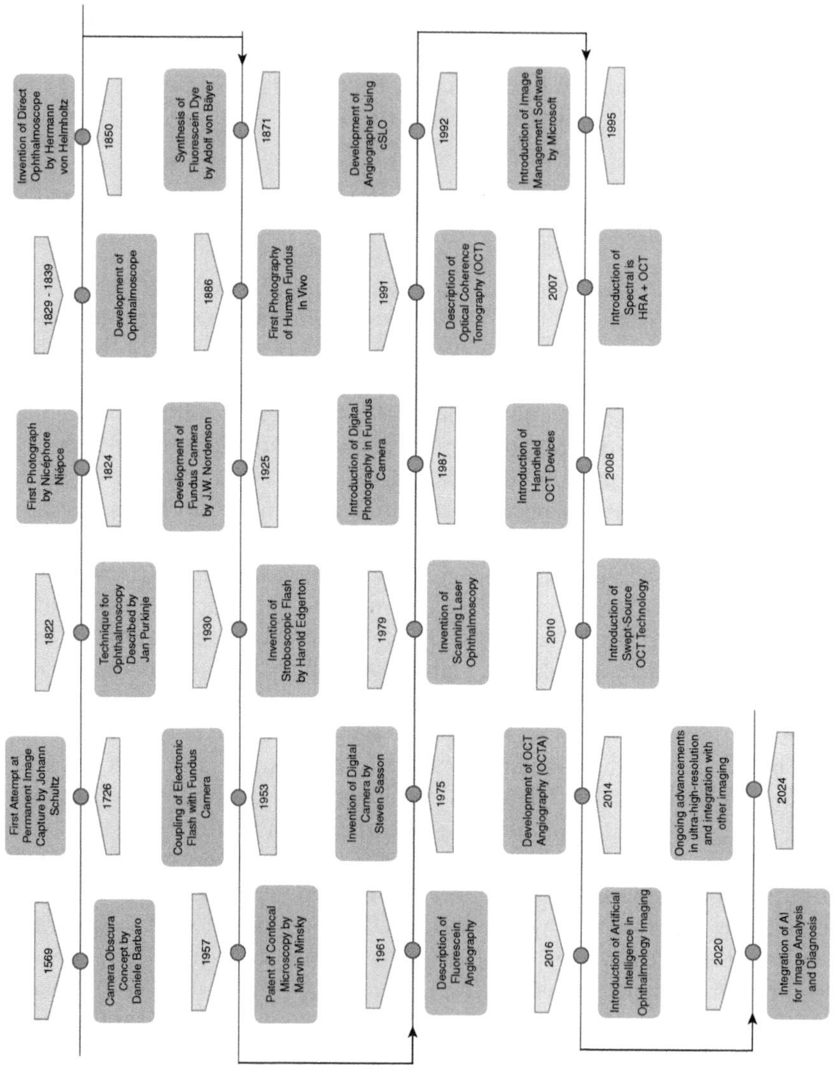

Figure 9.1 Timeline of evolution of ophthalmological imaging techniques.

advancements [4]. The 19th century witnessed the birth of photography alongside the development of the ophthalmoscope, revolutionizing the way we examine the eye. The 20th century saw the continuous refinement of imaging techniques with the introduction of fundus cameras, strobe flashes, and digital cameras. The invention of confocal microscopy and the discovery of OCT further expanded our ability to visualize the eye's intricate structures. The 21st century ushered in a new era of innovation with handheld OCT devices, swept-source technology, and the integration of AI, paving the way for increasingly sophisticated diagnostics and personalized care in ophthalmology.

9.1.2 Rationale for exploring thermal imaging

Exploring thermal imaging in ophthalmology offers numerous compelling reasons. Firstly, it's non-invasive and comfortable for patients, making it ideal for children or those with sensitive eyes [5]. Additionally, it provides real-time insights into temperature variations in the eye, aiding in rapid identification of issues like inflammation. Unlike traditional methods, thermal imaging reveals temperature-related physiological changes, offering deeper insights into ocular conditions. Its potential for early detection of pathologies such as dry eye and uveitis, before visible symptoms appear, can lead to timely intervention and better outcomes [6]. Moreover, thermal imaging complements existing techniques like OCT and fundus photography, providing unique information for accurate diagnosis. With ongoing technological advancements and its cost-effectiveness, thermal imaging holds promise for improving patient care and outcomes in ophthalmology, potentially unlocking new avenues for diagnosis, disease monitoring, and personalized treatment strategies.

9.1.3 Objectives of the comparative analysis

The objectives of the comparative analysis are to assess the strengths and limitations of thermal imaging in comparison to established ophthalmological imaging techniques such as fundus photography, OCT, OCTA, and ultrasound. The analysis seeks to explore how thermal imaging complements existing techniques, providing unique insights into ocular health and enhancing diagnostic accuracy. Notably, thermal imaging provides objective temperature data, unlike other techniques that rely on subjective assessment [7], offering valuable complementary information. Furthermore, the objectives include identifying potential applications and future directions for thermal imaging in ophthalmology, such as its integration into telemedicine, population-based screening programs, and personalized patient care initiatives. Ultimately, the comparative analysis aims to provide readers with a comprehensive understanding of thermal imaging's transformative potential in ophthalmology and its role in advancing the landscape of eye care.

9.2 PRINCIPLES OF THERMAL IMAGING IN OPHTHALMOLOGY

Thermal imaging in ophthalmology harnesses infrared technology to capture and depict the thermal patterns and temperature fluctuations present in ocular tissues and structures. Drawing from the foundational principles of thermography, this method utilizes the infrared segment of the electromagnetic spectrum to detect and record the thermal radiation emitted by objects, including biological tissues [8]. Subsequently, the captured infrared radiation undergoes processing to generate thermal images that elucidate temperature differentials, thereby offering valuable insights into ocular health and pathology. Operating on the concept of converting radiant heat into a visual representation based on temperature variations, thermal imaging utilizes the invisible infrared spectrum, detectable by specialized sensors [9]. The image formation process involves thermal cameras measuring emitted infrared radiation and assigning colors or gray scales to different temperature ranges. While thermal imaging boasts advantages such as non-invasiveness, real-time information, and sensitivity to subtle temperature changes, it also presents limitations, including lower image resolution compared to visible light imaging and the requirement for interpretation skills acquired through training.

9.2.1 Mechanisms of heat generation and transfer in the eye

Understanding how heat is produced and distributed within the eye is essential for interpreting thermal imaging data accurately. The eye generates heat through metabolic processes in its tissues and blood circulation, while factors like inflammation and external influences also contribute [10]. Heat travels within and around the eye through conduction, convection, and radiation. Different parts of the eye have unique temperature profiles, with the cornea typically being cooler and structures like the lacrimal glands warmer. Changes in blood flow, inflammation, or tissue damage can alter these heat patterns, which can be observed and analyzed using thermal imaging techniques.

9.2.2 Thermal imaging modalities and technologies

In ophthalmology, thermal imaging devices and equipment encompass a variety of infrared cameras, thermal sensors, and image-processing systems tailored specifically for capturing and analyzing thermal data from the eye. These devices are continuously evolving, with ongoing improvements in resolution, sensitivity, and integration with advanced image processing algorithms and artificial intelligence [11]. Infrared thermography, a key method, captures static thermal images of the eye and its surroundings, facilitating the detection of conditions such as dry eye, inflammation (uveitis), and neuro-ophthalmic disorders, as well as monitoring treatment responses. These thermal imaging devices come in various forms, including portable scanners, handheld devices, and integrated ophthalmic instruments. Advancements in technology have led to miniaturization, enhanced resolution, and the development of software for image analysis and AI integration. Factors to consider when selecting thermal imaging equipment include cost, portability, ease of use, image quality, and compatibility with other systems.

9.3 COMPARATIVE ANALYSIS OF IMAGING TECHNIQUES

This section compares various imaging techniques commonly used in ophthalmology, highlighting their respective strengths, limitations, and applications in detecting and monitoring ocular pathologies. Table 9.1 illustrates the detailed comparison of different imaging techniques in ophthalmology. The table outlines the imaging techniques, their focus, dilation requirements, distance from the patient, type of camera used, strengths, limitations, complementarity to thermal imaging, and applications. The comparative assessment of these imaging techniques reveals their complementary roles in ophthalmological diagnostics. While traditional imaging methods such as fundus photography, OCT, OCTA, and ultrasound excel in providing detailed anatomical and structural information, thermal imaging offers a distinct focus on revealing temperature variations associated with ocular pathologies. By combining these techniques, a comprehensive approach to ocular diagnostics is achieved, providing valuable insights into ocular health [12]. Each imaging modality contributes unique information, allowing for a more comprehensive understanding of ocular conditions and facilitating tailored treatment approaches [13]. Various imaging techniques offer distinct advantages and applications in ocular diagnostics. Fundus photography, utilizing specialized cameras, provides high-resolution images of retinal anatomy, complementing thermal imaging by offering detailed anatomical context crucial for detecting conditions like diabetic retinopathy. Optical coherence tomography (OCT), both standard and angiography, offers non-contact, high-resolution visualization of retinal layers and vasculature, aiding in structural assessment and monitoring diseases such as macular edema and glaucoma. Ultrasound, while limited in heat sensitivity, excels in visualizing deeper ocular structures like the vitreous cavity and optic nerve. Thermal imaging, on the other hand, specializes in detecting temperature variations associated with inflammation, providing unique insights crucial for examining conditions like uveitis and conjunctivitis. Each technique complements thermal imaging by offering specific diagnostic capabilities, enhancing the overall understanding and management of ocular diseases. Figure 9.2 displays standard images of the human eye captured through various imaging techniques.

Comparative tables are drawn using Google Data Studio. Google Data Studio is a user-friendly data visualization tool that enables users to create

Table 9.1 Comparison of imaging methods

Imaging techniques	*Focus*	*Dilation required*	*Distance from patient*	*Camera*	*Strength*	*Limitations*	*Complementary to thermal imaging?*	*Applications*
Fundus photography	Anatomy & Structure	Yes	Close Contact	Specialized fundus cameras	High-resolution visualization of retinal detail	Limited insight into thermal variations	Yes, for detailed anatomical context	Detecting retinal lesions (diabetic retinopathy, macular degeneration)
Optical coherence	Anatomy & Structure	No	Non-contact (uses Near-infrared (780–920 nm))	Scanning laser source & specialized detectors	High-resolution cross-sectional visualization of retinal layers	Limited insight into thermal variations	Yes, for detailed structural assessment	Diagnosing & monitoring conditions like macular edema, glaucoma
Optical coherence tomography angiography	Blood Flow	No	Non-contact (uses Near-infrared (780–920 nm))	Similar to OCT with specialized components for blood flow analysis	Non-invasive visualization of retinal vasculature	Newer technology, limited depth, requires interpretation training	Yes, for detailed structural assessment	Detecting & monitoring vascular disorders, assessing treatment efficacy
Ultrasound	Deeper Structures	No	Close contact or immersion probe	Piezoelectric transducers emitting & receiving sound waves	Good for visualizing deeper ocular structures	Not sensitive to heat variations	Limited, mainly for confirming ambiguous findings	Examining vitreous cavity, optic nerve, foreign bodies & tumors
Thermal imaging	Temperature	No	Non-contact (uses Specialized thermal cameras detecting infrared radiation)	Detects subtle temperature changes (inflammation)	Detects subtle temperature changes associated with pathologies	Lower resolution requires skilled interpretation	Yes, crucial for unique thermal insights	Examining dry eye, uveitis, conjunctivitis, and inflammations.

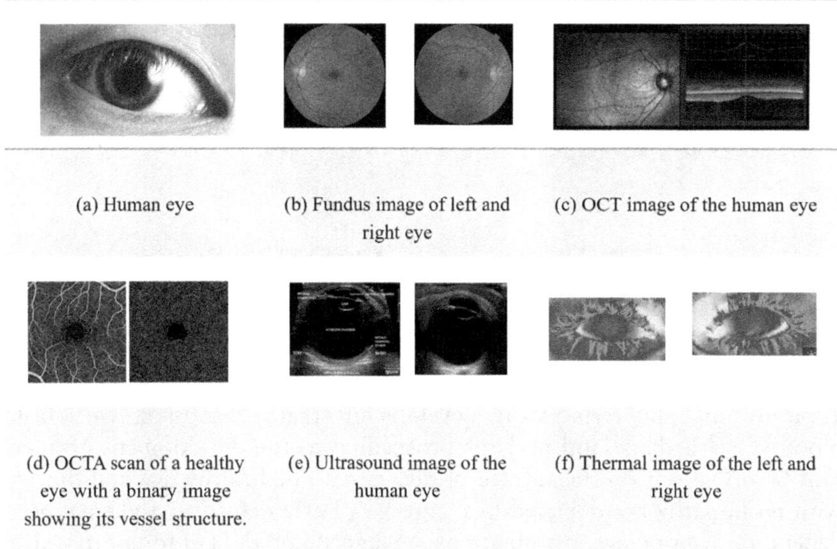

(a) Human eye
(b) Fundus image of left and right eye
(c) OCT image of the human eye
(d) OCTA scan of a healthy eye with a binary image showing its vessel structure.
(e) Ultrasound image of the human eye
(f) Thermal image of the left and right eye

Figure 9.2 Typical images of the human eye using different image modalities.

interactive and customizable reports and dashboards. Integrated seamlessly with other Google products like Google Sheets, Google Analytics, Data Studio offers drag-and-drop functionality for creating dynamic visualizations such as charts, graphs, and tables. With its collaborative features and ability to connect to various data sources, Google Data Studio empowers users to turn raw data into actionable insights effortlessly.

9.4 APPLICATIONS OF THERMAL IMAGING IN OPHTHALMOLOGY FOR EYE ANOMALY DETECTION

The use of ocular thermograms to spot abnormalities in the eyes is an emerging method in the field of ophthalmic diagnosis. This imaging technique can detect a variety of eye disorders and conditions by recording and examining the thermal patterns of the eye. These could be inflammatory reactions, dry eye syndrome, diabetic retinopathy, glaucoma, scleritis, conjunctivitis, and corneal conditions, all of which have been associated with changes in temperature on the ocular surface. Some of the eye diseases that can be detected using ocular thermograms are discussed below.

9.4.1 Diabetic retinopathy

A chronic eye condition that affects people with diabetes is diabetic retinopathy. It develops as a result of damage to the blood vessels in the retina,

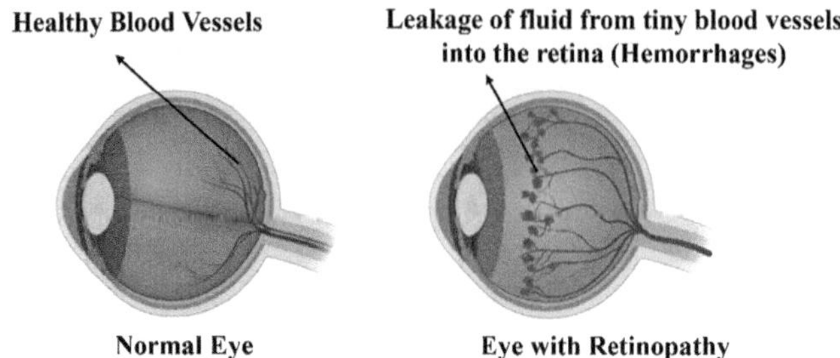

Figure 9.3 Diabetic retinopathy impact on the eye.

which can cause visual issues and, if ignored, could result in blindness. Typically, diabetic retinopathy develops in stages, beginning with little blood vessel leakage and perhaps proceeding to the development of aberrant blood vessels on the surface of the retina. The healthy eye and the eye with retinopathy are depicted in Figure 9.3. Early detection and treatment depend on routine eye examinations. Yashan Bu et al. [14] found that diabetes significantly affects the condition of the ocular surface in the human eye. In a related study, Bhuvaneswari Chandrasekar et al. [15], investigated alterations in ocular surface temperature resulting from diabetic retinopathy. Thermographic monitoring can assist in tracking the development of diabetic retinopathy over time [16], allowing for customized treatment strategies and improved control of the condition.

9.4.2 Glaucoma

Glaucoma is a serious eye disease characterized by increased intraocular pressure, which can damage the optic nerve and lead to irreversible vision loss if left untreated. It is often referred to as the "silent thief of sight" because it typically progresses slowly without noticeable symptoms in its early stages. Figure 9.4 shows a comparison between a normal eye and an eye affected by glaucoma. By capturing temperature variations in the

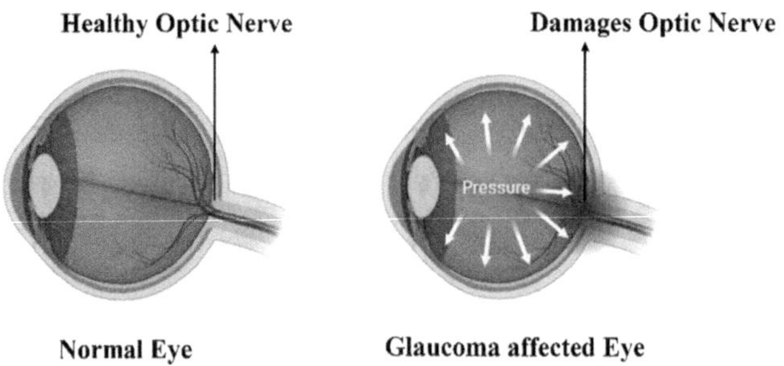

Figure 9.4 Glaucoma impact on the eye.

eye, especially around the optic nerve head, thermograms can help identify changes associated with increased intraocular pressure, a primary risk factor for glaucoma [17]. Thermographic imaging, in conjunction with traditional diagnostic methods, can assist ophthalmologists in identifying individuals at risk for or in the early stages of glaucoma, enabling timely intervention and better management of this sight-threatening condition.

9.4.3 Dry eye

Dry eye is a common eye condition characterized by a lack of sufficient moisture and lubrication on the eye's surface. It can result in discomfort, irritation, and a gritty sensation in the eyes. Dry eye can be caused by various factors, including environmental conditions, age, certain medications, and underlying medical conditions. Dry eye condition is illustrated in Figure 9.5. Meibomian Gland produces an oily substance called meibum. Meibum's role is to prevent the rapid evaporation of tears and maintain the stability of the tear film on the surface of the eye. Dysfunction of the meibomian glands can lead to conditions like meibomian gland dysfunction (MGD) and evaporative dry eye [18]. Lacrimal Glands is situated above the outer corner of each eye, and its primary function is to produce the watery component of tears. The watery tears produced by the lacrimal glands, contain enzymes and proteins that helps to keep the eye's surface clean and nourished. Tear ducts also known as lacrimal ducts are responsible for draining tears away from the eye's surface into the nasal passages. In cases of dry eye, improper tear drainage can exacerbate the condition, as tears don't adequately lubricate the eyes. Many invasive techniques are used to diagnose and treat dry eye [19, 20]. By capturing and analyzing these temperature variations, healthcare professionals can better understand the extent and severity of dry eye, aiding in its diagnosis and treatment planning [21]. Treatment options range from over-the-counter artificial tears to prescription medications, depending on the severity of the condition. Managing dry eye is essential to alleviate discomfort and maintain eye health.

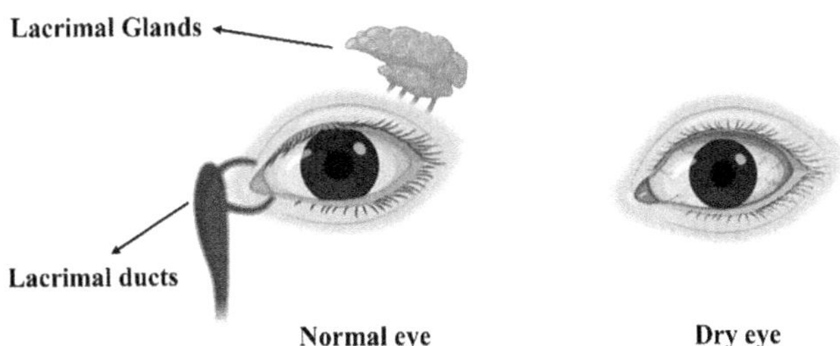

Figure 9.5 Dry eye impact on the eye.

9.4.4 Scleritis

Scleritis is a rare but serious condition characterized by inflammation of the white part of the eye (sclera). It often presents with severe eye pain, redness, and potential vision changes. Scleritis can be associated with underlying autoimmune diseases and requires prompt medical attention to prevent complications. In cases of scleritis, ocular thermography can detect increased temperature over the inflamed area of the sclera [22, 23]. Thermal imaging helps in quantifying the extent and severity of the inflammation, aiding in the diagnosis and monitoring of treatment response for this potentially serious eye condition.

Thermal imaging shows promise in detecting various ocular pathologies early by identifying subtle temperature changes indicative of underlying inflammatory processes. Its sensitivity to these changes makes it potentially valuable for detecting conditions in their early stages, before significant damage occurs. Detecting subclinical conditions such as pre-retinal inflammation or neuro-ophthalmological abnormalities early could allow for preventive measures and improve long-term outcomes. Additionally, thermal imaging can offer insights for guiding therapeutic interventions and surgical planning. For instance, identifying areas of inflammation during ocular surface reconstruction surgery can aid surgeons in targeting their efforts more precisely. Similarly, during laser procedures, thermal imaging could monitor tissue response and potential thermal damage, ensuring safer and more effective treatment delivery.

9.5 ADVANCEMENTS IN THERMAL IMAGING TECHNOLOGY

Advancements in thermal imaging technology are revolutionizing ophthalmic diagnostics, particularly through image processing algorithms and the integration of artificial intelligence (AI). Image processing algorithms are enhancing image quality by reducing noise, improving resolution through techniques like super-resolution, and fusing thermal images with data from other modalities for richer visualizations. Automated analysis and diagnosis are becoming more efficient with algorithms extracting features and recognizing patterns associated with specific pathologies, ensuring standardized and objective analysis. Meanwhile, AI models for image interpretation, particularly deep learning algorithms, offer high accuracy in analyzing thermal images, enabling personalized diagnostics and early disease detection. AI-assisted diagnostic decision support systems integrate thermal image analysis with clinical data, providing real-time feedback to clinicians and improving workflow efficiency. While these advancements show promise, ensuring high-quality data, retaining human expertize, and addressing ethical considerations are essential for realizing the full potential of thermal imaging in ophthalmology. Continued research and development hold significant promise for earlier diagnoses, personalized treatment strategies, and improved patient outcomes.

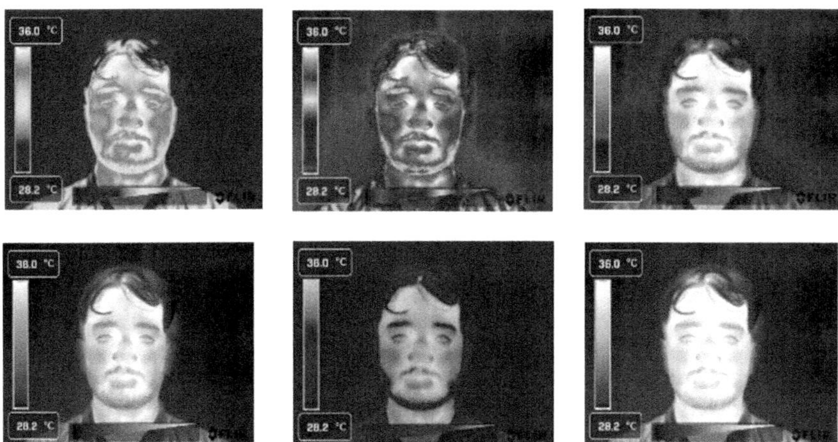

Figure 9.6 Different color palette representation of thermal images (rainbow, rainbow HC, lava, gray, cool, iron).

9.6 THERMAL IMAGING AND CAMERA

The FLIR E75 is an infrared thermal camera, featuring both far infrared imaging and a second camera capturing normal RGB images. To enhance detection capabilities, the images from both cameras are combined, enabling the calculation of absolute temperature estimates for specific points on the graph. Due to distinct temperature distribution characteristics in different detection materials, conventional single-image viewing is not sufficient. The FLIR E75 offers various image mode selections, each presenting heat sources distinctly with different colors and shades to quickly detect and highlight temperature variations. Figure 9.6 illustrates a variety of color palettes used in thermal imaging, providing a sample representation. The Rainbow palette is optimal for scenes with minimal heat changes, focusing on areas with similar heat energy. For facial thermal imaging in this chapter, the Rainbow mode is chosen, as it effectively captures relatively subtle regional temperature changes, allowing for observation of temperature variations in different areas of the face and creating new features based on these changes.

9.7 INTERACTIVE VISUALIZATION TECHNIQUES FOR THERMAL IMAGING

Various imaging modalities in ophthalmology offer unique features and capabilities for visualizing ocular structures and functions. Fundus photography provides 2D color images, offering a simple and familiar format for visualization and easy comparison of different images. Optical coherence tomography (OCT) offers 3D visualization with detailed cross-sectional

B-scans or en-face images, enabling quantitative measurements and assessment of retinal structures. OCT angiography (OCTA) provides composite images with flow information, facilitating visualization and analysis of blood flow and vessel structures. Ultrasound imaging offers real-time visualization of soft tissues and blood flow, with additional functional assessment capabilities using Doppler and elastography. Thermal imaging, while limited to 2D data, specializes in non-invasive assessment of temperature variations associated with inflammation and other abnormalities. Each modality has its strengths and limitations, offering complementary information crucial for comprehensive ocular diagnostics and management. Table 9.2 summarizes the comparison of different imaging techniques.

Dynamic visualization interfaces in ophthalmology are evolving to provide interactive ways of exploring thermal data. Interactive heatmaps, for instance, offer clinicians the ability to manipulate and delve into data in real-time, adjusting color palettes and zooming in on specific areas to identify subtle changes in temperature. These heatmaps can also be integrated with other imaging modalities like OCT or fundus photography for a comprehensive view of the eye. Additionally, temporal analysis tools enable visualization of thermal data changes over time, aiding in tracking disease progression and treatment efficacy. They can be particularly valuable for chronic conditions like dry eye or uveitis. These interactive visualization techniques offer benefits such as improved diagnostic accuracy, enhanced patient communication, and more precise guidance for interventions and surgical planning. However, challenges remain in developing user-friendly interfaces, integrating with existing diagnostic systems, and addressing cost and accessibility barriers. Despite these challenges, the potential of interactive visualization techniques in advancing thermal imaging in ophthalmology is substantial, promising improved patient outcomes through enhanced diagnostic and monitoring capabilities. The following sections discuss few visualization techniques.

9.7.1 FLIR thermal studio software

FLIR Thermal Studio Software is a comprehensive tool designed for analyzing and managing thermal imaging data captured by FLIR cameras [24]. It offers various features for processing, editing, and analyzing thermal images, including temperature analysis, image enhancement, and report generation. The software allows users to adjust color palettes, apply measurement tools, annotate images, and create customized reports. It is commonly used in research, industrial inspections, and building diagnostics for interpreting thermal data with precision and efficiency. Figure 9.7 provides a summary of various snapshots illustrating thermal analyses conducted using the FLIR Thermal Studio Software.

Different measurement shapes, such as spot meter, ellipse, line, and rectangle, are utilized for selecting the region of interest (ROI). Within the ROI,

Table 9.2 Comparative visualization of different imaging techniques

Feature	*Fundus photography*	*OCT*	*OCTA*	*Ultrasound*	*Thermal imaging*
Data dimensionality	2D	3D	3D	2D (multiple slices)	2D
Primary visualization	Color image	Cross-sectional B-scans or en-face images	Composite image with flow information	B-mode, Doppler, or elastography images	Color image with temperature overlay
Key interactive techniques	– Zoom & pan – Color adjustments – Side-by-side comparisons	– 3D volume rendering – Segmentation & measurement tools –VR visualization	– Flow visualization (color coding, animation) –Vessel analysis & quantification	– Multiplanar reconstruction – Doppler imaging – Elastography	– Color coding –Time-lapse visualization – ROI analysis
Strengths	– Simple and familiar visualization format – Easy to compare different images	– Detailed view of retinal structures – Quantitative measurements possible	–Visualization of blood flow & vessel analysis	– Real-time imaging of soft tissues & blood flow – Functional assessment with Doppler & elastography	– Non-invasive assessment of inflammation & other abnormalities
Limitations	– Limited depth information	– Requires specialized training for interpretation	– Requires specialized training for interpretation	– Limited penetration depth	– Limited anatomical detail compared to other modalities

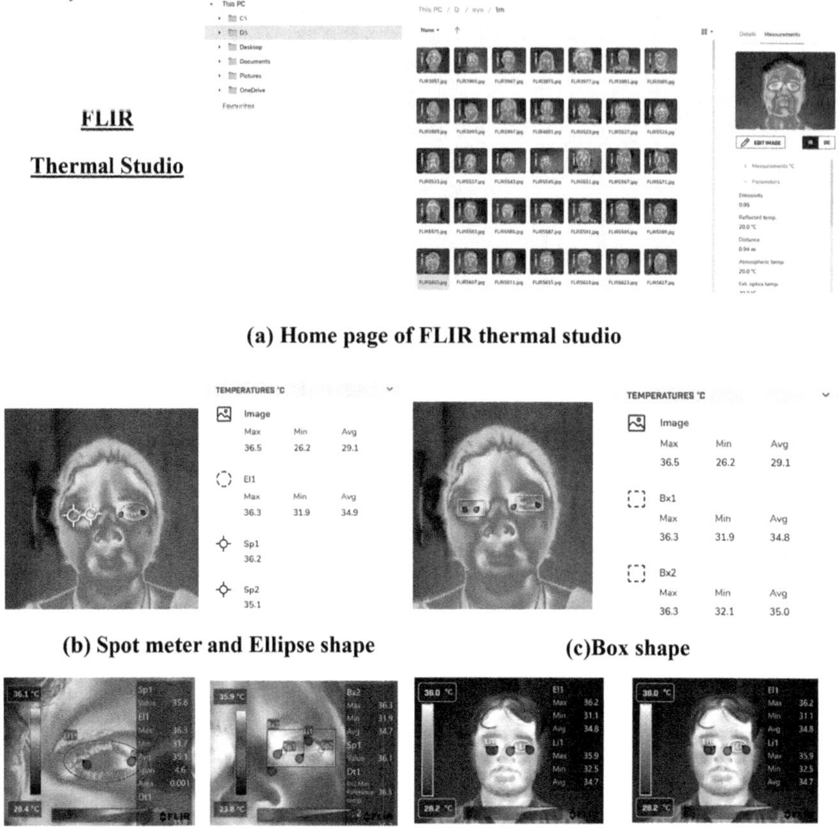

(d) Sample image analysis with minimum, maximum and average temperature.

Figure 9.7 FLIR thermal studio software analysis.

temperature distribution, histogram distribution, maximum temperature, minimum temperature, and average temperature can be computed. These temperature values can be graphically visualized and exported using appropriate options available in FLIR Thermal Studio. Ocular surface temperatures at the inner canthus, outer canthus, and central cornea are measured in 20 eye images and tabulated in Table 9.3 using the spot meter. Additionally, the ellipse shape is utilized to draw over the eye, measuring and comparing the minimum and maximum temperatures.

Figure 9.8 presents a comparison of temperature measurements taken at the inner canthus, outer canthus, and central cornea. The bar graph clearly illustrates that the inner canthus exhibits the highest temperature, while the central cornea displays the lowest temperature across the eye. This observation aligns with findings from prior research [25–27]. Analysis of 20 eye

Table 9.3 Ocular surface temperature analysis

Images	*Inner canthus*	*Outer canthus*	*Central cornea*	*Average temp*	*Min*	*Max*
1	35.7	35.1	34.6	34.8	32.8	36
2	35.6	35.1	34.5	34.9	31.9	36.3
3	35.9	34.7	34.1	34.7	32.5	35.9
4	34.6	34.2	33.3	33.7	31.3	34.9
5	35.4	34	33.6	34.1	30.8	35.5
6	35.1	34.5	33.8	33.8	30.5	35.4
7	34.9	34.6	34.2	34.1	31.4	35.7
8	35.9	35.1	34.4	34.6	31.7	36.2
9	35.7	35.5	34.7	35.2	32.7	36.7
10	36	35.8	35.1	35	31.7	36.3
11	36.1	35.3	35	35.1	31.7	36.7
12	35.3	34.5	34.3	34.6	31.4	35.8
13	35.2	35	34.6	34.6	32.8	35.5
14	36.2	35.1	34.1	35	31.9	36.5
15	35.5	35.3	35	34.9	31.5	36.5
16	36.7	36.1	35.9	35.8	33.5	36.9
17	36.6	35.2	35.1	35.5	33	36.6
18	35.9	34.9	34.8	34.9	31.8	36.1
19	35.6	35.3	34.6	34.8	33.2	35.7
20	36.2	35.6	35.2	35.3	32.7	36.7

images reveals an average eye temperature of 34.77 ± 0.51 degrees Celsius, consistent with previous studies conducted by researchers [28]. Figure 9.9 shows mean temperature value of sample thermal images.

FLIR Research Studio is also a software platform developed by FLIR Systems specifically for thermal imaging research applications [29]. It offers a range of advanced features tailored to meet the needs of scientists, engineers, and researchers working with FLIR thermal cameras. FLIR Research Studio provides powerful tools for analyzing thermal data, including advanced measurement and analysis capabilities, customizable image processing techniques, and scripting options for automation and customization. Figure 9.10 displays the temperature values generated for each pixel in the thermal images using FLIR Research Studio. Researchers can use the software to conduct precise temperature measurements, analyze thermal patterns, and generate detailed reports. FLIR Research Studio is widely used in fields such as medical research, materials science, aerospace, and environmental monitoring, where accurate thermal analysis is essential for understanding phenomena and solving complex problems.

The bar graph and analysis can be done using Microsoft Excell. Microsoft Excel is a versatile spreadsheet software that allows users to create impactful dashboards, including bar graphs, with ease. Using Excel's intuitive interface,

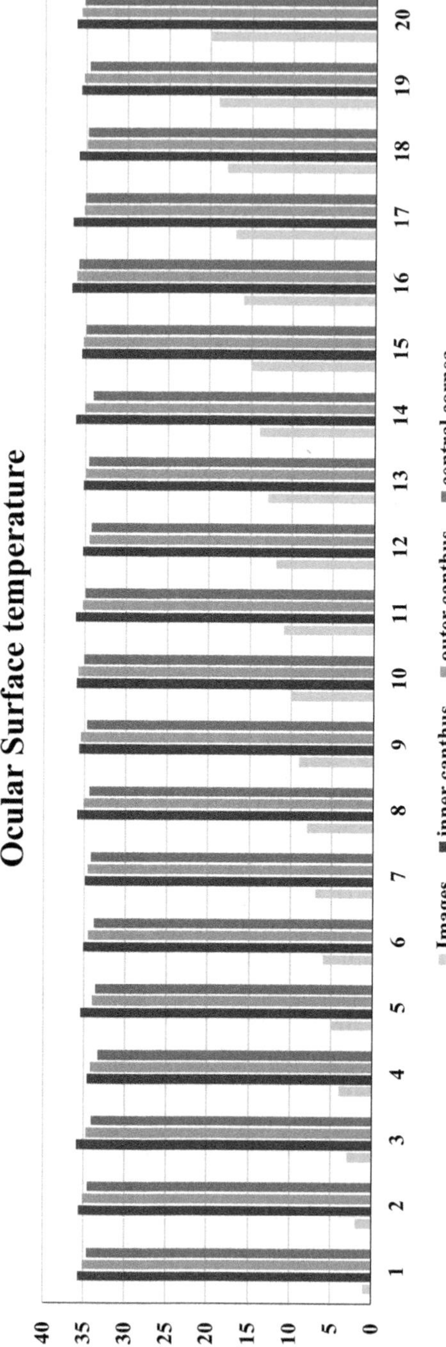

Figure 9.8 Comparison of temperature measured in human eye.

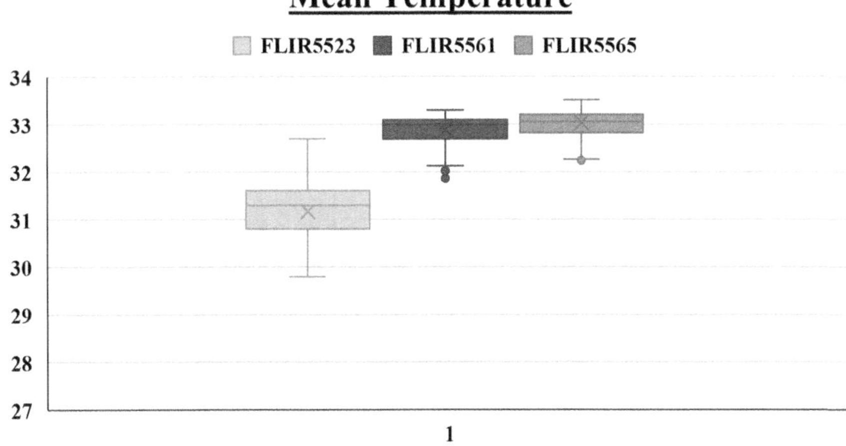

Figure 9.9 Mean temperature value of sample thermal images.

individuals can input and organize data in a spreadsheet and then effortlessly generate bar graphs through its charting tools. With a variety of customization options, users can adjust colors, labels, and axis settings to enhance the visual representation of their data. Excel dashboards facilitate data analysis and presentation, making it a widely-used tool for professionals across diverse fields who seek clear and concise visualization of their information using bar graphs.

9.7.2 Data exploration tools: flyr tools

Flyr is a Python library designed for extracting thermal data from FLIR images. Unlike other solutions that act as wrappers around ExifTool, Flyr offers a reimplementation of ExifTool's FLIR parser, providing several advantages. Firstly, it facilitates faster decoding as it avoids the need to start new processes and communicate in-memory data externally. Additionally, Flyr aims for enhanced accuracy by leveraging all available camera metadata to translate raw values into Kelvin, potentially yielding more precise results compared to methods with hardcoded conversion values. Moreover, Flyr boasts easier and more robust installation and deployment since it can be seamlessly installed from PyPI [30]. Its user-friendly design eliminates the need for creating additional extraction objects, simplifying the extraction process with a single call, such as thermogram = flyr.unpack(flir_file_path). With the Flyr Python package, temperature values in Celsius, Fahrenheit, or Kelvin can be extracted from radiometric thermal images. Additionally, it offers commands to retrieve embedded RGB images. Flyr includes built-in capabilities to convert thermal data to RGB images, utilizing either the embedded palette or one of the provided palettes. If optical imagery is

	A	B	C	D	E	F	G	H	I	J	K	L	M	N	O	P	Q	R	S	T	U	V	W
28	3.58E+01	3.58E+01	3.58E+01	3.58E+01	3.57E+01	3.57E+01	3.55E+01	3.55E+01	3.51E+01	3.51E+01	3.47E+01	3.36E+01	3.39E+01	3.45E+01	3.42E+01	3.41E+01	3.47E+01	3.43E+01	3.40E+01	3.39E+01	3.36E+01	3.33E+01	3.41E+01
29	3.58E+01	3.58E+01	3.57E+01	3.57E+01	3.57E+01	3.57E+01	3.56E+01	3.55E+01	3.55E+01	3.58E+01	3.59E+01	3.57E+01	3.47E+01	3.39E+01	3.35E+01	3.34E+01	3.33E+01	3.30E+01	3.32E+01	3.31E+01	3.30E+01	3.30E+01	3.30E+01
30	3.56E+01	3.56E+01	3.56E+01	3.56E+01	3.56E+01	3.55E+01	3.57E+01	3.56E+01	3.57E+01	3.55E+01	3.55E+01	3.57E+01	3.61E+01	3.64E+01	3.63E+01	3.54E+01	3.44E+01	3.35E+01	3.31E+01	3.29E+01	3.31E+01	3.33E+01	3.36E+01
31	3.56E+01	3.57E+01	3.56E+01	3.56E+01	3.54E+01	3.54E+01	3.55E+01	3.58E+01	3.58E+01	3.58E+01	3.55E+01	3.54E+01	3.50E+01	3.50E+01	3.49E+01	3.48E+01	3.54E+01	3.53E+01	3.47E+01	3.44E+01	3.45E+01	3.48E+01	3.57E+01
32	3.57E+01	3.56E+01	3.56E+01	3.56E+01	3.55E+01	3.54E+01	3.53E+01	3.51E+01	3.53E+01	3.52E+01	3.56E+01	3.61E+01	3.61E+01	3.61E+01	3.57E+01	3.55E+01	3.53E+01	3.53E+01	3.53E+01	3.55E+01	3.58E+01	3.59E+01	3.53E+01
33	3.57E+01	3.57E+01	3.57E+01	3.56E+01	3.56E+01	3.53E+01	3.50E+01	3.48E+01	3.50E+01	3.52E+01	3.53E+01	3.53E+01	3.52E+01	3.53E+01	3.57E+01	3.54E+01	3.57E+01	3.57E+01	3.55E+01	3.52E+01	3.53E+01	3.51E+01	3.52E+01
34	3.58E+01	3.58E+01	3.57E+01	3.57E+01	3.59E+01	3.57E+01	3.57E+01	3.60E+01	3.56E+01	3.53E+01	3.53E+01	3.52E+01	3.57E+01	3.59E+01	3.58E+01	3.59E+01	3.54E+01	3.52E+01	3.51E+01	3.49E+01	3.45E+01	3.52E+01	3.50E+01
35	3.58E+01	3.58E+01	3.54E+01	3.56E+01	3.57E+01	3.60E+01	3.61E+01	3.60E+01	3.63E+01	3.62E+01	3.63E+01	3.59E+01	3.55E+01	3.56E+01	3.56E+01	3.54E+01	3.48E+01	3.45E+01	3.45E+01	3.46E+01	3.47E+01	3.44E+01	3.49E+01
36	3.58E+01	3.59E+01	3.54E+01	3.56E+01	3.58E+01	3.60E+01	3.61E+01	3.61E+01	3.61E+01	3.61E+01	3.61E+01	3.58E+01	3.54E+01	3.51E+01	3.47E+01	3.43E+01	3.45E+01	3.46E+01	3.46E+01	3.48E+01	3.47E+01	3.49E+01	3.61E+01
37	3.58E+01	3.58E+01	3.59E+01	3.59E+01	3.59E+01	3.60E+01	3.60E+01	3.57E+01	3.58E+01	3.57E+01	3.55E+01	3.54E+01	3.52E+01	3.49E+01	3.51E+01	3.55E+01	3.55E+01	3.55E+01	3.59E+01	3.53E+01	3.54E+01	3.66E+01	3.64E+01
38	3.58E+01	3.59E+01	3.59E+01	3.59E+01	3.60E+01	3.60E+01	3.60E+01	3.60E+01	3.60E+01	3.60E+01	3.55E+01	3.57E+01	3.58E+01	3.59E+01	3.61E+01	3.61E+01	3.58E+01	3.58E+01	3.55E+01	3.53E+01	3.65E+01	3.66E+01	3.65E+01
39	3.58E+01	3.59E+01	3.59E+01	3.59E+01	3.60E+01	3.61E+01	3.61E+01	3.61E+01	3.63E+01	3.63E+01	3.61E+01	3.61E+01	3.62E+01	3.59E+01	3.62E+01	3.57E+01	3.55E+01	3.55E+01	3.59E+01	3.60E+01	3.67E+01	3.65E+01	3.66E+01
40	3.59E+01	3.59E+01	3.59E+01	3.60E+01	3.60E+01	3.62E+01	3.62E+01	3.62E+01	3.63E+01	3.60E+01	3.60E+01	3.62E+01	3.60E+01	3.56E+01	3.57E+01	3.55E+01	3.58E+01	3.60E+01	3.60E+01	3.63E+01	3.64E+01	3.65E+01	3.65E+01
41	3.59E+01	3.60E+01	3.60E+01	3.60E+01	3.60E+01	3.62E+01	3.61E+01	3.61E+01	3.61E+01	3.58E+01	3.60E+01	3.58E+01	3.56E+01	3.62E+01	3.60E+01	3.59E+01	3.61E+01	3.58E+01	3.59E+01	3.63E+01	3.66E+01	3.65E+01	3.65E+01
42	3.59E+01	3.59E+01	3.60E+01	3.60E+01	3.60E+01	3.60E+01	3.60E+01	3.61E+01	3.59E+01	3.59E+01	3.62E+01	3.62E+01	3.63E+01	3.63E+01	3.59E+01	3.60E+01	3.59E+01	3.58E+01	3.62E+01	3.63E+01	3.63E+01	3.63E+01	3.63E+01
43	3.58E+01	3.59E+01	3.59E+01	3.60E+01	3.60E+01	3.60E+01	3.61E+01	3.61E+01	3.61E+01	3.62E+01	3.61E+01	3.62E+01	3.62E+01	3.62E+01	3.59E+01	3.58E+01	3.56E+01	3.62E+01	3.61E+01	3.62E+01	3.62E+01	3.62E+01	3.56E+01
44	3.58E+01	3.58E+01	3.59E+01	3.59E+01	3.60E+01	3.60E+01	3.60E+01	3.60E+01	3.61E+01	3.61E+01	3.61E+01	3.61E+01	3.61E+01	3.61E+01	3.59E+01	3.55E+01	3.61E+01	3.60E+01	3.61E+01	3.61E+01	3.61E+01	3.57E+01	3.58E+01
45	3.58E+01	3.58E+01	3.59E+01	3.59E+01	3.60E+01	3.60E+01	3.60E+01	3.60E+01	3.60E+01	3.61E+01	3.60E+01	3.61E+01	3.61E+01	3.60E+01	3.58E+01	3.58E+01	3.60E+01	3.60E+01	3.61E+01	3.60E+01	3.58E+01	3.58E+01	3.60E+01
46	3.57E+01	3.58E+01	3.58E+01	3.59E+01	3.60E+01	3.60E+01	3.60E+01	3.60E+01	3.60E+01	3.60E+01	3.60E+01	3.61E+01	3.60E+01	3.58E+01	3.57E+01	3.58E+01	3.58E+01	3.59E+01	3.59E+01	3.57E+01	3.59E+01	3.60E+01	3.60E+01
47	3.57E+01	3.57E+01	3.58E+01	3.58E+01	3.59E+01	3.59E+01	3.59E+01	3.59E+01	3.59E+01	3.60E+01	3.59E+01	3.59E+01	3.59E+01	3.58E+01	3.58E+01	3.56E+01	3.57E+01	3.59E+01	3.55E+01	3.58E+01	3.59E+01	3.59E+01	3.59E+01
48	3.56E+01	3.57E+01	3.57E+01	3.58E+01	3.57E+01	3.59E+01	3.59E+01	3.59E+01	3.59E+01	3.59E+01	3.59E+01	3.59E+01	3.58E+01	3.58E+01	3.57E+01	3.56E+01	3.56E+01	3.56E+01	3.58E+01	3.59E+01	3.59E+01	3.59E+01	3.58E+01
49	3.55E+01	3.56E+01	3.56E+01	3.57E+01	3.57E+01	3.58E+01	3.58E+01	3.58E+01	3.59E+01	3.59E+01	3.59E+01	3.58E+01	3.58E+01	3.58E+01	3.57E+01	3.54E+01	3.55E+01	3.58E+01	3.58E+01	3.58E+01	3.58E+01	3.59E+01	3.55E+01
50	3.55E+01	3.56E+01	3.55E+01	3.56E+01	3.57E+01	3.57E+01	3.57E+01	3.58E+01	3.58E+01	3.58E+01	3.58E+01	3.58E+01	3.58E+01	3.58E+01	3.57E+01	3.58E+01	3.58E+01	3.58E+01	3.58E+01	3.58E+01	3.58E+01	3.57E+01	3.50E+01
51	3.54E+01	3.55E+01	3.55E+01	3.55E+01	3.56E+01	3.57E+01	3.57E+01	3.57E+01	3.58E+01	3.58E+01	3.58E+01	3.58E+01	3.58E+01	3.58E+01	3.58E+01	3.58E+01	3.58E+01	3.58E+01	3.57E+01	3.58E+01	3.58E+01	3.54E+01	3.51E+01
52	3.54E+01	3.54E+01	3.55E+01	3.55E+01	3.55E+01	3.55E+01	3.56E+01	3.57E+01	3.57E+01	3.57E+01	3.57E+01	3.58E+01	3.57E+01	3.57E+01	3.57E+01	3.57E+01	3.57E+01	3.57E+01	3.57E+01	3.57E+01	3.55E+01	3.53E+01	3.57E+01
53	3.53E+01	3.53E+01	3.54E+01	3.54E+01	3.55E+01	3.55E+01	3.56E+01	3.55E+01	3.57E+01	3.57E+01	3.57E+01	3.57E+01	3.57E+01	3.57E+01	3.57E+01	3.57E+01	3.57E+01	3.57E+01	3.57E+01	3.56E+01	3.56E+01	3.58E+01	3.57E+01
54	3.52E+01	3.53E+01	3.53E+01	3.53E+01	3.54E+01	3.55E+01	3.55E+01	3.56E+01	3.56E+01	3.56E+01	3.57E+01	3.57E+01	3.57E+01	3.57E+01	3.57E+01	3.57E+01	3.57E+01	3.57E+01	3.56E+01	3.56E+01	3.56E+01	3.57E+01	3.57E+01

Figure 9.10 Temperature value generated for each pixel in the thermal images.

available, it can aid in edge detection and improve delineation. Furthermore, common EXIF metadata can be conveniently accessed through thermogram.camera_metadata. Furthermore, Flyr offers extra features including different units, built-in rendering, and adjustable thermal data. For installation, Flyr can be easily installed via pip using the command pip install flyr. Recent updates to Flyr include features such as accessing EXIF metadata, masking renders, rendering grayscale-inverted images, and emphasizing edges in renders using optical data. While it supports various FLIR models, complete compatibility with all FLIR image formats cannot be guaranteed. Nonetheless, for those working with FLIR images and requiring thermal data extraction, Flyr presents a promising option. Users are advised to consult the library's documentation for comprehensive usage guidance and compatibility details.

9.7.3 Bokeh

Bokeh is a Python library for creating interactive data visualizations in web browsers. It provides a flexible and powerful framework for building interactive plots and applications with rich, interactive features. With Bokeh, users can create interactive plots with tools like zooming, panning, hovering, and selecting data points. Bokeh applications can be deployed as standalone HTML documents or as part of larger web applications using Bokeh server. This allows users to create highly customizable and interactive data-driven applications that can be easily shared and accessed through web browsers. Bokeh's intuitive syntax and extensive documentation make it accessible to users with varying levels of programming experience, enabling the creation of dynamic and engaging visualizations for data exploration and presentation. Figure 9.11 illustrates the output achieved using the Bokeh

OUTPUT:

```
ROI mean temperature: 34.758923169546314
ROI minimum temperature: 33.22870919738549
ROI maximum temperature: 36.0206666858037
```

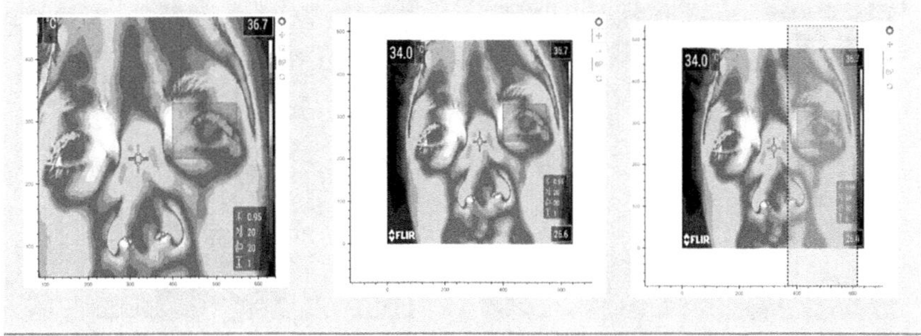

Figure 9.11 Sample outputs obtained using Bokeh's python library.

Python library. It demonstrates the ability to select the Region of Interest (ROI) from input thermal images, zoom in, and specify the desired area. Additionally, it showcases the computation of mean, minimum, and maximum temperatures within the selected ROI, utilizing both the thermal image and the corresponding temperature data.

9.7.4 Matplotlib and ipywidgets

Matplotlib and ipywidgets are used to create an interactive plot where you can adjust the region of interest (ROI) in real-time using sliders. This is a powerful feature for exploratory data analysis. However, you seem to be using two different libraries for creating the interactive plot (Bokeh and Matplotlib). Both of these libraries have their own unique advantages. Bokeh is generally more powerful for creating highly interactive, web-ready plots, while Matplotlib, especially when combined with ipywidgets, is great for creating simple interactive plots in Jupyter notebooks. Figure 9.12 depicts the interactive interface facilitated by ipywidgets for manual selection of the Region of Interest (ROI). Once the ROI is chosen, the thermal images and corresponding temperature values are utilized to compute, display, and annotate the mean, minimum, and maximum temperature values. This outcome is showcased in Figure 9.13. As illustrated in the figure, the minimum temperature is highlighted in red, indicating 31.1 degrees Celsius, while the maximum temperature is indicated in blue, representing 36.9 degrees Celsius.

Utilizing interactive visualization methods facilitates the analysis of temperature values across different regions of interest. By capturing thermal images of both normal and abnormal human eyes affected by various eye diseases, temperature values can be extracted using suitable tools and recorded for subsequent analysis. Once a dataset of labeled thermal images,

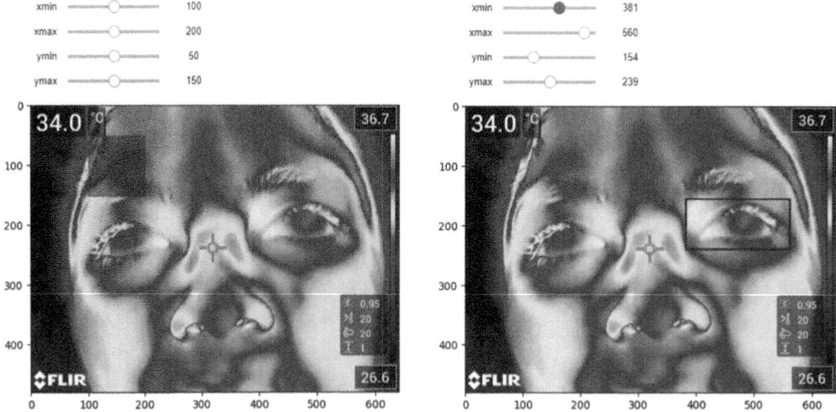

Figure 9.12 Interactive interface for updating the region of interest (ROI).

OUTPUT:

```
ROI mean temperature: 35.312254114863514
ROI minimum temperature: 31.0923140904392
ROI maximum temperature: 36.89571075675649
```

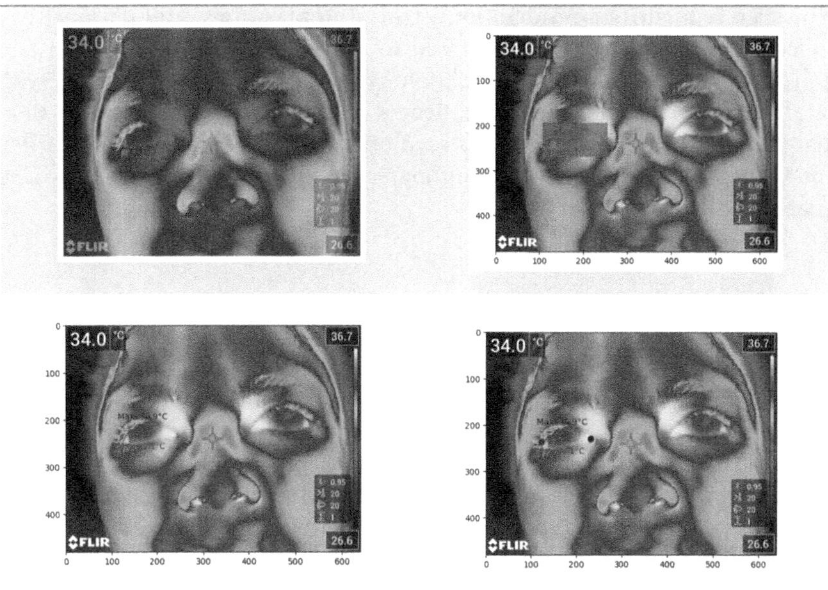

Figure 9.13 Sample output obtained using Matplotlib and ipywidgets.

along with the extracted temperatures, is prepared, a range of image processing and artificial intelligence techniques can be employed and compared to achieve highly accurate and automated diagnosis and inference.

9.8 FUTURE DIRECTIONS AND EMERGING APPLICATIONS

The integration of thermal imaging with established techniques such as OCT, OCTA, and ultrasound provides a more comprehensive understanding of ocular health, enhancing diagnostic accuracy and offering insights into disease pathophysiology. For example, combining thermal imaging with OCT can assess inflammatory activity in macular edema, while integrating it with ultrasound improves visualization of deeper ocular structures. Furthermore, the development of wearable thermal imaging devices holds promise for home monitoring and continuous data collection, empowering patients to track their eye health and detect early signs of inflammation.

Population-based screening programs leveraging thermal imaging's non-invasive nature could enable early detection of ocular pathologies in high-risk groups, potentially reducing the burden of vision loss. Additionally, integrating thermal imaging with teleophthalmology platforms facilitates remote consultations and monitoring, particularly for underserved areas. However, regulatory considerations, ethical implications, and data privacy concerns must be carefully addressed to ensure responsible adoption and utilization of thermal imaging technology in ophthalmology. Overall, with careful navigation of existing challenges and ethical considerations, thermal imaging has the potential to revolutionize the diagnosis, monitoring, and management of eye diseases, ultimately enhancing vision and improving quality of life for millions.

9.9 CONCLUSION

The analysis has delved into the potential of thermal imaging in ophthalmology, delineating its strengths and weaknesses in comparison to established techniques. Noteworthy findings encompass its sensitivity to minute temperature changes, particularly in detecting inflammatory processes, and its promising applications in early disease detection, disease monitoring, and therapeutic guidance. Integration with other modalities like OCT and ultrasound offers a more comprehensive understanding of ocular health, while emerging technologies such as wearable devices and AI-powered analysis indicate future avenues for innovation and accessibility. In terms of clinical practice, thermal imaging could enhance diagnostic accuracy, facilitate personalized medicine, optimize workflow efficiencies, and raise ethical considerations regarding data privacy and AI usage. Recommendations for further research include standardizing protocols and image analysis, conducting large-scale clinical trials, developing and validating AI algorithms, and assessing cost-effectiveness to support wider adoption. Ultimately, by addressing existing challenges, fostering responsible development, and advancing research, thermal imaging stands poised to revolutionize ophthalmic care, promising improved diagnosis, personalized treatment, and enhanced vision outcomes.

REFERENCES

[1] T. Ilginis, J. Clarke, and P. J. Patel, "Ophthalmic imaging," *Br. Med. Bull.*, vol. 111, no. 1, pp. 77–88, 2014, https://doi.org/10.1093/bmb/ldu022
[2] M. Tkáčová, J. Živčák, and P. Foffová, "A Reference for Human Eye Surface Temperature Measurements in Diagnostic Process of Ophthalmologic Diseases," *Meas. 8th Int. Conf*, pp. 406–409, 2011, [Online]. Available: http://www.measurement.sk/M2011/doc/proceedings/406_Tkacova-2.pdf

[3] R. Gulias-Cañizo, M. E. Rodríguez-Malagón, L. Botello-González, V. Belden-Reyes, F. Amparo, and M. Garza-Leon, "Applications of infrared thermography in ophthalmology," *Lifestyles*, vol. 13, no. 3, 2023, https://doi.org/10.3390/life13030723

[4] M. Alvise and L. F. Von, "The evolution of confocal scanning laser ophthalmoscopy," *Retina Today*, pp. 63–65, 2012. https://retinatoday.com/articles/2012-sept/the-evolution-of-confocal-scanning-laser-ophthalmoscopy

[5] A. Modrzejewska, "The role of thermography in ophthalmology," *OphthaTherapy. Ther. Ophthalmol.*, vol. 9, no. 1, pp. 14–21, 2022, https://doi.org/10.24292/01.OT.291221

[6] P. D. Abusharha Ali, "Analysis of ocular surface temperature in patients with dry eye," *Med. J. Cairo Univ.*, vol. 89, no. 12, pp. 2549–2553, 2021, https://doi.org/10.21608/mjcu.2021.217392

[7] S. B. Han, Y. C. Liu, K. Mohamed-Noriega, and J. S. Mehta, "Advances in imaging technology of anterior segment of the eye," *J. Ophthalmol.*, vol. 2021, 2021, https://doi.org/10.1155/2021/9539765

[8] E. F. J. Ring, "The historical development of temperature measurement in medicine," *Infrared Phys. Technol.*, vol. 49, no. 3 SPEC. ISS., pp. 297–301, 2007, https://doi.org/10.1016/j.infrared.2006.06.029

[9] O. S. Zadorozhnyy, A. R. Korol, V. O. Naumenko, N. V. Pasyechnikova, and L. L. Butenko, "Heat exchange in the human eye: A review," *Oftalmol. Zh.*, vol. 101, no. 6, pp. 50–58, 2022, https://doi.org/10.31288/OFTALMOLZH202265058

[10] E. F. J. Ring and K. Ammer, "The technique of infrared imaging in medicine," *Infrared Imaging A Caseb. Clin. Med.*, no. September, 2015, https://doi.org/10.1088/978-0-7503-1143-4ch1

[11] L. Pinto-Coelho, "How artificial intelligence is shaping medical imaging technology: A survey of innovations and applications," *Bioengineering*, vol. 10, no. 12, 2023, https://doi.org/10.3390/bioengineering10121435

[12] T. Y. Y. Lai, "Ocular imaging at the cutting-edge," *Eye*, vol. 35, no. 1, pp. 1–3, 2021, https://doi.org/10.1038/s41433-020-01268-1

[13] P. Alexopoulos, C. Madu, G. Wollstein, and J. S. Schuman, "The development and clinical application of innovative optical ophthalmic imaging techniques," *Front. Med.*, vol. 9, no. June, pp. 1–33, 2022, https://doi.org/10.3389/fmed.2022.891369

[14] Y. Bu, K. C. Shih, and L. Tong, "The ocular surface and diabetes, the other 21st Century epidemic," *Exp. Eye Res.*, vol. 220, no. May, p. 109099, 2022, https://doi.org/10.1016/j.exer.2022.109099

[15] B. Chandrasekar et al., "Ocular surface temperature measurement in diabetic retinopathy," *Exp. Eye Res.*, vol. 211, no. July, p. 108749, 2021, https://doi.org/10.1016/j.exer.2021.108749

[16] R. Raut, A. Sapkal, V. Ingale, P. Borkar, and P. Bhanarkar, "Assessment of diabetic retinopathy progression using CNN from ocular thermal images," *Soft. Comput.*, vol. 1, 2023, https://doi.org/10.1007/s00500-023-08238-1

[17] F. Galassi, B. Giambene, A. Corvi, and G. Falaschi, "Evaluation of ocular surface temperature and retrobulbar haemodynamics by infrared thermography and colour Doppler imaging in patients with glaucoma," *Br. J. Ophthalmol.*, vol. 91, no. 7, pp. 878–881, 2007, https://doi.org/10.1136/bjo.2007.114397

[18] A. S. Hura, A. T. Epitropoulos, C. N. Czyz, and E. D. Rosenberg, "Visible meibomian gland structure increases after vectored thermal pulsation treatment in dry eye disease patients with meibomian gland dysfunction," *Clin. Ophthalmol.*, vol. 14, pp. 4287–4296, 2020, https://doi.org/10.2147/OPTH.S282081

[19] N. A. B. S. Amri, M. H. Alkawaz, K. L. Teow Aik, and M. G. Md Johar, "n Overview of Dry Eye Analysis Algorithms for Tear Film Break-Up Time Detection," *2021 IEEE Symp. Ind. Electron. Appl. ISIEA 2021*, 2021, https://doi.org/10.1109/ISIEA51897.2021.9510004

[20] U. R. Acharya et al., "Automated diagnosis of dry eye using infrared thermography images," *Infrared Phys. Technol.*, vol. 71, pp. 263–271, 2015, https://doi.org/10.1016/j.infrared.2015.04.007

[21] J. Persiya, A. Sasithradevi, and S. Mohamed Mansoor Roomi, "Infrared Thermograms for Diagnosis of Dry Eye: A Review," *Proc. 9th Int. Conf. Biosignals, Images, Instrumentation, ICBSII 2023, no. Icbsii 23*, pp. 1–6, 2023, https://doi.org/10.1109/ICBSII58188.2023.10181092

[22] A. Kawali, "Thermography in ocular inflammation," *Indian J. Radiol. Imaging*, vol. 23, no. 3, pp. 281–283, 2013, https://doi.org/10.1055/s-0041-1734381

[23] A. A. Kawali, S. Sanjay, P. Mahendradas, and R. Shetty, "Thermography in posterior scleritis," *Russ. J. Clin. Ophthalmol.*, vol. 20, no. 4, pp. 204–208, 2020, https://doi.org/10.32364/2311-7729-2020-20-4-204-208

[24] "FLIR Thermal Studio." https://www.flir.com/products/flir-thermal-studio-suite/

[25] E. Micheletti, N. W. El-Nimri, R. N. Weinreb, and J. H. K. Liu, "Relative stability of regional facial and ocular temperature measurements in healthy individuals," *Transl. Vis. Sci. Technol.*, vol. 11, no. 12, pp. 1–6, 2022, https://doi.org/10.1167/tvst.11.12.15

[26] A. A. Fernandes et al., "Validity of inner canthus temperature recorded by infrared thermography as a non-invasive surrogate measure for core temperature at rest, during exercise and recovery," *J. Therm. Biol.*, vol. 62, no. September, pp. 50–55, 2016, https://doi.org/10.1016/j.jtherbio.2016.09.010

[27] R. Vardasca, A. R. Marques, J. Diz, A. Seixas, J. Mendes, and E. F. J. Ring, "The influence of angles and distance on assessing inner-canthi of the eye skin temperature," *Thermol. Int.*, vol. 27, no. 4, pp. 130–135, 2017.

[28] J. H. Tan, E. Y. K. Ng, and Rajendra Acharya U, "An efficient automated algorithm to detect ocular surface temperature on sequence of thermograms using snake and target tracing function," *J. Med. Syst.*, vol. 35, no. 5, pp. 949–958, 2011, https://doi.org/10.1007/s10916-010-9552-6

[29] "FLIR Research Studio." https://www.flir.eu/products/flir-research-studio?vertical=rd+science&segment=solutions

[30] "PYPI." https://pypi.org/project/flyr/

Chapter 10

Mind scan

Dynamic brain cancer detection dashboard with MRI imaging

J Suneetha and Smita Darandale
GITAM Deemed to be University, Bengaluru, Karnataka, India

Niranjan L
CMR Institute of Technology, Bengaluru, Karnataka, India

Tanvir H Sardar
GITAM Deemed to be University, Bengaluru, Karnataka, India

10.1 INTRODUCTION

The brain structure pertains to the soft, pliable tissue enclosed inside the skull bones. Within the realm of medicine, there is a complex arrangement known as the meninges, which consists of three intricate layers and is accompanied by the existence of cerebrospinal liquid. The brain tumor characterized by an aberrant development of cells inside cerebral tissue [1]. The two distinct categories of tumor are with the primary classification being referred to as Benign Brain Tumor Cells. The extraction of these entities from the body of a person is a straightforward process. However, the presence of benign tumors may lead to significant health complications due to their interference with the delicate structures inside the brain [2]. The second condition is referred to as Malignant Brain Tumor or brain cancer. The observed behavior exhibits a significant rate of expansion and has been shown to have detrimental effects on the integrity of brain tissue.

Magnetic Resonance Imaging (MRI) is a technique that illustrates pathological changes in living tissues. This imaging method utilizes magnetic waves, allowing us to employ conventional software to detect and classify tumors (malignant and benign) [3]. This characteristic makes MRI a valuable tool for visualizing many anatomical structures, including the brain, muscles, heart, and cancerous tissues [4]. A collection of magnetic resonance imaging (MRI) images, including both normal and pathological cases, including benign and malignant instances, were used, as seen in Figure 10.1.

Texture analysis, a key topic in processing images, is used for categorization and segmentation. This texture analysis approach is based on the degree

DOI: 10.1201/9781003542735-10

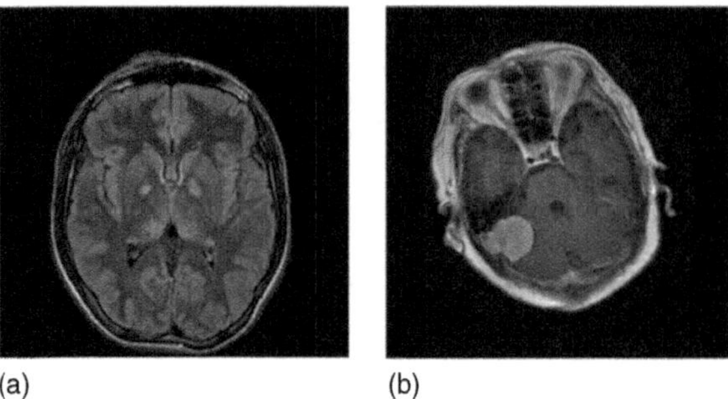

(a) (b)

Figure 10.1 MRI images from various phases were examined: (a) cancer-free brain, and (b) cancerous brain.

of intensity or color variation within a picture. Texture analysis approaches are classified into two types: structural and statistical techniques [5].

The structural technique investigated the texture field, whether it is steady or not. If the texture's local attribute is extreme or sluggish, it provides a change in the pixel's action [6–8]. The statistical technique addresses and evaluates pixel difference in imaging studies. It is critical for all texture pictures that the attributes studied in texture analysis have the same perspective, which means they have the same direction and scale. This strategy, however, is regarded a shortcoming since it is difficult to maintain the exact same rotation or scale across all photos [9]. Maintaining an appropriate viewpoint is important for all photographs shot. The textural properties of a tumor frequently match those of regular tissue, making reliable tumor identification difficult. This might result in inaccurate reporting and needs further procedures to address the problem. Segmentation and classification are proposed strategies for detecting and identifying cancers [10].

Numerous research on the identification and detection of cancers in medical imaging have been done. MRI scans to detect brain tumors in a two stages multi-model deep learning-based technique. The basic idea is to accurately localize the tumor by categorizing pictures into normal and malignant categories. ECOC model and Convolutional neural network (CNN) techniques are used to do this, approaches used for extracting topographies and brain tumor classification. Based on a prototype, Alam Shahariar et al. created a method for detecting brain cancers in MRI scans using average and fuzzy median algorithms [11]. The grey level value is the first variable for the prototype-centered K-means algorithm. For good results, the fuzzy C-mean technique depends on cluster data membership. These approaches were used to localize the tumor site by employing a function with membership that considers several tumor picture attributes like contrast, capacity, difference, uniformity, entropy, and relationship [12]. In 2018, Ergena and Baykara performed research [13] whereby they used spatial methodologies to extract characteristics from medical photographs, with a particular emphasis on evaluating the rate of retrieval efficiency. As analytic tools, the investigation employed the Gabor wavelet, grey level co-occurrence matrix and grey-level run length matrix. However, a substantial quantity of medical photos was used as an extensive repository [14]. The findings indicate that the GLCM exhibits favorable efficiency, yet the Gabor Wavelet technique appears as the most efficacious and precise strategy The fuzzy C-means approach, which relies on cluster data membership, has been successfully utilized for detecting tumor locations [15]. The membership function, which is based on numerous tumor picture features contrast, capacity, difference, uniformity, entropy, and relationship is critical in this finding [16]. The simulation findings demonstrate that the suggested strategy effectively distinguishes between normal and diseased brain tissues by imposing limitations on grey-level intensity detachment. Ahasan Kabir suggested this automated technique for brain tumor identification and categorization. The suggested method consists of five phases. The MRI input picture is first preprocessed to eliminate unwanted artefacts using an approach that combines primary component-based conversion of grayscale and an anisotropic diffusion filter. Following that, a Contrast Limited Adapt Histogram Equalization (CLAHE) algorithm is used to improve picture contrast. The tumor is then segmented via the Chan-Vese approach and a multi-variant thresholding technique. Finally, statistical traits, texture qualities, and wavelet properties are measured to separate the segmented objects.

Tunnelmole provides a public URL for a web server operating on your local machine, allowing access from any Internet-connected device. It functions as a reverse proxy that directs internet traffic to your local workstation and can manage many sorts of HTTP/HTTPS requests including text, HTML& pictures. Tmole bypasses typical internet firewalls and routers by routing all requests via the client app without initiating its own server. It should function properly even whether connected to a corporate network like VPN. No specialized network configuration or networking expertize is required. The software is compatible with several types of web servers such as Nginx, Native NodeJS applications, and servers operating within containerized environments like Docker, Kubernetes, or a Virtual Machine.

10.2 METHODOLOGY

The process of image segmentation entails the partitioning of an image into several categories or areas that correspond to various items or features [17]. Every pixel inside the input picture is assigned to a certain category or location. The maximum level of segmentation is equivalent to a value of one. The phrase "highest-value" denotes a grayscale value that is identical to the grayscale values of adjacent pixels. Segmentation is an essential factor in the field of image analysis, since it entails the meticulous evaluation of individual pixels and their subsequent grouping into discernible objects within the image. Consequently, segmentation is measured a critical phase in image investigation with the aim of simplicity [18]. The method of threshold is critical in the field of picture segmentation since it involves identifying a particular range of pixels that are less than a predetermined threshold value. This category includes the histogram thresholding approach. The information was divided into different groups. Object distinction within an image can be accomplished by studying the color levels of the pixel that comprise the image. Pixels exhibiting greater grayscale values are often correlated with a certain entity, whereas pixels displaying lower grayscale values are commonly linked to a distinct item. The achievement of this technique may be accomplished by the use of a threshold value, referred to as "T" [19]. Therefore, an object point is identified when the pixel value, represented as f(r, c), above a certain threshold value, T. In contrast, when the value of f(r, c) is less than or equal to T, it is deemed that the pixel is part of the background. The terminology in question has been established and documented in the reference provided [20].

In this research, the threshold value was determined by a manual process including the iterative testing of many values for "T" in order to identify the optimal choice. In the present context, the process of identifying objects of interest is aided by two morphological tactics [7, 8], namely cleaning by reconstruction and fill via reconstruction. These strategies are graphically shown in Figure 10.2.

The goal of performing extra study is to collect data in a given topic. The data presented in this study is obtained via the use of the structured Grey Level Run Length Matrix (GLRLM), The co-occurrence Matrix for Grey

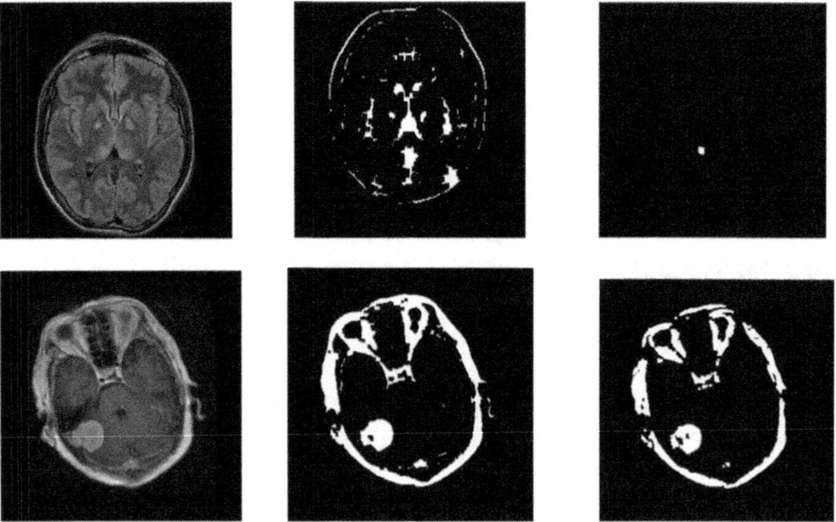

Figure 10.2 Displays the grayscale pictures that have been used, using the optimum threshold cost and morphological methods such as cleaning and filling.

Levels (SGLCM) and methodologies. The SGLCM, an empirical texture analysis technique, is used to extract the second-order statistical data from pairs of pixels. The SGLCM (Spatial Grey Level Co-occurrence Matrix) is a statistical method used to analyze the distribution of pixel brightness levels in a picture. It characterizes the relationship between pixel values based on their spatial proximity and calculates various statistical properties [13]. A computational process is executed to build a matrix that reflects the durations of runs in grey levels, with a specific emphasis on runs that demonstrate a particular direction. The visual representation of the image matrix, denoted as (v, w), signifies the frequency of run length (w) for each grayscale value (v). The matric provided in this study provide quantitative evaluations of texture, as used by Haralick [19, 20]. The run-length matrices, as shown in Table 10.1, may be represented by different equations depending on the orientation of the texture run. The run-length matrix is often represented as B (v, w) in academic literature [20]. The variable Ng is used to denote the overall quantity of grey levels that are present inside the picture, whereas the variable Nr is utilized to represent the count of viable run lengths. The extent of grey-level runs is determined based on the identification and quantification of grey levels in numerous directions.

The SGLCM Algorithm is given below.

1. Load the Image
2. Convert the image to grayscale

Table 10.1 Texture features and its mathematical equations

Texture feature	*Equation*
High Grey level Run Emphasis (HGLRE)	$\frac{\sum_{v=1}^{N_g}\sum_{w=1}^{N_r} B(v,w)}{N_\circ}$
Low Grey Level Run Emphasis (LGLRE)	$\frac{\sum_{v=1}^{N_g}\sum_{w=1}^{N_r} \frac{B(v,w)}{w^2}}{N_\circ}$
Long Run Emphasis (LGS)	$\frac{\sum_{v=1}^{N_g}\sum_{w=1}^{N_r} B(v,w).w^2}{N_\circ}$
Long Run High Grey Level Run Emphasis (LRHGLRE)	$\frac{\sum_{v=1}^{N_g}\sum_{w=1}^{N_r} B(v,w).v^2w^2}{N_\circ}$
Short Run Emphasis (SRE)	$\frac{\sum_{v=1}^{N_g}\sum_{w=1}^{N_r} \frac{B(v,w)}{w^2}}{N_\circ}$
Run Percentage (RP)	$\frac{N_\circ}{\sum_{v=1}^{N_g}\sum_{w=1}^{N_r} B(v,w).w}$

3. Apply a filter on the grayscale image
4. For each pixel in the image: Create a co-occurrence matrix capturing the frequency of pairs of pixel brightness values at specific distances and angles
5. Calculate Haralick parameters from the co-occurrence matrix and represent them in matrices
6. Detect edges from the extracted images using a gradient operation[7, 10]
7. Segment the image using a method for joining edge and region growing
8. Apply a thresholding operation to create a binary image, completing the segmentation process.

The GLRLM method involves quantifying the frequency of successive pixels that possess identical grey levels. The orientation of the texture specification is unrestricted, except for the constraint imposed by the total length of the sample run. The determination of the run duration of an image's texture involves the examination of a single row consisting of consecutive pixels with similar grey values. This data can be used to identify the location and length of a texture's grey level inside the image. As a result, the GLRLM is labelled as such.

$$\Phi(d,\theta) = x(a, b/d, \theta), 0 < a \leq Qa, 0 < a \leq Pmax \tag{10.1}$$

In this instance from equation 10.1, Qa means the greatest grey level, while P max specifies the longest possible run duration. The matrix "x" holds the final details about the run length "b" at a specified grey level "a" for a specific angle direction It considers four angle orientations: 0°, 45°, 90°, and 135°. Using the extraction function, methods like LRHGLE, SRE, and RP are retrieved for individually period matrix within the grey-level path.

Proposed Algorithm

1. Load the Image
2. Convert the image to grayscale
3. For each pixel in the image: – Create a run length matrix capturing the aggregate count of neighboring pixels that possess identical grey levels.
4. Calculate quality descriptors from the run length matrix
5. Apply these descriptors for further image analysis tasks such as segmentation or classification

10.2.1 Morphological process

This study analyzed three distinct MRI cases of brain images (Figure 10.3). The first case involved a healthy brain image, which had a size of 121 × 128

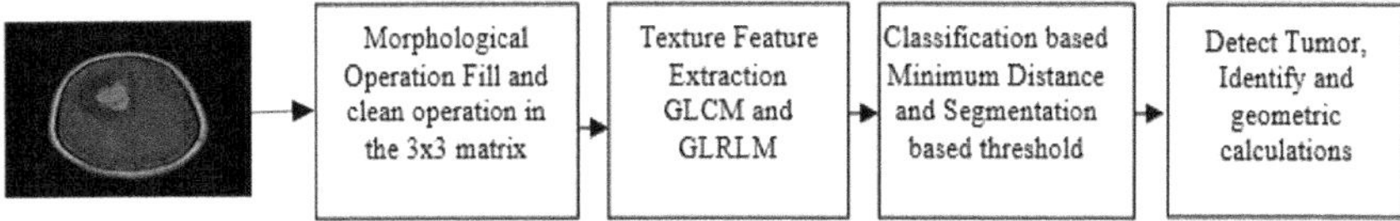

Figure 10.3 Brain cancer process of detection and identification.

pixel sizes and a bit depth of 24 [14]. The second case examined a malignant cognitive image gathered from a cancer care medical center, specifically for a tumor. This image had a resolution of 320 × 180 pixels and a bit depth of 22 [14]. Lastly, the third case focused on a brain image depicting a benign tumor, with a resolution of 124 × 146 pixel values and a bit depth of 24 [14].

10.3 EXPECTED RESULTS

The statistical performance of the algorithms, which illustrates the distinction and effectiveness in tumor extraction, can be encapsulated in Table 10.2. These features are exclusive to the tumor region and elucidate the actions employed in classification and segmentation.

The results of performance are pre-processed before classification or segmentation begins. Cleaning and filling via reconstruction is a morphological approach used in this step, and the outcomes are visible in the data. The malignant tumor in the first image and the characteristic image were automatically separated using the defined area in the original image. When compared with the photo features, the algorithm results show that the histogram successfully diagnosed the cancer. The image properties differ in both approaches. As a consequence, based on histogram data, there is clear separation between all features. Displaying the results implemented an interactive dashboard for MRI data analysis using Tunnel mole techniques. The first step in utilizing Tunnel mole to build a dynamic, interactive dashboard is to install the Timm package using pip. Then, create a "static" directory to hold

Table 10.2 Using the new techniques, we provide a statistical comparison of cancerous and healthy images

Benign image	*GLRLM*	*Malignant image*	*GLRLM*
HGLRE	**35.85**	**HGLRE**	**51.575**
LRE	**36.831**	**LRE**	**36.831**
LRHGLE	**68.54**	**LRHGLE**	**49.586**
LRGLE	**64.975**	**LRGLE**	**75.017**
SRE	**68.542**	**SRE**	**49.486**
RP	**0**	**RP**	**0**

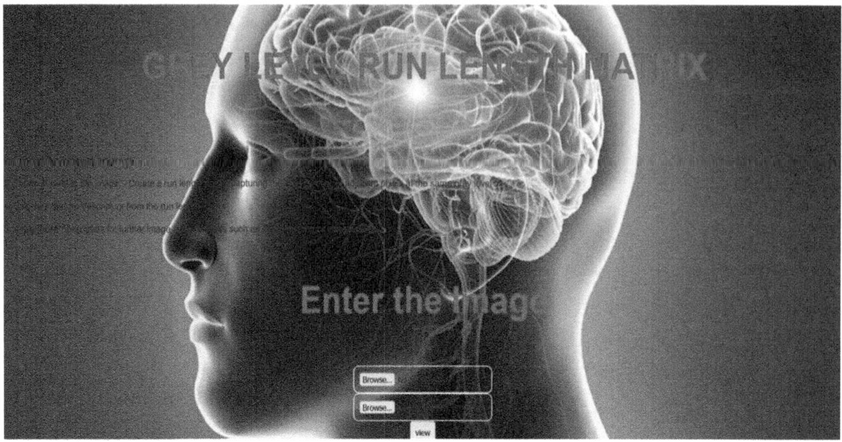

Figure 10.4 Interactive dynamic dashboard frontend application.

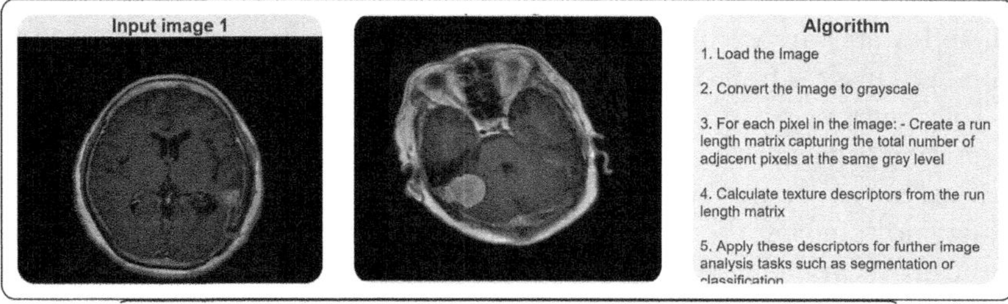

Figure 10.5 Input images for calculating features.

the input photos and produce graphs, and a "Dynamic" directory to hold the HTML file that will be used for webpage interactivity. After extracting features using the GREY LEVEL RUN LENGTH MATRIX technique, present them in a graph manner. To run the program and see its results on the dashboard, use HTTP requests with the GET and POST methods. The HTML website displaying the dynamic, interactive dashboard is shown in Figure 10.4. The dashboard interface displays the two uploaded photos, as seen in Figure 10.5. In Figure 10.6, bar charts are used to display the GLRLM characteristics that are produced.

Two figures, one showing malignant and the other benign samples (Figures 10.7 and 10.8), are tested using the Dynamic Dashboard.

In Figure 10.8, displaying the two images outputs along with the results. In Figure 10.9, you can see all the statistics related to the original, complement, and tumor images. Important measurements like index, dimensions, region, radius, and volume are part of it, allowing for a thorough examination of the pictures' properties.

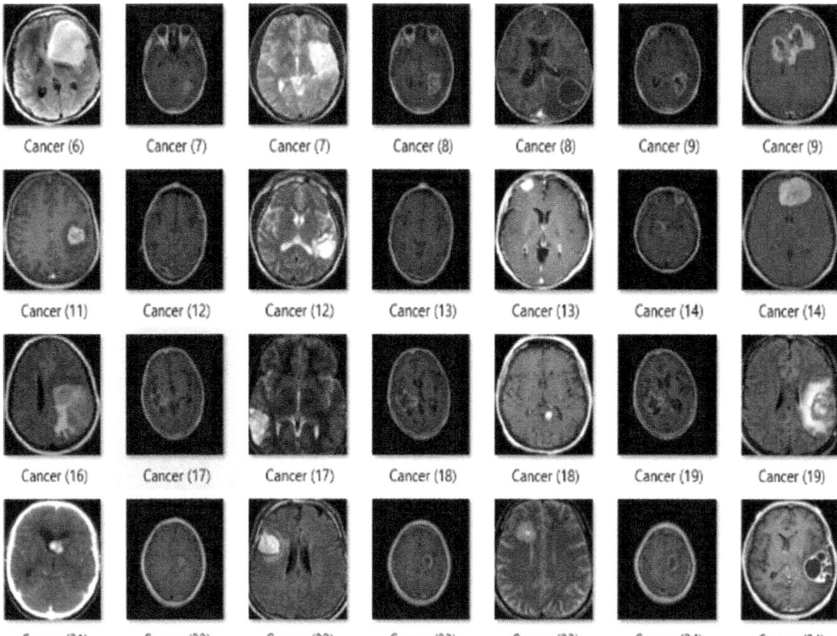

Figure 10.6 Sampled malignant images.

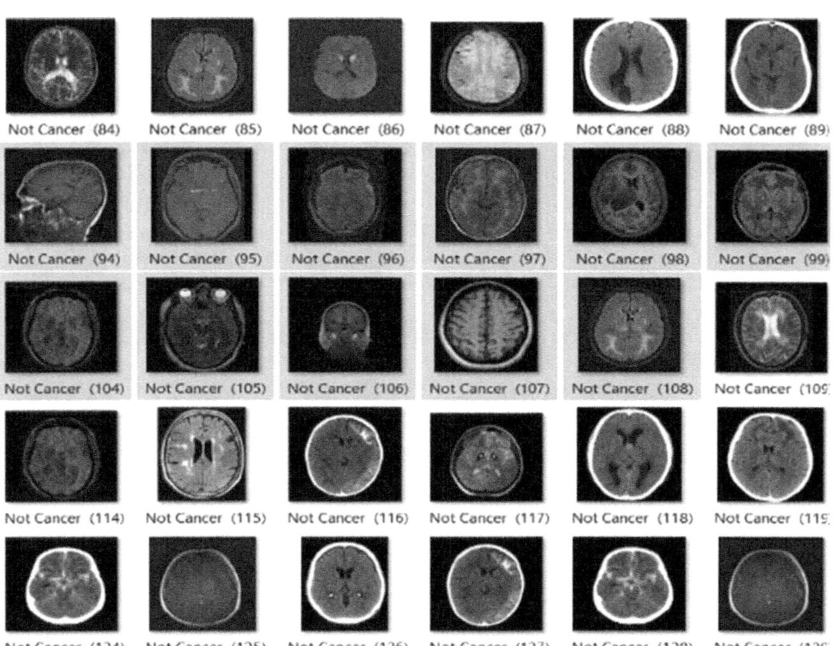

Figure 10.7 Sampled benign images.

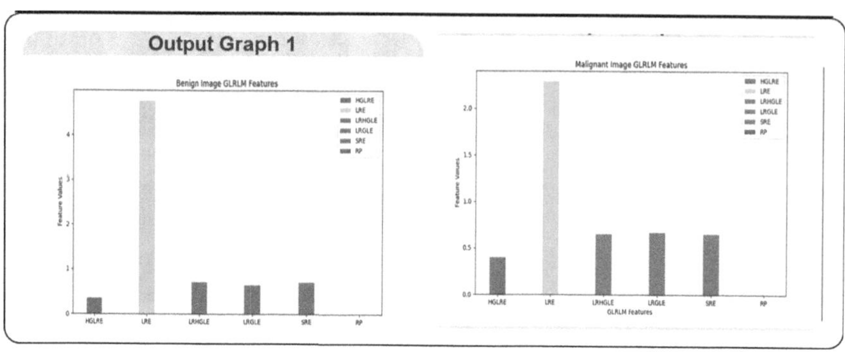

Figure 10.8 Result graphs for benign & malignant images.

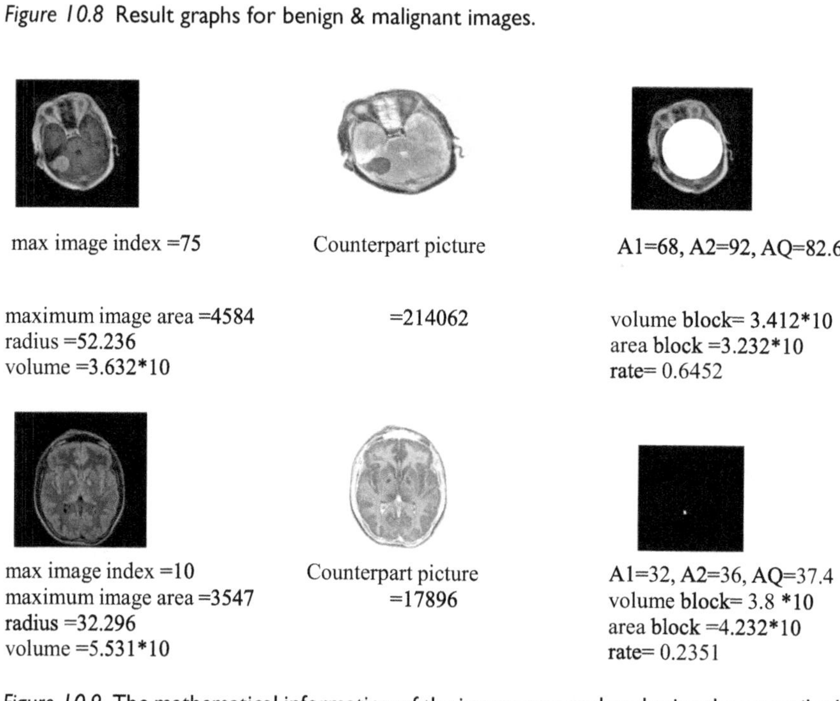

Figure 10.9 The mathematical information of the images was analyzed using the prescribed methodologies.

10.4 CONCLUSION

Tumor identification inside an MRI picture is a difficult task. As a result, two approaches were offered to assess the efficacy and quality of tumor extraction. Three MRI scans with natural, benign & malignant tumors were segmented and classified using segmentation and classification methods. The thresholding-based segmentation approach proved to be very efficient in identifying tumor with amazing precision. The histogram graphs clearly show the contrast between the tumor region and the complimentary area.

The histogram of both the tumor or the complimentary picture displays the textural feature pattern in each image. The proposed algorithm, which enhances the performance and reliability of automated diagnosis, is recommended for tumor detection in medical images. The dimension of the tumor might assist determine what kind of tumor When the A1 and A2 levels exhibit a high degree of similarity, it indicates that the tumor is harmless; conversely, if they differ significantly, it suggests cancer. The results obtained from both procedures indicate the GLRLM method outperforms the original image in terms of providing more comprehensive and precise histogram pixel counts, even if it registers a value of 0 at the corresponding location. The efficacy of the Grey Level Run Length Matrix (GLRLM) is contingent upon the selection of the grey level choice. The identification, localization, and separation of a tumor in an image are influenced by its characteristics, such as its kind, dimensions, and spatial position. The distinguishing capabilities of the traits vary, with SRE demonstrating an advantage in identifying cancerous cases. The GLRLM methodology enhances the GLCM method by using the complement picture, which aligns with the initial image and feature discoveries. The complimentary image is seen as a unique image with a negative value. An interactive MRI data analysis dashboard allows for easy exploration and visualization of intricate medical imaging data, enabling users to make well-informed decisions and progress in medical research and diagnosis.

REFERENCES

[1] Zhang, W., & Ge, Y. R. (2012, April 9). Texture feature extraction by incomplete tree-structed wavelet based on morphology pre-processing. *Journal of Computer Applications*, 31(6), 1592–1594. https://doi.org/10.3724/sp.j.1087.2011.01592

[2] Abbas, H. K., Fatah, N. A., Mohamad, H. J., & Alzuky, A. A. (2021, April 1). Brain tumor classification using texture feature extraction. *Journal of Physics: Conference Series*, 1892(1), 012012. https://doi.org/10.1088/1742-6596/1892/1/012012

[3] Vadivel, A., Sural, S., & Majumdar, A. (2007). An integrated color and intensity co-occurrence matrix. *Pattern Recognition Letters*, 28(8), 974–983. https://doi.org/10.1016/j.patrec.2007.01.004

[4] Vadivel, A., & Surendiran, B. (2013). A fuzzy rule-based approach for characterization of mammogram masses into BI-RADS shape categories. *Computers in Biology and Medicine*, 43(4), 259–267. https://doi.org/10.1016/j.compbiomed.2013.01.004

[5] Wady, S. H., Yousif, R. Z., & Hasan, H. R. (2020, July 11). A novel intelligent system for brain tumor diagnosis based on a composite neutrosophic-slantlet transform domain for statistical texture feature extraction. *BioMed Research International*, 2020, 1–21. https://doi.org/10.1155/2020/8125392

[6] Vidyarthi, A., & Mittal, N. (2017, March 29). Texture based feature extraction method for classification of brain tumor MRI. *Journal of Intelligent & Fuzzy Systems*, 32(4), 2807–2818. https://doi.org/10.3233/jifs-169223

[7] Hu, Y., Wang, Z., & AlRegib, G. (2020, April). Texture classification using block intensity and gradient difference (BIGD) descriptor. *Signal Processing: Image Communication*, 83, 115770. https://doi.org/10.1016/j.image.2019.115770
[8] Borvornvitchotikarn, T., & Kurutach, W. (2020, December 15). miRID: Multi-modal image registration using modality-independent and rotation-invariant descriptor. *Symmetry*, 12(12), 2078. https://doi.org/10.3390/sym12122078
[9] Song, T., Xin, L., Gao, C., Zhang, G., & Zhang, T. (2018, May). Grayscale-inversion and rotation invariant texture description using sorted local gradient pattern. *IEEE Signal Processing Letters*, 25(5), 625–629. https://doi.org/10.1109/lsp.2018.2809607
[10] Ashwathi, S., Swamy Goud, A., Niranjan, L., Sreekantha, B., & Suneetha, J. (2023). A novel approach to prognosticate CKD using a supervised and unsupervised learning algorithms. *Intelligent Manufacturing and Energy Sustainability*, 107–116. https://doi.org/10.1007/978-981-19-8497-6_11
[11] Kou, Q., Cheng, D., Chen, L., & Zhao, K. (2018). A multiresolution gray-scale and rotation invariant descriptor for texture classification. *IEEE Access*, 6, 30691–30701. https://doi.org/10.1109/access.2018.2842078
[12] Soni M, A. Bhat, S. Aralikatti, A. Pasha, Niranjan L and Yousuf Madar R (2023). An Efficient Digital Cluster Scheme to Improve the Lifetime Ratio of Wireless Sensor Networks, *2023 International Conference on Smart Systems for applications in Electrical Sciences (ICSSES), Tumakuru, India*, pp. 1–5, https://doi.org/10.1109/ICSSES58299.2023.10199240
[13] Szychot E, Youssef A, Ganeshan B, Endozo R, Hyare H, Gains J, Mankad K, Shankar A. (2021, June) Predicting outcome in childhood diffuse midline gliomas using magnetic resonance imaging based texture analysis. *Journal of Neuroradiology*, 48(4), 243–247. https://doi.org/10.1016/j.neurad.2020.02.005. Epub 2020 Mar 14. PMID: 32184119.
[14] Kunimatsu, A., Yasaka, K., Akai, H., Sugawara, H., Kunimatsu, N., & Abe, O. (2022, January 1). *Texture Analysis in Brain Tumor MR Imaging. Magnetic Resonance in Medical Sciences*. Japan Society of Magnetic Resonancein Medicine. https://doi.org/10.2463/mrms.rev.2020-0159
[15] Yang, Y., Yan, L., Zhang, X., Nan, H. Y., Hu, Y. C., Han, Y., Zeng, J., Liu, Z., Sun, Y. Z., Tian, Q., Yu, Y., Sun, Q., Wang, S., Zhang, X., Wang, W., & Cui, G. (2019, January 9). Optimizing texture retrieving model for multimodal MR image-based support vector machine for classifying glioma. *Journal of Magnetic Resonance Imaging; Wiley-Blackwell*. https://doi.org/10.1002/jmri.26524
[16] Zhang, J., Wang, J., Lyu, W., and Yin, Z. (2023) "Local Texture Complexity Guided Adversarial Attack," *2023 IEEE International Conference on Image Processing (ICIP), Kuala Lumpur, Malaysia*, pp. 2065–2069, https://doi.org/10.1109/ICIP49359.2023.10222176
[17] J. Suneetha, Niranjan L., H. Tabasum, S. Goud, R. Taseen, and M. B. Neelagar, (2023) A Wireless Detector Network for Three-Dimensional Positioning Using Artificial Neural Networks, *2023 International Conference on Recent Trends in Electronics and Communication (ICRTEC), Mysore, India*, pp. 01–05, https://doi.org/10.1109/ICRTEC56977.2023.10111934
[18] Srivastava, D., Singh, S. S., Rajitha, B., Verma, M., Kaur, M., and Lee, H.-N. (2023). Content-based image retrieval: A survey on local and global features selection, extraction, representation, and evaluation parameters. *IEEE Access*, 11, 95410–95431, https://doi.org/10.1109/ACCESS.2023.3308911

[19] Devareddi, R. B., and Srikrishna, A. (2022). Review on Content-based Image Retrieval Models for Efficient Feature Extraction for Data Analysis, *2022 International Conference on Electronics and Renewable Systems (ICEARS), Tuticorin, India*, pp. 969–980, https://doi.org/10.1109/ICEARS53579.2022.9752281

[20] Haralick, R. M., Shanmugam, K., and Dinstein, I., (1973, November). Textural features for image classification. *IEEE Transactions on Systems, Man, and Cybernetics*, SMC-3(6), 610–621, https://doi.org/10.1109/TSMC.1973.4309314

Chapter 11

Interactive and dynamic stock market dashboard

Prasan Mittal and Kanimozhi S
Vellore Institute of Technology, Chennai, India

Sairamesh L
St. Joseph's Institute of Technology, Chennai, India

11.1 INTRODUCTION

In the fast-paced digital era, financial markets experience rapid fluctuations and generate an overwhelming amount of data. This environment demands advanced tools for analysts and investors to sift through data, identifying valuable insights to remain competitive. Microsoft Power BI has emerged as a frontrunner in the realm of data visualization and analysis, offering unparalleled support for creating interactive and dynamic stock market dashboards [1]. These dashboards not only facilitate a deeper understanding of stock movements through intuitive visualizations but also empower users with machine learning algorithms and statistical models for advanced analytics. The crafting of an interactive and dynamic stock market dashboard using Power BI integrates the principles of effective data integration, analysis, and visualization. Power BI's robust data modeling capabilities allow for efficient organization and management of complex stock data, enabling users to uncover hidden patterns and insights that inform better investment decisions [2]. The ability to connect to live data sources, integrate various data sets, and collaborate among team members further enhances the tool's utility, making it an indispensable asset for anticipating market trends and optimizing profit forecasting. This research paper explores the application of Power BI in developing a sophisticated stock market dashboard, highlighting its significant benefits for business intelligence (BI) and analytics. By analyzing Power BI's functionality and advantages, along with a review of previous research and methodologies, the paper aims to elucidate the transformative potential of BI tools in financial analysis. It underscores the logic behind the adoption of Power BI by different sectors, from cement companies focusing on profit forecasting to the automotive industry seeking competitive advantages through innovative strategy and effective decision-making [3]. By blending insights from various sources, including case studies on the use of Power BI in corporate settings and academic research on its analytical capabilities, this introduction sets the stage for a comprehensive discussion on the development and utilization of interactive and dynamic

 DOI: 10.1201/9781003542735-11

stock market dashboards. Through this exploration, the paper aims to contribute valuable perspectives on the future of stock analysis and investment strategy, powered by advanced BI tools like Power BI.

11.2 LITERATURE REVIEW

11.2.1 Power BI

Power BI, a business analytics service by Microsoft, enables users to visualize and share insights drawn from their data across their organization or embed them in an app or website. It's known for its user-friendly, intuitive interface that allows users to create and publish dashboards and reports easily. Power BI's strength lies in its ability to provide real-time updates, integrate various data sources, and apply advanced analytics and machine learning to unearth deeper insights [4].

11.2.2 Benefits of business intelligence (BI)

BI technologies provide historical, current, and predictive views of business operations, often using data collected into a data warehouse or a data mart and small subsets of data in a real-time dashboard. Power BI stands out amongst BI tools for its ease of use, interactive visualizations [5], and ability to handle large volumes of data efficiently, thus empowering organizations to make well-informed decisions.

11.2.3 The importance of dashboards

Dashboards are crucial for displaying critical metrics and KPIs through data visualization. With Power BI, dashboards become not just informative but interactive and dynamic, allowing users to explore and analyze data in various dimensions [6]. This interactivity enhances the decision-making process, making it possible to respond quickly to market changes.

11.2.4 Previous research

Previous research illuminates the evolution from traditional profit forecasting methods to more sophisticated techniques facilitated by Microsoft Power BI [7]. This transformation is significantly marked in sectors with complex data management needs, such as the cement industry, where precise forecasting and visualization techniques are critical for informed decision-making.

The application of Power BI in the cement industry serves as a testament to its transformative impact on profit forecasting [11]. Through a meticulous process of data collection, relationship modeling between tables in data warehouses, and the creation of customized dashboards, cement companies have been able to gain clearer insights into profit forecasts [8]. This process not only refines the raw data into actionable intelligence but also encapsulates it in a user-friendly visual format, enhancing interpretability and decision-making efficiency.

The development of performance dashboards in Power BI goes beyond traditional visualization, adopting an advanced analytical stance. By incorporating Azure Machine Learning Studio, Power BI elevates business analysis to a more predictive and prescriptive level [10]. This capability empowers users to quickly forge prediction models and uncover complex modeling and data combination scenarios, thereby enabling organizations to make data-driven decisions across various business facets [9]. Such advanced analytics engender a more nuanced understanding of the factors influencing stock movements and profit margins, thereby portraying Power BI as a pivotal tool in the arsenal of modern businesses striving for data-informed excellence.

11.3 METHODOLOGY

This research employs a mixed-methods approach, combining qualitative analysis through literature review and quantitative analysis by implementing a prototype stock market dashboard in Power BI. The focal point of this research is the development and evaluation of an interactive and dynamic stock market dashboard, designed to empower investors and analysts with advanced analytical tools and intuitive data visualizations.

11.3.1 Development steps

The initial step involves setting up Power BI and connecting it to relevant data sources. Power BI's extensive data connectors facilitate the integration of various financial databases, spreadsheets, and online platforms, ensuring a comprehensive dataset for analysis [12]. This connectivity is pivotal for importing stock data into Power BI for subsequent analysis as shown in Figure 11.1.

11.3.1.1 Data collection and preparation

Following data import, Power BI's data modeling capabilities come into play, transforming the raw data into a format suitable for detailed analysis and visualization. This process involves Extract Transform Load (ETL) operations [13] to harmonize data from diverse sources, crucial for ensuring accuracy and relevance in the ensuing analysis. Our Sample is shown in Figures 11.2 and 11.3. our Data has 15 features, which are Date of format date, Symbol of data type string representing the symbol of the company, Series of data type String, pre, close, high, low, last, close of data type double

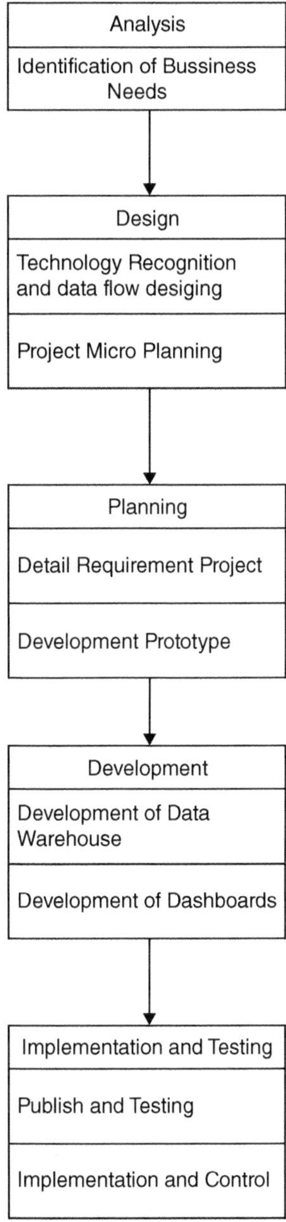

Figure 11.1 Development steps involved in making dashboard from the stock data.

giving us the information about the intraday trading information at the end of the day, VWAP of data type double, and volume of data type int representing the total volume traded that day, Turnover is of data type double, Trades of data type int giving total number of trades performed on the given companies stock, and we have deliverable of data type int and %delivered

1	Date	Symbol	Serie	Prev Clo	Open	High	Low	Last	Close	VWAP	Volume	Turnove	Trades	Delivera	%Delive
3244	01-01-2021	ADANIPORTS	EQ	483.75	485	508	482.55	505	503.85	495.57	7815730	3.9E+14	79171	834159	0.1067
8550	01-01-2021	ASIANPAINT	EQ	2764.5	2759	2792	2750	2770.1	2775.55	2774.59	1246698	3.5E+14	44528	95779	0.0768
13856	01-01-2021	AXISBANK	EQ	620.45	620.25	625.45	617.55	623.4	623.8	622.15	6047062	3.8E+14	84795	1473702	0.2437
17058	01-01-2021	BAJAJ-AUTO	EQ	3444.05	3446	3494	3446	3479	3481.25	3468.09	421643	1.5E+14	23426	69494	0.1648
20259	01-01-2021	BAJAJFINSV	EQ	8906.35	8930	8958.7	8840	8870.2	8870.45	8900.4	288101	2.6E+14	23360	59708	0.2072
25494	01-01-2021	BAJFINANCE	EQ	5295.2	5310.2	5338	5250	5272	5280.15	5302.83	1447187	7.7E+14	74896	247821	0.1712
30268	01-01-2021	BHARTIARTL	EQ	509.7	512.25	516.4	508.2	515	515.15	511.92	1E+07	5.2E+14	115803	2857212	0.2805
35574	01-01-2021	BPCL	EQ	381.1	381.1	384	380.4	382.5	381.95	382.47	3861740	1.1E+14	84002	900773	0.3146
40829	01-01-2021	BRITANNIA	EQ	3576.35	3575	3605	3563.05	3573	3567.8	3578.02	453083	1.6E+14	27548	189208	0.4176
46185	01-01-2021	CIPLA	EQ	819.95	822.8	828.95	820.65	827.3	826.6	826.17	2474916	2E+14	44615	359704	0.1453
48783	01-01-2021	COALINDIA	EQ	135.45	135.4	136.25	135.05	135.5	135.35	135.51	6995084	9.5E+13	42330	2455945	0.3511
54089	01-01-2021	DRREDDY	EQ	5205.1	5217.25	5254.85	5200	5233	5241.35	5232.78	583043	3.1E+14	34946	75016	0.1287
59390	01-01-2021	EICHERMOT	EQ	2530.9	2531	2555	2515.8	2551.9	2542.7	2534.15	914818	2.3E+14	37867	132246	0.1446
64375	01-01-2021	GAIL	EQ	123.25	123.9	124.4	122.55	123.35	123.65	123.44	8751786	1.1E+14	42637	1655957	0.1892
69681	01-01-2021	GRASIM	EQ	927.85	924.5	938	920.25	932.95	933.4	929.67	1151605	1.1E+14	23247	208640	0.1812
74981	01-01-2021	HCLTECH	EQ	946.15	942	955.5	942	950.25	950.5	950.65	3142822	3E+14	52922	650483	0.207
80287	01-01-2021	HDFC	EQ	2558.65	2549	2593.3	2541.15	2565	2568.75	2574.95	2054108	5.3E+14	77982	564766	0.2749
85593	01-01-2021	HDFCBANK	EQ	1436.3	1440	1443	1420.6	1423.45	1425.05	1433.57	4405469	6.3E+14	85415	1426213	0.3237
90899	01-01-2021	HEROMOTOCO	EQ	3110	3115	3120.8	3093	3105	3102.65	3106.98	408952	1.3E+14	20269	97066	0.2374
96205	01-01-2021	HINDALCO	EQ	240.55	239	240	237.6	238.25	238.35	238.52	5091899	1.2E+14	40872	1011170	0.1986
101511	01-01-2021	HINDUNILVR	EQ	2395.4	2395.4	2404	2382	2393	2387.55	2388.16	830096	2E+14	53330	263382	0.3173
106817	01-01-2021	ICICIBANK	EQ	535.05	535.55	537	526.1	527.8	527.5	529.68	1.4E+07	7.2E+14	156898	4502751	0.3313
111802	01-01-2021	INDUSINDBK	EQ	894.95	895	904	893.05	901.15	900.15	899.82	4632855	4.2E+14	64207	681808	0.1472
117108	01-01-2021	INFY	EQ	1255.8	1257.9	1265.5	1255.8	1259.5	1260.45	1261.66	4253550	5.4E+14	97445	1089987	0.2563
122414	01-01-2021	IOC	EQ	90.95	91.3	91.75	91.1	91.65	91.5	91.42	7144142	6.5E+13	24893	1903745	0.2665
127720	01-01-2021	ITC	EQ	209	209.9	214.2	209.3	213.9	213.85	212.11	2E+07	4.1E+14	97216	6064448	0.3106
131714	01-01-2021	JSWSTEEL	EQ	387.2	387.25	391.5	384.6	388.5	389.7	388.46	4461761	1.7E+14	37517	361794	0.0811
136699	01-01-2021	KOTAKBANK	EQ	1995.6	1996.9	2007.95	1990.7	1993.2	1994.05	1998.31	1106099	2.2E+14	34245	286907	0.2594
140883	01-01-2021	LT	EQ	1287.6	1283	1299.95	1283	1296.65	1297	1294.82	2058419	2.7E+14	60674	512462	0.249
146189	01-01-2021	M&M	EQ	720.6	725	744.75	723	731.8	732.45	737.58	9543128	7E+14	141701	1418987	0.1487
150616	01-01-2021	MARUTI	EQ	7649.6	7654	7748.5	7650	7687	7691.3	7704.92	767574	5.9E+14	55693	93385	0.1217
153422	01-01-2021	NESTLEIND	EQ	18390.3	18389.7	18520	18306.1	18425	18450.7	18434.3	50286	9.3E+13	12539	9592	0.1907
157510	01-01-2021	NTPC	EQ	99.35	99.95	100.2	98.8	99.1	99.05	99.19	1.3E+07	1.3E+14	47342	4506952	0.334
162816	01-01-2021	ONGC	EQ	93.05	93.75	94.45	93	93.35	93.2	93.39	1.5E+07	1.4E+14	67116	4668969	0.3086
166175	01-01-2021	POWERGRID	EQ	189.85	189	189.9	188.5	189.75	189.5	189.37	1729475	3.3E+13	20133	502554	0.2906
171481	01-01-2021	RELIANCE	EQ	1985.3	1988	1997	1982	1988	1987.5	1989.5	4622002	9.2E+14	139680	1013314	0.2192

NIFTY50_all

Figure 11.2 Sample data of each day.

1	Date	Symbol	Serie	Prev Clo	Open	High	Low	Last	Close	VWAP	Volume	Turnove	Trades	Delivera	%Delive
1025	17-01-2012	ADANIPORTS	EQ	135.5	137.1	141	135	140.1	140	138.13	1636196	2.3E+13	18374	1004327	0.6138
1026	18-01-2012	ADANIPORTS	EQ	140	142	143.8	138.7	143	141.7	141.25	890591	1.3E+13	15615	404925	0.4547
1027	19-01-2012	ADANIPORTS	EQ	141.7	144	150.55	143.15	149.5	149.4	146.72	1456077	2.1E+13	31299	721545	0.4955
1028	20-01-2012	ADANIPORTS	EQ	149.4	151.9	157.6	150.25	155.4	155.4	153.76	1634070	2.5E+13	23335	861145	0.527
1029	23-01-2012	ADANIPORTS	EQ	155.4	155.4	155.4	145.1	146.4	146.75	149.54	1657609	2.5E+13	12400	820653	0.4951
1030	24-01-2012	ADANIPORTS	EQ	146.75	147.05	152.9	145.6	149.8	150.05	150.29	1337362	2E+13	15441	703939	0.5264
1031	25-01-2012	ADANIPORTS	EQ	150.05	150.95	150.95	142.25	144.35	143.2	144.8	1859617	2.7E+13	14803	906254	0.4873
1032	27-01-2012	ADANIPORTS	EQ	143.2	145.8	149.65	144.8	146.75	147.1	147.69	1264483	1.9E+13	17693	711931	0.563
1033	30-01-2012	ADANIPORTS	EQ	147.1	147.1	147.4	137.35	138.4	138.4	142.13	757694	1.1E+13	12020	406470	0.5365
1034	31-01-2012	ADANIPORTS	EQ	138.4	138.95	148.5	137	147.1	146.25	143.17	1291344	1.8E+13	14640	683754	0.5295
1035	01-02-2012	ADANIPORTS	EQ	146.25	146.55	147.75	142.15	143.4	143.9	144.33	899183	1.3E+13	12697	350195	0.3895
1036	02-02-2012	ADANIPORTS	EQ	143.9	144.5	151.8	144.5	151	151	148.82	2355597	3.5E+13	26053	1072287	0.4552
1037	03-02-2012	ADANIPORTS	EQ	151	151	152.7	148.25	149.6	150.25	150.11	814359	1.2E+13	14137	351641	0.4318
1038	06-02-2012	ADANIPORTS	EQ	150.25	152.85	152.85	146.3	148.15	148.1	150.34	1518103	2.3E+13	17353	550880	0.3629
1039	07-02-2012	ADANIPORTS	EQ	148.1	149.15	149.45	137.85	137.85	139.15	142.49	1735290	2.5E+13	26089	970202	0.5591
1040	08-02-2012	ADANIPORTS	EQ	139.15	139	143.3	136.6	140	140.3	140.87	1532614	2.2E+13	16856	705330	0.4602
1041	09-02-2012	ADANIPORTS	EQ	140.3	139.65	141.8	135.2	138	137.95	138.09	1907732	2.6E+13	16753	1172113	0.6144
1042	10-02-2012	ADANIPORTS	EQ	137.95	138.15	144.15	137.2	142.95	142.9	140.28	1780060	2.5E+13	21630	1017352	0.5715
1043	13-02-2012	ADANIPORTS	EQ	142.9	143.95	144.3	141	142.75	143.05	142.63	718239	1E+13	7886	265151	0.3692
1044	14-02-2012	ADANIPORTS	EQ	143.05	143.3	145.55	142	143.8	144.15	143.2	1580236	2.3E+13	9737	1243143	0.7867
1045	15-02-2012	ADANIPORTS	EQ	144.15	145	147.95	144.15	146.6	147.15	145.97	886834	1.3E+13	11877	276880	0.3122
1046	16-02-2012	ADANIPORTS	EQ	147.15	147.45	149.9	145.75	147.25	147.6	147.81	856476	1.3E+13	10284	276539	0.3229
1047	17-02-2012	ADANIPORTS	EQ	147.6	149	152.9	145.75	151.5	150.5	150.36	1749511	2.6E+13	22821	660950	0.3778
1048	21-02-2012	ADANIPORTS	EQ	150.5	151.2	155.75	150.65	152.35	152.65	153.44	1744030	2.7E+13	19690	693099	0.3974
1049	22-02-2012	ADANIPORTS	EQ	152.65	155.75	157.75	142.4	143.5	143.85	151.21	1863666	2.8E+13	26638	628868	0.3374
1050	23-02-2012	ADANIPORTS	EQ	143.85	142	145.4	140.5	144.95	144.05	143.69	1533912	2.2E+13	11593	886359	0.5778
1051	24-02-2012	ADANIPORTS	EQ	144.05	145.05	146.5	141	142.25	142.7	142.67	527338	7.5E+12	7730	244104	0.4629
1052	27-02-2012	ADANIPORTS	EQ	142.7	144.4	145.15	137.25	142.7	143.65	141.87	947560	1.3E+13	13999	433571	0.4576
1053	28-02-2012	ADANIPORTS	EQ	143.65	144.8	149.45	144.1	148.65	148.1	146.92	629276	9.2E+12	8877	244844	0.3891
1054	29-02-2012	ADANIPORTS	EQ	148.1	149.75	152	145.25	147.3	148.65	148.8	627017	9.3E+12	9602	240577	0.3837
1055	01-03-2012	ADANIPORTS	EQ	148.65	148	152	144	145.5	146.85	146.38	370864	5.4E+12	5672	121763	0.3283
1056	02-03-2012	ADANIPORTS	EQ	146.85	147.5	150	144	146.3	147.35	146.3	497666	7.3E+12	9330	169822	0.3412
1057	03-03-2012	ADANIPORTS	EQ	147.35	147.6	147.75	145.4	145.6	146.15	146.45	22092	3.2E+11	417	9831	0.445
1058	05-03-2012	ADANIPORTS	EQ	146.15	147	147.05	142.4	142.9	143.2	144.31	264293	3.8E+12	3683	115981	0.4388
1059	06-03-2012	ADANIPORTS	EQ	143.2	144.65	147	140.65	144.9	143.55	144	502716	7.2E+12	7452	277511	0.552
1060	07-03-2012	ADANIPORTS	EQ	143.55	141.05	146.65	141.05	145.45	145.25	143.83	389292	5.6E+12	6600	148706	0.382

NIFTY50_all

Figure 11.3 Sample data of each company.

and data type double [15]. Our data has 235193 rows of data including different company's different days stock information.

In Figure 11.2, we present the structured format of daily stock data. Each row corresponds to a specific day, and the columns represent various attributes related to different companies' stocks. These attributes may include stock prices (open, close, high, low), trading volume, market capitalization, and other relevant metrics. By organizing the data in this tabular format, we facilitate efficient analysis and comparison across multiple companies [16].

Figure 11.3 illustrates the time-series behavior of a single company's stock over a specified period. We focus on one company at a time, tracking its stock price fluctuations, trends, and potential patterns. By isolating individual companies, we gain insights into their performance dynamics, volatility, and long-term trends. This approach allows us to delve deeper into the nuances of each company's stock behavior.

11.3.1.2 Dashboard development

Leveraging Power BI's drag-and-drop interface and advanced visualization options, the creation of customized dashboards begins. This step involves designing interactive visualizations that allow users to explore stock data from various perspectives, uncovering correlations and patterns critical for understanding market dynamics [17]. The integration of predictive analytics and machine learning algorithms further enhances the dashboard's capability to provide actionable insights into stock movements and trends. Our Dashboard's UI is shown in Figure 11.4 which represent the data visualization of the NIFTY-50 Stock Data.

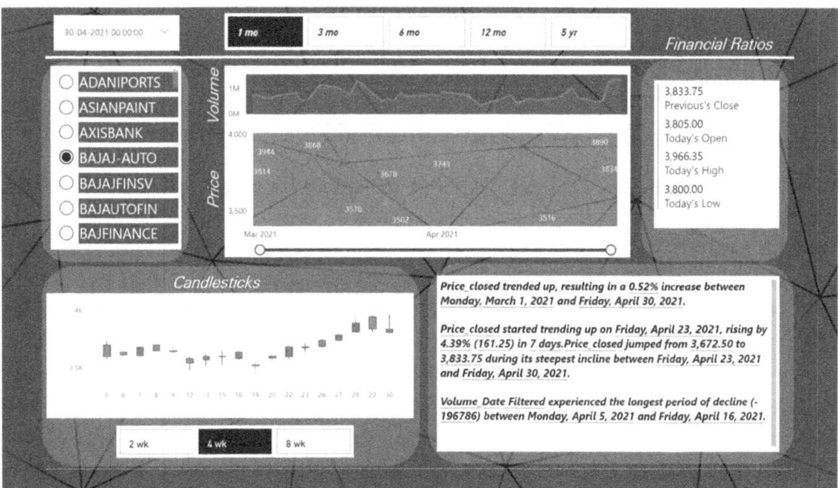

Figure 11.4 Power BI stock monitoring dashboard.

11.3.2 Data analysis method

The research adopts a three-part data analysis method, focusing on literature review, quantitative analysis through the prototype dashboard, and feedback collection for refinement. This multifaceted approach ensures a thorough understanding of Power BI's capabilities and the effectiveness of the developed dashboard in real-world scenarios. The final evaluation of the dashboard pivots on its usability, the precision of insights generated, and its impact on decision-making processes, reflecting the careful consideration of user involvement from the initial analysis and design phases through to result validation. By meticulously following these development steps and employing a comprehensive data analysis method, this research endeavors to bridge the gap between complex data analytics and user-friendly visualizations through Power BI [18]. The ultimate goal is to equip investors and analysts with a dynamic tool that enables them to navigate the volatility of the stock market with greater foresight and precision, thereby fostering more informed investment decisions.

Content shown in our dashboard Figure 11.4. gives the visual representation of change volume and price over a specific period of time for the selected stock, it also gives us a candlestick chart giving us a chance to gain insight into the intraday trading information and is also using Artificial Intelligence to analyze this data, and these analyzed result can be seen in the text box next to candlestick chart. For a brief overview our dashboard also provides financial ratio information of the stock for the selected date [19].

By bridging the gap between complex data analytics and user-friendly visualizations, Power BI enables investors and analysts to navigate the volatility of the stock market with greater foresight and precision [20].

11.4 RESULTS AND DISCUSSIONS

11.4.1 Business requirement analysis

Critical to the creation of an interactive and dynamic stock market dashboard is the meticulous analysis of business requirements. Such analysis ensures that the dashboard delivers the functionalities that match the specific needs of stock market analysis, which can range from real-time data processing to the provision of detailed historical data analysis capabilities [21]. This facet underpins not only what data is visualized but also how it is presented to support quick yet informed decision-making. In particular, business requirements for stock market dashboards typically include the ability to handle high-frequency data updates, integration with various data sources, and advanced analytical capacities to detect trends and anomalies. Utilizing Power Query [22] within Power BI greatly facilitates this, enabling seamless data transformation and enrichment, which is crucial for designing dashboards that are not only interactive but also provide expansive capabilities for generating insightful reports.

11.4.2 Planning

The initial phase of constructing the interactive and dynamic stock market dashboard revolves around meticulous planning to ensure coherent development. Initially, the project requirements were detailed, specifying critical functionalities such as real-time data updates, user interactivity, and comprehensive stock market analytics. Subsequently, a mockup design of the dashboard was developed to provide a visual blueprint, guiding the aesthetic and practical aspects of the user interface. In parallel, the ETL (Extract, Transform, Load) architecture [23] was meticulously designed to streamline the process of data extraction from various stock market sources, transformation into a coherent format, and loading into a data repository. We also delineated the SSAS (SQL Server Analysis Services) architecture to support complex analytical operations, such as predictive modeling and trend analysis. Finally, the query design was carefully crafted to optimize the retrieval of stock market data, ensuring that the dashboard reflects the most accurate and current information.

11.4.3 Design of the stock market dashboard

The design phase of our stock market dashboard was a meticulous process that involved multiple critical steps aimed at creating an effective and user-centric tool. The journey began with the development of a data warehouse that housed the various data elements essential for stock analysis, organized for optimal access and utilization [24]. To ensure relevance and ease of use, a relationship model was constructed, linking the disparate data sets and establishing meaningful connections between them. This model underpinned the dashboard's foundation, allowing for seamless interaction with the data. Beyond the technical architecture, we placed significant emphasis on the dashboard's aesthetics and functionality, incorporating filtering options to curate data views to specific user needs. In orchestrating the dashboard's interface, the design team prioritized clarity and accessibility to make it both attractive and simple for end-users to interpret the wealth of market data presented. As shown in Figure 11.5 these are all the existing and Derived fields that are used in our Dashboard.

11.4.4 Implementation and control

In the critical phase of Implementation and Control of the stock market dashboard, meticulous steps were undertaken to ensure the accuracy and effectiveness of the platform. The sequence began with the creation of a

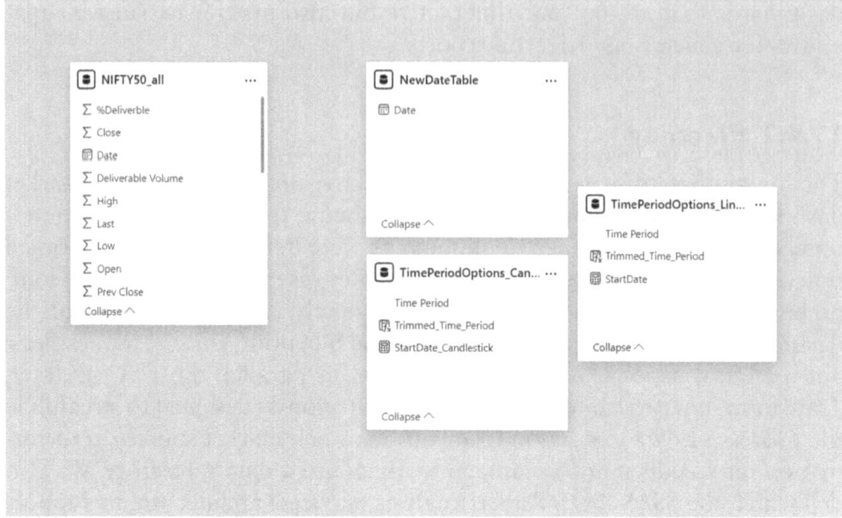

Figure 11.5 Represents data model view.

suite of visualizations tailored to highlight key stock market metrics, utilizing Power BI's extensive array of chart and graph capabilities. This was followed by a rigorous process of data cleansing, an indispensable step that entailed weeding out inconsistencies and refining the dataset to maintain high-quality, reliable information [25]. Thereupon, testing was methodically conducted to verify that the dashboard's performance was faultless and user interactions precipitated the expected outcomes. The testing phase not only sought to iron out technical abnormalities but also focused on evaluating the overall user experience, ensuring that the dashboard was both intuitive and insightful for stakeholders engaging with the dynamic stock analysis tool. As can be seen in Figure 11.6 we have our working dashboard which is interactive and can give you insight into the trading trends by analyzing the candlestick graph of intraday trading using Power BI's AI capabilities, and gives you a freedom to switch between different time periods.

11.5 CONCLUSION

The proposed research work has elucidated the pivotal role of Power BI in revolutionizing the analysis of stock market data with its interactive and dynamic dashboard capabilities. The key findings indicate that through meticulous business requirement analysis, thoughtful design, and strategic implementation, Power BI can substantially augment the efficiency of stock market monitoring and analysis. Importantly, the utilization of Power BI

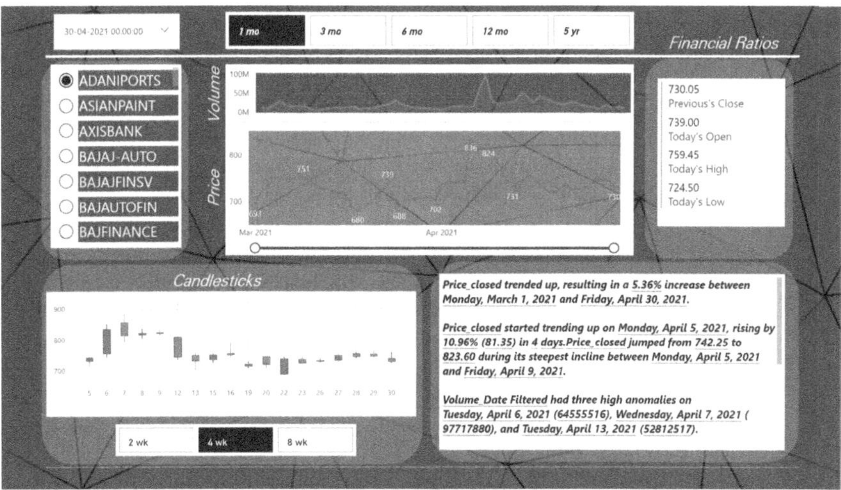

Figure 11.6 Power BI stock monitoring dashboard (showing Bajaj-Auto stock price and volume chart for previous six month).

enables a more agile and informed decision making process, underpinned by its advanced data transformation and visualization tools. The implications of this study suggest a beneficial impact for financial analysts, investors, and business strategists who seek to harness the power of data-driven insights for competitive advantage in the stock market arena.

The proposed work can be enhanced by the evolution of the interactive and dynamic stock market dashboard is poised to continue, potentially ingraining even greater depths of analysis through the integration of external data sources and the expansive analytics capabilities of Power BI. These developments will not only streamline the infusion of real-time economic indicators and ancillary financial data into the decision-making process but also pave the way for incorporating cutting-edge machine learning and AI-driven insights to finely tune investment strategies. As such, the dashboard's evolution will extend its utility beyond mere visualization, becoming a vital tool endowed with predictive power, capable of navigating the complexities of market dynamics with informed precision.

REFERENCES

1. Kumar, N. B., and Mohapatra, S. *The Use of Technical and Fundamental Analysis in the Stock Market in Emerging and Developed Economies*. United Kingdom: Emerald Group Publishing Limited, 2015.
2. Nafiisa, Birra Lailatul, et al. "Profit Forecasting Analysis and Visualization of Cement Companies Listed in the Indonesia Stock Exchange." *Jurnal Akuntansi Universitas Jember*, vol. 21, no. 1, 2023, pp. 40.

3. Widjaja, Surlisa, and Tuga Mauritsius. "The Development of Performance Dashboard Visualization with Power BI as Platform." *International Journal of Mechanical Engineering and Technology (IJMET)*, vol. 10, no. 5, May 2019, pp. 235–249, Article ID: IJMET_10_05_024. ISSN Print: 0976-6340, ISSN Online: 0976-6359.
4. PT. Indomobil Sukses Internasional. "Annual Report 2017 PT Indomobil Sukses International Tbk." (http://www.indomobil.com), 2017.
5. Moss, L., and Atre, S. *Business Intelligence Roadmap*. Boston, MA: Addison-Wesley, 2003.
6. Delen, D., and Demirkan, H. "Data, Information and Analytics as Services." *Decision Support Systems*, vol. 55, no. 1, 2013, pp. 359–363.
7. Appelbaum, D., Kogan, A., Vasarhelyi, M., and Yan, Z. "Impact of Business Analytics and Enterprise Systems on Managerial Accounting." *International Journal of Accounting Information Systems*, vol. 25, March 2017, pp. 29–44.
8. Alexandru, A. "Cloud Computing and Business Intelligence." *Database Systems Journal*, vol. V, no. 4, 2014, pp. 49–58.
9. Andersson, D. "Business Intelligence - The Impact on Decision Support and Decision Making Processes." *Analysis*, vol. 10, January 2008, pp. 40–50.
10. Sharda, R., Delen, D., Turban, E., Aronson, J. E., Liang, T.-P., King, D., York, N., Francisco, S., and Kong, H. *Business Intelligence, Analytics, and Data Science: A Managerial Perspective*, 4th ed. New York, NY: Pearson, 2018.
11. Microsoft Power BI. "Bring Your Data to Life with Power BI - Power BI Whitepaper." (http://powerbi.com), July 2015.
12. Krishnan, V. "Research Data Analysis with Power BI." *11th International CALIBER-2017, Anna University, Tamil Nadu*, 02–04 August 2017.
13. Tounsi, M. I. "Application and Survey of Business Intelligence (BI) Tools within the Context of Military Decision Making." Thesis, Naval Postgraduate School, June 2012.
14. Williams, S. *Business Intelligence Strategy and Big Data Analytics*. New York, NY: Morgan Kaufmann, 2016.
15. Al-Aqrabi, H., Liu, L., Hill, R., and Antonopoulos, N. "Cloud BI: Future of Business Intelligence in the Cloud." *Journal of Computer and System Sciences*, vol. 81, no. 1, 2015, pp. 85–96.
16. IBM. "Descriptive, Predictive, Prescriptive: Transforming Asset and Facilities Management with Analytics." IBM Software Thought Leadership White Paper, October 2013.
17. Mithas, S., Lee, M. R., Earley, S., Murugesan, S., and Djavanshir, R. "Leveraging Big Data and Business Analytics." *IT Professional*, vol. 15, no. 6, 2013, pp. 18–20.
18. Visinescu, L. L., Jones, M. C., and Sidorova, A. "Improving Decision Quality: The Role of Business Intelligence." *Journal of Computer Information Systems*, vol. 57, no. 1, 2017, pp. 58–66.
19. Giesen, E., Riddleberger, E., Christner, R., and Bell, R. "When and How to Innovate Your Business Model." *Strategic Leadership*, vol. 38, no. 4, 2010, pp. 17–26.
20. Rouhani, S., Ashrafi, A., Zare Ravasan, A., and Afshari, S. "The Impact Model of Business Intelligence on Decision Support and Organizational Benefits." *Journal of Enterprise Information Management*, vol. 29, no. 1, 2016, pp. 19–50.
21. Eckerson, W. W. *Performance Dashboards: Measuring, Monitoring, and Managing Your Business*. 2nd ed. Hoboken, NJ: John Wiley & Sons, 2010.

22. Vercellis, C. *Business Intelligence: Data Mining and Optimization for Decision Making*. TJ International, Padstow, Cornwall, 2009.
23. Ali, S. M., Gupta, N., Nayak, G. K., and Lenka, R. K. "Big Data Visualization: Tools and Challenges." *2016 2nd International Conference on Contemporary Computing and Informatics (IC3I)*, Noida, India 2016. https://doi.org/10.1109/IC3I.2016.7918044
24. Stefanovic, N. "Collaborative Predictive Business Intelligence Model for Spare Parts Inventory Replenishment." *Computer Science and Information Systems*, vol. 12, no. 3, 2015, pp. 911–930.
25. Murugesan, R., Shanmugaraja, V. & Vadivel, A. Forecasting Bitcoin Price Using Interval Graph and ANN Model: A Novel Approach. *SN Computer Science* vol. 3, 2022, p. 411.

Chapter 12

Performance analysis of hierarchical clustering and high-dimensional clustering algorithms on network IDS benchmark datasets using interactive dynamic dashboard

S. Srivarshinee and Menaka Pushpa A
Vellore Institute of Technology, Chennai, India

12.1 INTRODUCTION

Earlier, businesses and individuals often struggled with static reports that quickly became outdated, hindering decision-making and obscuring trends. Data analysis involves time-consuming manual updates and a lack of interactivity, limiting the ability to drill down into details or explore specific data segments. Dynamic dashboards [1, 2] have revolutionized this process by offering real-time data updates, empowering users to immediately spot emerging trends and react to changes. A dynamic dashboard is a real-time, personalized platform that allows users to add and update data. Dynamic dashboards are typically web-based applications that can be accessed from any device with internet connectivity. Their interactive features have led to detailed exploration, and their customizable layouts and no-code options provide easy accessibility to relevant information. This shift has helped make timely insights, improve decision making, and promote a data-driven approach to problem-solving. By connecting to a database, dynamic dashboards offer real-time updates, ensuring that changes in the database are reflected automatically. This feature is invaluable for the timely identification of new patterns through cluster analysis, particularly beneficial in scenarios like detecting new attacks or tracking real-time data in areas such as health and fitness, finance management, and business performance monitoring.

Many dynamic dashboard environments such as PowerBi, Tableau, Klipfolio, IBM Cognos Analytics, etc., While commonly utilize in business analytics, dynamic dashboards prove valuable in any scenario with access to real-time data, facilitating rapid analysis and informed decision-making processes. Amongst the many dynamic dashboard tools that exist, Tableau offers a wider range of advanced visualization options and greater customization

DOI: 10.1201/9781003542735-12

capabilities. Tableau also provides a Tableau Python Server, to run Python scripts directly within it, and a Tableau Software Development Kit, which allows users to build custom extensions for Tableau using various programming languages. Hence, machine learning algorithms can be executed as Python scripts within Tableau, with the resulting outcomes seamlessly visualized on the same platform. The two most prevalent machine learning techniques in the realm of data analysis are classification and clustering. While classification is a supervised learning approach that relies on labeled data for training, clustering is an unsupervised learning method that operates without the need for labeled data, identifying patterns or groupings within the dataset based on inherent similarities or distances between data points. Although both clustering and classification involve assigning labels to data points, clustering is employed when the labels or categories are not predefined, and the goal is to discover intrinsic patterns or latent groupings within the data based on metrics like distance or density. The process of grouping a set of samples into two similar subsets is called clustering. A cluster is a collection of points in a vector space such that the similarity index between any two points within the cluster is more than that of any two points in the cluster and any point not on it. Clustering algorithms are broadly categorized into partition methods, density-based methods, and advanced methods. Partition methods include algorithms like that of k-Means [3] and k-Medoids which partition a dataset of N elements into k partitions based on a distance metric, with each partition representing a cluster. Both k-Means and k-Medoids iteratively assign data points to clusters based on distance metrics and update cluster centroids until convergence, aiming to minimize intra-cluster distance and maximize inter-cluster dissimilarity. While k-Means selects the centroids after calculating the mean of all the data points in the cluster, iteratively, k-Medoids assign a data point within the cluster as the centroid. Thus, k-Medoids are better equipped to deal with outliers than kMeans. Density-based methods include Density-Based Clustering based on Connected region with high density (DBSCAN) or Ordering Points To Identify the Cluster Structure (OPTICS). DBSCAN focuses on grouping closely packed points based on a density criterion, requiring two parameters—epsilon (ε) and minPts—to define the neighborhood of points and the minimum number of points needed to form a dense region, respectively. It excels at discovering clusters of arbitrary shapes and sizes while also identifying outliers. In contrast, OPTICS extends DBSCAN by creating a reachability plot that orders points based on their density and distance to other points, providing a hierarchical clustering structure. Although OPTICS does not require explicit specification of parameters like ε and minPts, it introduces xi (ξ) to control the minimum steepness of the reachability plot. DBSCAN is well-suited for datasets featuring non-linear and irregularly shaped clusters, offering efficient identification of clusters with varying densities and shapes. Conversely, OPTICS excels when datasets contain clusters with differing densities or when a hierarchical understanding of the data structure is desired. Normal clustering algorithms, such as k-means and DBSCAN, provide foundational techniques for grouping data points based on similarity or density. These methods offer effective means of partitioning datasets and identifying clusters, albeit with certain limitations in handling complex data structures or high dimensional datasets. In contrast, advanced clustering algorithms, like hierarchical clustering and high-dimensional clustering techniques, represent a significant evolution in clustering methodologies. Hierarchical clustering, for instance, enables the exploration of hierarchical relationships within data, offering insights into nested cluster structures that traditional methods may overlook. Meanwhile, high-dimensional clustering algorithms are specifically tailored to address the challenges posed by datasets with numerous dimensions, mitigating issues such as the curse of dimensionality. By incorporating more sophisticated approaches and leveraging advanced techniques, these algorithms provide enhanced capabilities for uncovering intricate patterns and structures in data, thus expanding the scope and depth of clustering analysis across diverse domains. Clustering is widely used to find inherent patterns in the data and thus, is employed for mining and analysis of patterns in data, research in market and processing of images. Another inevitable aspect of clustering a dataset is outliers extremely high or extremely low values of a data such that it doesn't lie close to any of the clusters formed. Network administrators use outlier detection to identify abnormal network traffic patterns that might indicate cyberattacks, malware intrusions, or unusual system activity. Outlier analysis is also especially useful in the case of fraud detection in transactions banks and other financial institutions utilise outlier detection to identify potentially fraudulent transactions. The paper is organized as follows; Section 2 describes the characteristics of Network Intrusion Detection System benchmark datasets used for this analysis. Section 3 and Section 4 explain hierarchical and high-dimensional clustering algorithms. Section 5 shows the implementation results obtained from the dynamic dashboard tool tableau. Section 6 concludes this clustering analysis work with the help of an interactive dynamic dashboard.

12.2 DATASET SELECTION

Intrusion detection, which entails vigilant monitoring of network or system activities to uncover unauthorized access or malicious behaviors, relies on specialized systems like NIDS for comprehensive network surveillance and HIDS for nuanced host-level scrutiny, enabling swift threat detection and response. To enhance the efficacy and precision of these systems, the development of robust datasets becomes imperative, serving as crucial resources for training, validating, and refining intrusion detection algorithms. These datasets typically consist of labeled network traffic data captured from real-world environments or simulated scenarios, encompassing various types of benign activities, as well as different forms of attacks and anomalies. Widely recognized benchmark datasets such as UNSW-NB15 [4, 5], CIC-IDS2017 [6–8], and NSL-KDD serve as invaluable resources for cybersecurity research, fostering innovation and advancements in the field of intrusion detection and network security. Out of these benchmark IDS datasets, UNSW-NB15 and CIC-IDS2017 datasets have been selected to perform clustering. The UNSW-NB15 dataset stands as a cornerstone in the field of cybersecurity research, offering invaluable insights into network traffic behavior and facilitating the development of robust intrusion detection systems (IDS). This dataset, curated by the University of New South Wales (UNSW), comprises a comprehensive collection of labeled network traffic data captured from a real-world environment. It presents about 10 types of attacks including Fuzzers, Backdoors, Generic, Worms, Analysis, DoS, Exploits, Reconnaissance, and Shellcode. It has 2540044 samples with 175341 in the training set, and 82332 in the test set. The CIC-IDS2017 dataset offers an extensive collection of labeled network traffic data essential for the development and evaluation of intrusion detection systems (IDS) and related security solutions. Curated by the Canadian Institute for Cybersecurity (CIC), this dataset encompasses a wide range of network traffic scenarios and attack types, captured from diverse environments and network configurations. It categorizes the attacks into 14 types including Heartbleed, Web Attack—Sql Injection, DDoS, Infiltration, PortScan, Bot, Web Attack-Brute Force, SSH-Patator, DoS Hulk, Web Attack-XSS, FTP-Patator, DoS slowloris, DoS Slowhttptest, and DoS GoldenEye. It has 2830540 samples with 83.1% of it being Benign samples and the rest are attacks. The smaller scale of the UNSW-NB15 makes it more manageable for research tasks and experiments and provides a rich set of features, including basic flow features, content features, and statistical attributes, enabling detailed analysis and modelling of network behavior. However, CIC-IDS2017 offers a larger dataset with a diverse range of instances, suitable for testing and validating intrusion detection algorithms on a larger scale. It also offers detailed feature sets, including packet-level information, flow-based characteristics, and statistical metrics, allowing for in-depth exploration of network traffic patterns and security threats. For the purpose of this paper, the training set of UNSW-NB15 containing 174341 samples and random samples of the CIC-IDS2017 (distribution of samples based on the attack categories maintained) dataset constituting around 283054 samples have been used. The distribution of attack categories of CIC-IDS2017 and NB-15 datasets are shown in Figures 12.1 and 12.2.

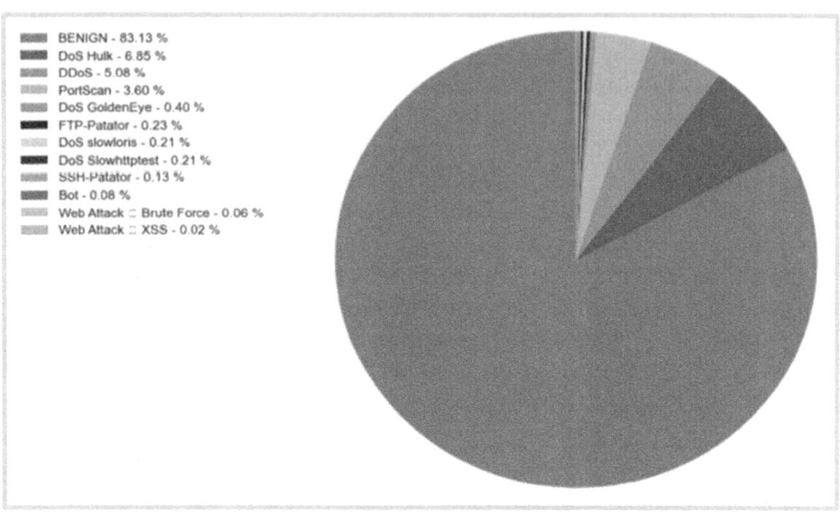

Figure 12.1 Distribution of various attack categories in the CIC-IDS2017 dataset.

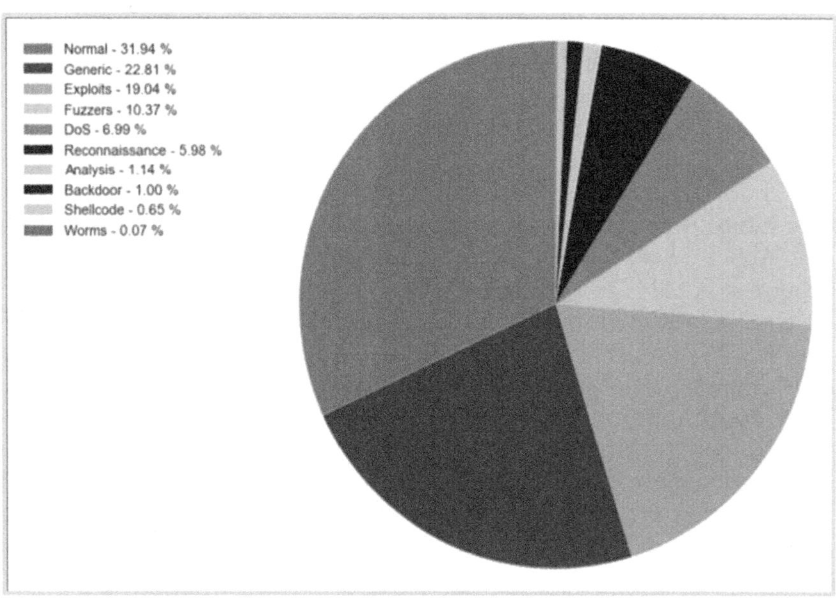

Figure 12.2 Distribution of various attack categories in the UNSW-NB15 dataset.

12.3 HIERARCHICAL CLUSTERING

Hierarchical clustering groups data samples into a cascading tree of clusters, where the clusters of one level are derived from the clusters of the previous level. This establishes a hierarchy amongst the clusters which is useful in studying datasets with an underlying taxonomy. Hierarchical clustering can be classified into two categories on the basis of the strategy employed- divisive clustering and agglomerative clustering. While the agglomerative [9] uses a bottom-up approach, divisive uses a top-down method. Divisive clustering method begins with a single cluster, the root cluster, consisting of all the data samples, which are recursively split into sub-clusters. This continues until the each cluster either contains a single object, or the objects within each cluster are adequately similar. Agglomerative clustering, on the other hand, associate a cluster with every data sample which are then progressively combined to form higher-level clusters. Usually a single cluster is formed at the end step of agglomerative clustering. This cluster acts as the root cluster and decision on which clusters to merge is made depending upon similarity measures usually distance measures like Euclidean or Manhattan Measures. The algorithm of agglomerative clustering is as follow:

If the number of samples in the distance matrix is N, and the specified number of clusters to be formed at the last level be k, then

```
1. Compute the distance matrix i.e,
        for i=1 to N:
                for j=1 to i:
                        find dist[i][j]=distance(i,j)
2. Repeat until only k clusters remain:
      • merge cluster[I] and cluster[j] if
      • dist[i][j]<=dist[m][n] for all m,n such that m<N and n<N
      • update the dist[I][j] with the new distance
```

As for divisive clustering, for N number of data samples

1. Compute the proximity matrix
2. Repeat until only 1 cluster remains:
   - split into clusters using any flat clustering algorithm, like k-Means
   - choose cluster with largest Sum Squared Error (SSE)

The linkages between the different clusters—how they are grouped together or partitioned from each other is a given by a tree data structure called a dendrogram. On an average the time complexity of both the above clustering algorithms is $O(n^3)$ and space complexity is $O(n^2)$. However, in agglomerative clustering, in the rth, iteration, N-r clusters are formed, and hence it requires more iterations as compared to divisive algorithm where the numbers of clusters formed at each stage is dependent on the specific flat clustering algorithm used, and thus can take less number of iterations. This can have an effect on the running time of the algorithm especially for large datasets. However, this also becomes a challenge as there would be $2^{N-1} - 1$ ways to partition the cluster, and examining all these possibilities for high values of N becomes computationally expensive. In order to overcome challenges related to choosing of the split or merge points, an other hierarchical clustering methodology called multiple-phase clustering was introduced. Balanced Iterative Reducing and Clustering using Hierarchies (BIRCH) [10] is a type of multiphase clustering algorithm, which like agglomerative and divisive clustering utilizes a tree data structure, with the leaf nodes representing microclusters. Different clustering algorithms can then be performed on these microclusters to form macroclusters. BIRCH algorithm uses a clustering feature which is a 3D-vector to summarize information about the clusters and organized in a hierarchical structure using a Clustering feature tree (CF-Tree). Chameleon [11], also a multi-phase hierarchical clustering method models the similarity between every two clusters, dynamically. Chameleon algorithm assesses the cluster similarity on the basis of how interconnected the samples within a cluster are and inter-cluster proximity. This algorithm optimizes the edge-cut whilst partitioning the k-nearest-neighbor graph into sub clusters using a graph partitioning algorithm. While BIRCH algorithm performs promisingly for clustering data with interference, Chameleon is widely used for identifying arbitrarily shaped clusters.

12.4 HIGH-DIMENSIONAL CLUSTERING

As the dimensionality of the data grows, data becomes more sparse, as a result, it gives the notion of all the samples being equidistant from each other which makes the distance measure pointless. High dimensionality can also add extraneous noise to the data due to the presence of irrelevant features [12, 13]. In order to deal with these challenges, techniques like feature selection and dimensionality reduction have been developed. Techniques like Principal Component Analysis (PCA) and Singular Value Decomposition (SVD) express the high dimensional data by forming linear combinations of the features, thereby helping find latent patterns in the data. However, the irrelevant features may infiltrate this combination. Thus feature selection becomes crucial while dealing with high dimensional data. Attribute subset selection is a technique employed to reduce data size by identifying and discarding redundant or irrelevant features that don't contribute significantly to the analysis. Feature selection can be performed by either supervised learning or unsupervised learning. The former method would involve dropping features whose correlation with the target variable is low, unsupervised methods perform entropy analysis on the clusters to find out the subspace where cluster entropy is minimum. Subspace clustering algorithms have been developed to identify meaningful clusters in different subspaces, and have proven to be highly effective with high-dimensional data. These algorithms are categorized into dimension-growth and dimension reduction projected clustering. CLIQUE (Clustering In QUEst) [14] is a dimension-growth subspace clustering method which integrates both grid-based and density-based clustering methods. The CLIQUE clustering method addresses the curse of dimensionality by employing a grid-based approach to partition the data space. By dividing the dataset into grid cells of equal size, CLIQUE reduces the effective dimensionality of the problem. This grid partitioning allows CLIQUE to focus on identifying dense regions in each grid cell, rather than considering every possible combination of data points in the original high-dimensional space. Additionally, CLIQUE utilizes a minimum cell density parameter to filter out grid cells that do not contain enough data points to form meaningful clusters. By setting the minimum points per grid cell, CLIQUE ensures that only regions with sufficient density are considered during the clustering process. This helps mitigate the effects of sparsity in high-dimensional spaces, where data points may be scattered across a large number of dimensions.

Furthermore, CLIQUE employs a refinement step to merge adjacent dense units with similar densities, effectively reducing redundancy and eliminating noise in the clustering results.

This refinement process helps improve the quality of the identified clusters while also reducing the computational complexity associated with processing high-dimensional data Let grid Size refer to the size of each cell that the dataset needs to be divided into, minCellDensity refer to the threshold density that a cells needs to have in order to be dense, and significanceThreshold refer to the minimum density a cell needs to possess in order to be considered significant.

CLIQUE:

```
1. Divide dataset into grid cells of size gridSize.
2. Calculate density of each grid cell.
3. Find dense units:
        for each grid cell:
                if density of cell > minCellDensity: mark cell as dense
4. Identify significant dense units: for each dense grid cell: if density of cell >
     significanceThreshold:
                mark cell as significant dense unit
5. Refine dense units:
        for each significant dense unit:
                merge adjacent units with similar density
6. Form clusters:
        for each refined dense unit:
                assign data points to cluster
7. Output clusters and corresponding data points
```

CLIQUE algorithm is oblivious to the order of input dataset and automatically finds the highest dimensionality containing high density clusters. It has good scalability since it linearly increases with the size of input. PROCLUS (PROjected CLUStering) on the other hand is a dimensionality reduction subspace clustering algorithm. It tackles the challenges of high-dimensional data by leveraging the concept of subspaces, which are lower-dimensional projections of the original feature space. Its methodology involves a two-step process aimed at identifying meaningful clusters amidst the vast dimensionality of the data. Initially, PROCLUS randomly selects candidate subspaces using projections and then refines these selections based on a cost function that balances cluster compactness and coverage. This iterative approach allows PROCLUS to explore various subspaces efficiently, focusing on those that best capture the inherent structure of the data while disregarding irrelevant dimensions, thus alleviating the curse of dimensionality. Let k_clusters be the number of clusters to be formed, max_iterations be the maximum number of iterations to be performed by the algorithm if the desired convergence isn't achieved, and threshold be the minimum cost allowed within the clusters.

PROCLUS:

```
    Initialize best_cost to infinity
    Initialize best_clusters to an empty list
        for iteration = 1 to max_iterations:
        1. Randomly selection a subspace in the dataset. random_subspace =
        RandomlyGenerateSubspace(dataset)
        2. Perform K-means clustering in the subspace
        clusters = KMeans(dataset, k_clusters, random_subspace)
        3. Calculate the cost of the current clustering cost =
        CalculateCost(clusters)
        if cost < best_cost:
            best_cost = cost
            best_clusters = clusters
        if cost < threshold:
            Stop
```

By operating within lower-dimensional subspaces, PROCLUS not only reduces computational complexity but also addresses the curse of dimensionality, a common issue in high-dimensional data analysis. The algorithm's selective exploration of subspaces enables it to uncover clusters that may be obscured or fragmented in the original space.

12.5 ANALYSIS OF IMPLEMENTATION RESULTS

With the aim of studying clustering algorithms applied to network intrusion detection datasets, namely UNSW-NB15 and CIC-IDS2017, an analysis involving k-Means, Hierarchical clustering, and PROCLUS has been conducted. These algorithms were employed to discover underlying patterns and structures within the network traffic data. Subsequently, the clustering results were visualized using Tableau through integration with the TabPy server, facilitating comprehensive exploration and interpretation of the clustering outcomes.

The elbow method was used in determining the optimal number of clusters to be formed and the value was found to be 5 for both UNSW-NB15 and CIC-IDS2017 as can be inferred from Figures 12.3 and 12.6. The inter cluster plot map as presented in Figures 12.4 and 12.7, was plotted using the python library, Yellowbricks and it can be inferred that cluster 4 (Figure 12.4) is completely isolated from the rest of the 4 clusters and is located far away from the

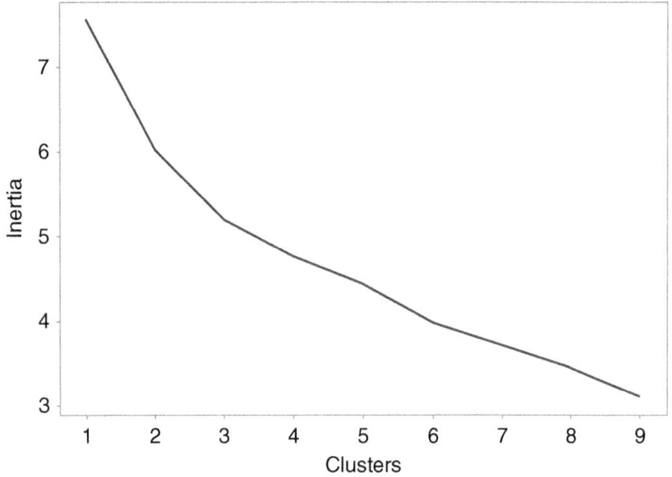

Figure 12.3 Inertia vs number of clusters for the UNSW-NB15 dataset.

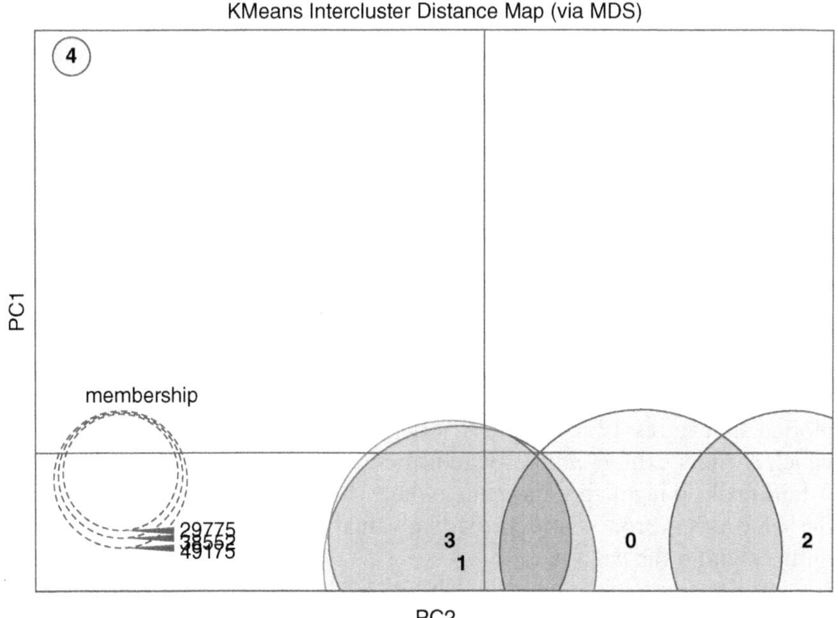

Figure 12.4 K-means inter cluster distance map (UNSW-NB15).

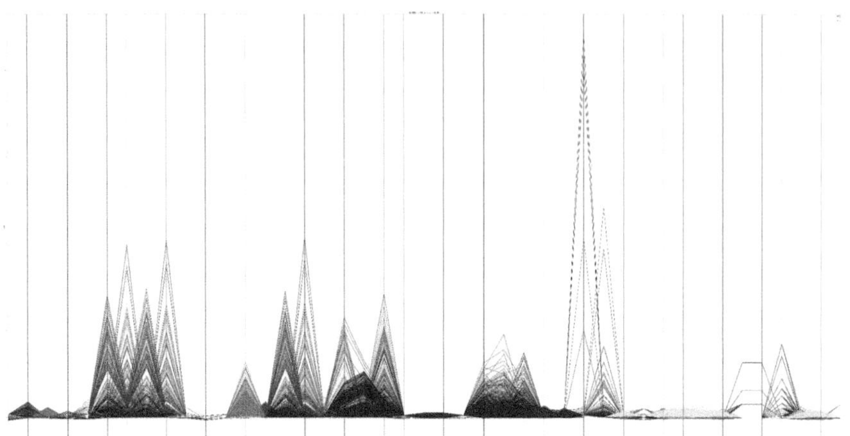

Figure 12.5 Parallel coordinate plot after applying k-means on UNSW-NB15.

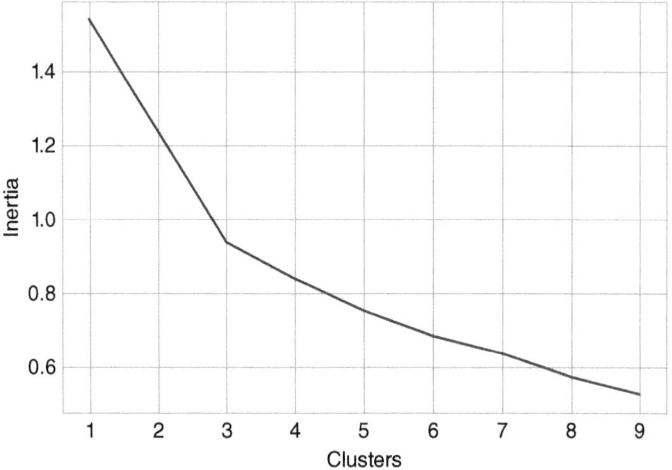

Figure 12.6 Inertia vs number of clusters for the CIC-IDS2017 dataset.

former four, for the UNSW-NB15 dataset, and cluster 1 (Figure 12.7) takes up the place of cluster 4 in the CIC-IDS2017 in the aspect of being isolated from the rest of the clusters. The parallel coordinate plot for K-Means clustering plotted in Figures 12.5 and 12.8, offers insights into the attributes along which clusters exhibit similarity, indicated by areas with denser col-oration. Additionally, it highlights anomalies where the variance between two lines on the same axis is pronounced, providing valuable information about potential outliers within the dataset.

On the basis of metrics, for the UNSW-NB15 dataset, hierarchical clustering has demonstrated the lowest V-Measure, indicating less agreement with

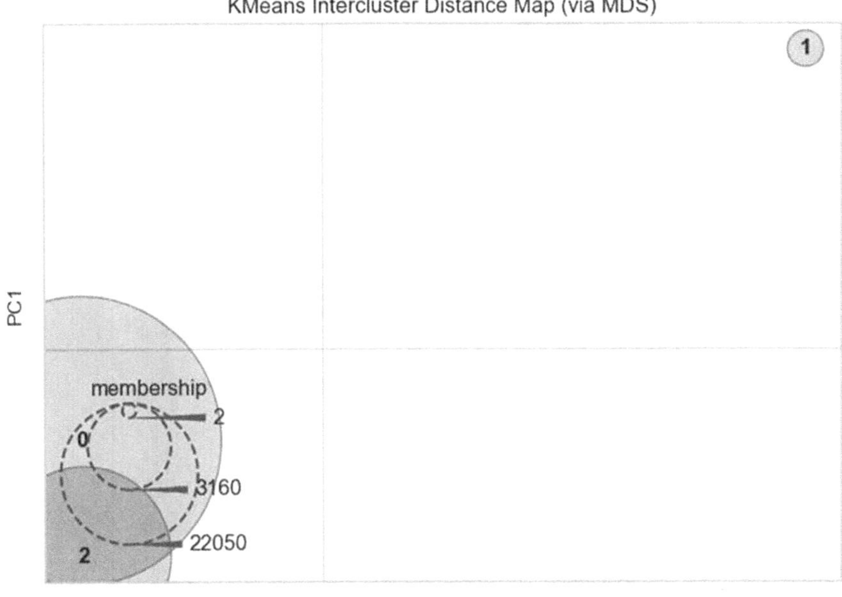

Figure 12.7 K-means inter cluster distance map (CIC-IDS2017).

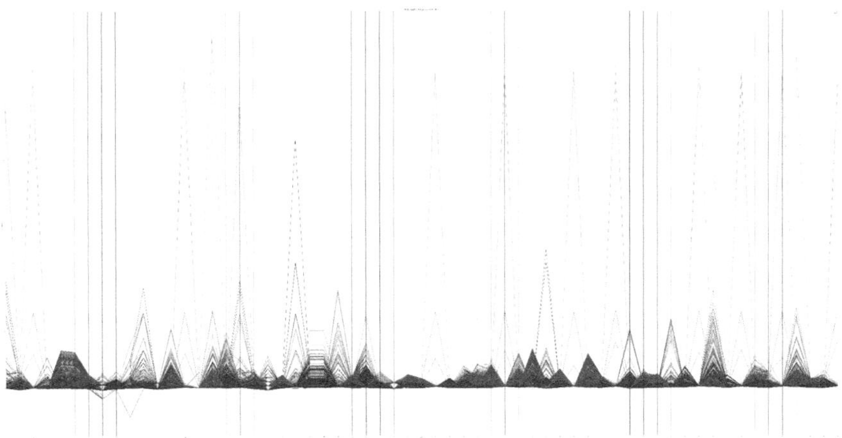

Figure 12.8 Parallel coordinate plot after applying k-means on CIC-IDS2017.

the true class labels compared to k-means and PROCLUS. While k-means and PROCLUS (in Figures 12.9 and 12.11) have exhibited higher V-Measures of approximately 0.4034 and 0.4013, respectively, suggesting better alignment with the ground truth, Hierarchical clustering (in Figure 12.10) has fell short with a V-Measure of 0.1332. Moreover, k-means and PROCLUS have showcases superior purity metrics, with values around 0.5776 and

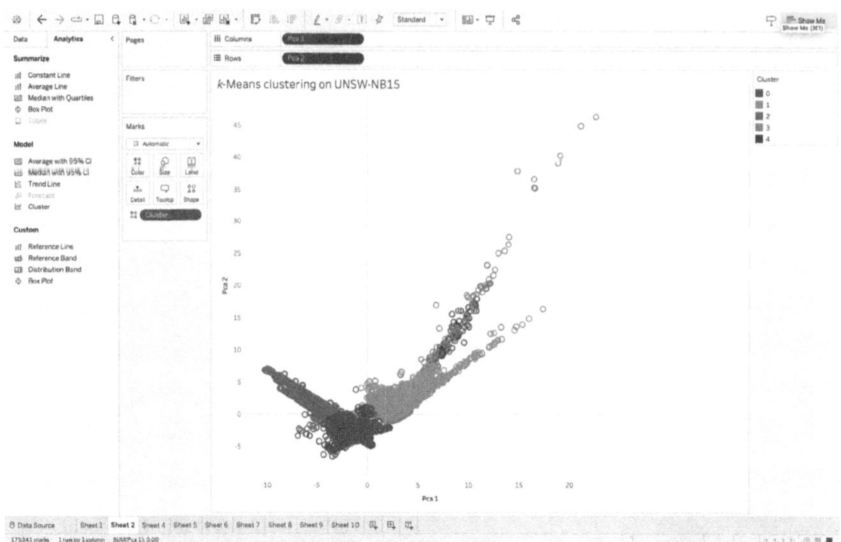

Figure 12.9 PCA projection of k-means clustering on UNSW-NB15 (on Tableau).

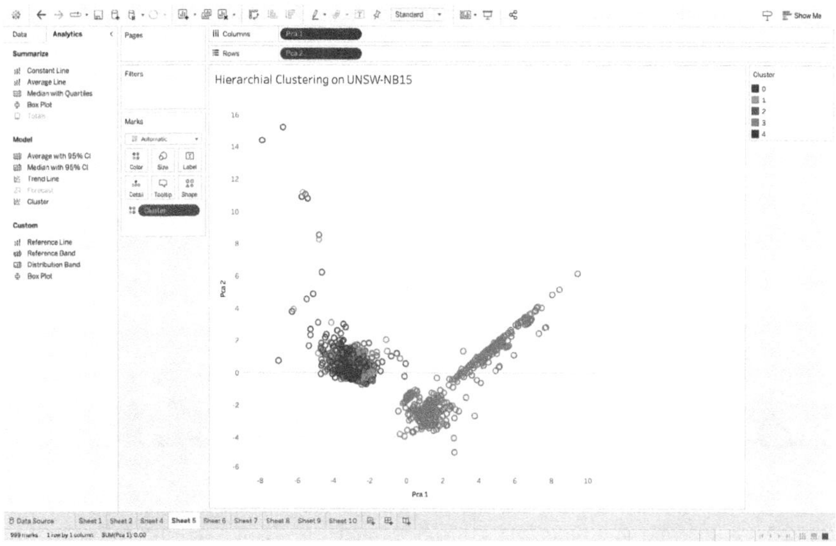

Figure 12.10 PCA projection of hierarchical clustering on UNSW-NB15 (on Tableau).

0.5786, indicating more consistent class assignments within clusters. In contrast, Hierarchical clustering has achieved a lower purity of 0.4374, suggesting less coherence in cluster formation.

Additionally, k-means and PROCLUS have demonstrated higher completeness and homogeneity scores compared to Hierarchical clustering, indicating

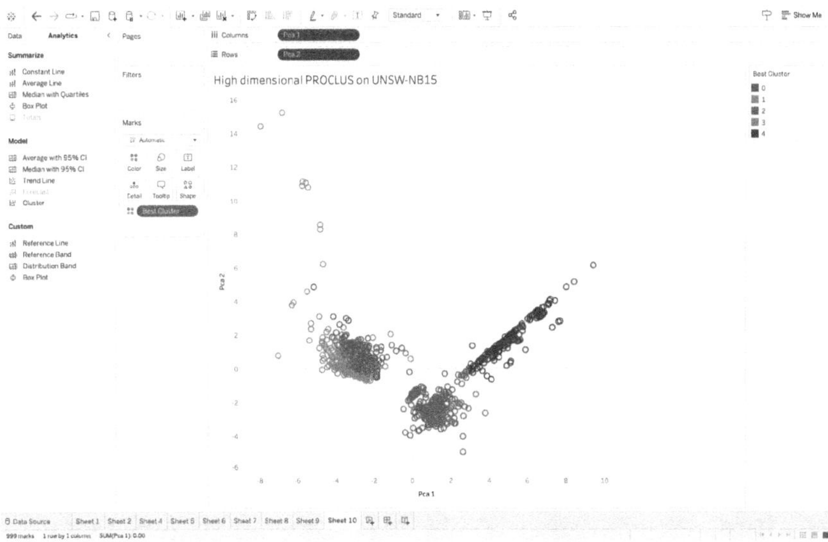

Figure 12.11 PCA projection of high-dimensional clustering on UNSW-NB15 (on Tableau).

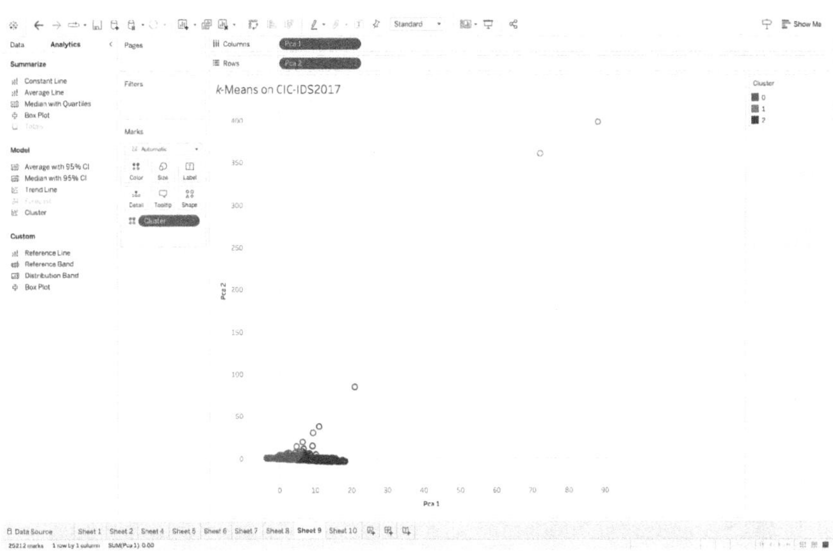

Figure 12.12 PCA projection of k-means clustering on CIC-IDS2017 (on Tableau).

more uniform clusters in terms of class composition. With completeness scores around 0.4600 and 0.4526, and homogeneity scores approximately at 0.3592 and 0.3605, respectively, kmeans and PROCLUS have outperformed. This analysis suggests that k-means and PROCLUS are better suited for capturing meaningful structures within the dataset, as evidenced by their superior

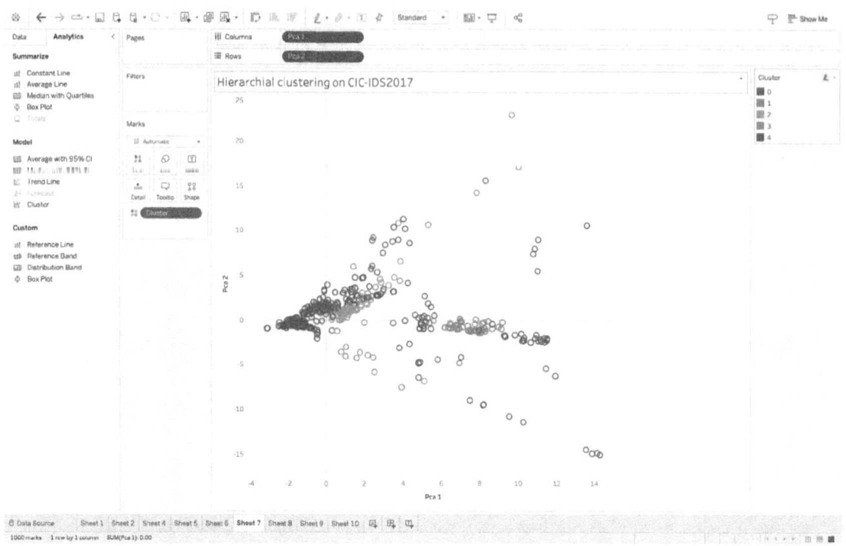

Figure 12.13 PCA projection of hierarchical clustering on CIC-IDS2017 (on Tableau).

performance across multiple metrics. Both k-means and high-dimensional clustering PROCLUS, have exhibited efficacy in partitioning the dataset into discernible and non-overlapping clusters. The resultant clusters demonstrate clear delineation, characterised by well-defined boundaries and minimal inter-cluster ambiguity. This observation underscores the capability of both PROCLUS and k-means to capture underlying structures and patterns inherent in the data, thereby facilitating insightful data segmentation and analysis.

Hierarchical clustering has performed poorly on this basis, with cluster 3 in Figure 12.10 being the only distinct cluster, which when compared to the ground truth is dominated by the attack category "Generic" (40.89%). Comparing the performance of k-means (Figure 12.12), hierarchical clustering (Figure 12.13), and PROCLUS (Figure 12.14) on the basis of metrics like V-measure, purity, completeness and homogeneity, on the CIC-IDS2017 dataset has yielded valuable insights. K-means clustering has demonstrated the highest purity of 0.8466, indicating a significant proportion of instances within each cluster belonging to the same true class. However, its V-Measure of 0.2615 suggests a moderate agreement with the ground truth labels, implying room for improvement in capturing the underlying data structure.

Hierarchical clustering has achieved a slightly lower purity of 0.868, indicating similar cluster consistency as k-means, albeit with a marginally better VMeasure of 0.2539. This suggests that hierarchical clustering may achieve slightly better alignment with the true class labels, but it still falls short of providing a robust clustering solution. PROCLUS, on the other hand, has exhibited the lowest V-Measure of 0.2254, implying a lesser degree of agreement with the ground truth labels compared to k-means and hierarchical

Figure 12.14 PCA projection of high-dimensional clustering on CIC-IDS2017 (on Tableau).

clustering. While PROCLUS has demonstrated relatively high purity (0.863), its completeness and homogeneity scores (0.1987 and 0.2604, respectively) indicate suboptimal cluster uniformity and coherence. In the context of k-means clustering, the presence of a cluster with a notably small number of samples, such as cluster 3 in Figure 12.12, signifies the potential existence of outliers within the dataset. However, despite this outlier presence, all three clustering methods, including k-means, have demonstrated the ability to generate relatively distinct clusters, indicating their effectiveness in partitioning the dataset into cohesive groups. In the case of hierarchical clustering, cluster 4 in Figure 12.13, emerges as the most densely populated cluster, characterised by a majority of samples classified as BENIGN, accounting for 88.23% of the cluster's composition when compared with the ground truth labels. Notably, the prevalence of BENIGN samples across three out of the five clusters under-scores the dominance of normal network traffic patterns within the dataset, followed by various types of Denial-of-Service (DoS) attacks. In inference, while k-means and hierarchical clustering has shown strengths in achieving high purity and relatively better alignment with the ground truth labels, they have also exhibited limitations in fully capturing the underlying data structure as indicated by moderate V-Measure scores. PROCLUS, despite demonstrating high purity, has fell short in terms of completeness and homogeneity, suggesting challenges in forming clusters that are both coherent and representative of the true class labels. In terms of CPU runtimes, k-means generally exhibits faster execution times compared to PROCLUS and hierarchical clustering. K-means algorithm's computational complexity is relatively lower, making it more efficient for processing large datasets with a significant number of dimensions. However, PROCLUS and hierarchical clustering may require longer CPU runtimes due to their iterative nature and higher computational demands, especially when dealing with high-dimensional data or complex clustering structures. K-means typically requires minimal additional memory overhead compared to PROCLUS and hierarchical clustering. K-means operates by maintaining centroids and cluster assignments, resulting in relatively lower memory usage. On the other hand, PROCLUS and hierarchical clustering may require additional memory for storing hierarchical structures, proximity matrices, or cluster prototypes, leading to higher space requirements. While k-means can be seamlessly integrated and implemented in the TabPy server, additional configuration and optimization efforts would be required for executing hierarchical and high-dimensional clustering algorithms.

12.6 CONCLUSION

In the analysis of the UNSW-NB15 dataset, both the k-means and PROCLUS algorithms demonstrate comparable performance across various evaluation metrics. Despite this similarity, k-means holds an advantage due to its ease of integration and execution within Tableau, making it the preferred choice for clustering tasks in this context. Conversely, when examining the CICIDS2017 dataset, both PROCLUS and hierarchical clustering algorithms exhibit similar performance levels. Hence, either of these algorithms can be effectively employed for clustering purposes in this scenario. PROCLUS, with its adeptness in handling large high-dimensional datasets, stands out as a suitable option. On the other hand, hierarchical clustering may be favored in situations requiring a detailed dataset analysis or anomaly detection, owing to its ability to capture intricate hierarchical relationships within the data.

REFERENCES

1. E. Juliano, C. Thakkar, C. Taber, M. Raval, T. Kaya and S. Senbel, "A Dynamic Online Dashboard for Tracking the Performance of Division 1 Basketball Athletic Performance," *2023 IEEE 28th Pacific Rim International Symposium on Dependable Computing (PRDC)*, Singapore, 2023, pp. 314–318. https://doi.org/10.1109/PRDC59308.2023.00050
2. R. Toasa, M. Maximiano, C. Reis and D. Guevara, "Data visualization techniques for realtime information – A custom and dynamic dashboard for analyzing surveys' results," *2018 13th Iberian Conference on Information Systems and Technologies (CISTI), Caceres, Spain*, 2018, pp. 1–7. https://doi.org/10.23919/CISTI.2018.8398641
3. V. S. Parekh and M. A. Jacobs, "A multidimensional data visualization and clustering method: Consensus similarity mapping," *2016 IEEE 13th International Symposium on Biomedical Imaging (ISBI), Prague, Czech Republic*, 2016, pp. 420–423, https://doi.org/10.1109/ISBI.2016.7493297

4. U. C. Akuthota and L. Bhargava, "Evaluation of Machine Learning Models for Intrusion Detection with the UNSW-NB15 Dataset," *2023 IEEE Silchar Subsection Conference (SILCON), Silchar, India*, 2023, pp. 1–5, https://doi.org/10.1109/SILCON59133.2023.10404204
5. N. A. Putri, D. Stiawan, A. Heryanto, T. W. Septian, L. Siregar and R. Budiarto, "Denial of service attack visualization with clustering using K-means algorithm," *2017 International Conference on Electrical Engineering and Computer Science (ICECOS), Palembang, Indonesia*, 2017, pp. 177–183, https://doi.org/10.1109/ICECOS.2017.8167129
6. S. S. Panwar, Y. P. Raiwani and L. S. Panwar, "An Intrusion Detection Model for CICIDS2017 Dataset Using Machine Learning Algorithms," *2022 International Conference on Advances in Computing, Communication and Materials (ICACCM), Dehradun, India*, 2022, pp. 1–10. https://doi.org10.1109/ICACCM 56405.2022.10009400
7. Kurniabudi, D. Stiawan, Darmawijoyo, M. Y. Bin Idris, A. M. Bamhdi and R. Budiarto, "CICIDS-2017 Dataset Feature Analysis With Information Gain for Anomaly Detection," in *IEEE Access*, vol. 8, pp. 132911–132921, 2020, https://doi.org/10.1109/ACCESS.2020.3009843
8. S. Arshad, W. Ashraf, S. Ashraf, I. Hassan and F. S. Masoodi, "Comparative Study of Machine Learning Techniques for Intrusion Detection on CICIDS-2017 Dataset," *2023 10th International Conference on Computing for Sustainable Global Development (INDIACom), New Delhi, India*, 2023, pp. 929–934.
9. Y. Diao, K.-Y. Liu, L. Hu, D. Jia and W. Dong, "Classification of Massive User Load Characteristics in Distribution Network Based on Agglomerative Hierarchical Algorithm," *2016 International Conference on Cyber-Enabled Distributed Computing and Knowledge Discovery (CyberC), Chengdu, China*, 2016, pp. 169–172, https://doi.org/10.1109/CyberC.2016.41
10. Zhang, Tian, Ramakrishnan Ramakrishnan, and Mirek Ostrowski. "BIRCH: An Efficient Data Clustering Method for Very Large Databases." In *Proceedings of the 1996 ACM SIG-MOD International Conference on Management of Data*, New York, NY, USA, 1996, pp. 103–114.
11. Karypis, George. "Chameleon: Hierarchical Clustering Using Dynamic Modeling." *IEEE Transactions on Knowledge and Data Engineering*, vol. 32, no. 8, pp. 352–371, 1999.
12. Anastasiou, A., & Karypis, M. "A Projective Clustering Algorithm for High Dimensional Data." *Proceedings of the 2000 SIAM International Conference on Data Mining (SDM)*, Cincinnati, Ohio, U.S., 2000, pp. 261–265.
13. P. Patel, B. Sivaiah and R. Patel, "Approaches for finding Optimal Number of Clusters using K-Means and Agglomerative Hierarchical Clustering Techniques," *2022 International Conference on Intelligent Controller and Computing for Smart Power (ICICCSP), Hyderabad, India*, 2022, pp. 1–6, https://doi.org/10.1109/ICICCSP53532.2022.9862439
14. Agrawal, Rakesh, Jiansheng Ge, Prabhakar Gupta, and Vipin Kumar. "Clustering in Quest (CLIQUE) of Dense Subspaces: A Grid-Based Approach." In *Proceedings of the 1998 ACM SIGKDD International Conference on Knowledge Discovery and Data Mining*, New York, NY, USA, 1998, pp. 144–152.

Chapter 13

Breaking boundaries

The next frontier in skin cancer diagnosis combining transfer learning and multi-scale deep learning

S. A. Sahaaya Arul Mary, Sameer Chauhan, S. Sanjith Surya, and Luv Sachdeva
VIT University, Vellore, India

1 INTRODUCTION

Background

Given that the skin constitutes the body's largest organ, it should come as no surprise that skin cancer is one of the most often diagnosed cancer forms [1, 2]. It can be divided into two groups: non-melanoma and melanoma[3]. For the best results, early detection is essential, especially in light of the fact that melanoma is the most lethal type of skin cancer[4], arises from uncontrolled proliferation of melanocyte cells. While it can affect any part of the body, it's more prevalent in areas exposed to sunlight[5]. Currently, biopsy remains the primary method for skin cancer detection, involving the sampling of skin tissue to determine malignancy. To aid in timely detection, leveraging automated systems can be beneficial, given occasional medical errors[6]. Machine learning algorithm, particularly deep learning methods, have emerged as effective tools for detection and recognition in the past decade[5].

Rationale and knowledge gap

When it comes to machine learning techniques, data is everything. To improve learning model training and lessen overfitting, more data may be used. Overfitting occurs when a model fit all or almost all of the training data but is unable to generalize well. Typically, learning problems have insufficient data because of challenges in building the data, particularly when the unfiltered data consists of pictures. categorizing the data is necessary, and for large datasets, this can take a while. Because of this, one of the primary goals of every learning problem is to figure out how to add more photographs to the dataset. The research has covered a variety of techniques for increasing the images.

A classification system's second crucial component is feature extraction which seeks to extract key information from data. The extraction of features

DOI: 10.1201/9781003542735-13

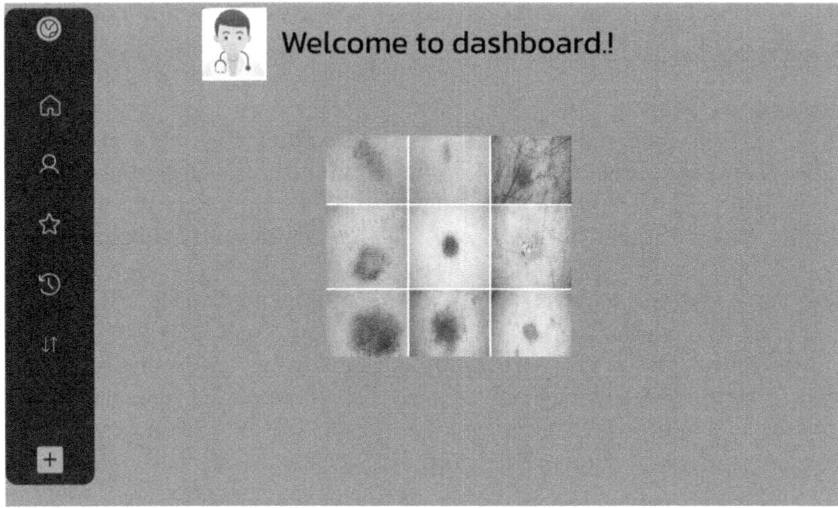

Figure 13.1 Nine images from ISIC 2020 dataset, original size [7]. The licensed material is licensed under this public license: https://creativecommons.org/licenses/by-nc/4.0/legalcode.txt

and classification process is completed in a single step in deep learning structures. When working with photos, it is crucial to extract features from the places where the information we are seeking is most likely to be present. The majority of the skin areas that require care are seen in the image's center in cancer of the skin datasets (see Figure 13.1). The present research makes use of this knowledge to enhance the feature extraction section. After training three separate convolutional neural networks (CNNs), the outputs from each are merged and fed into a network that is fully connected. Considering this combination, the outcome indicates a boost in the diagnosis rate.

This research paper is organized into total 4 sections. Deep learning and the transfer of learning are concepts that are thoroughly explained in the "Methods" section, along with the approach that was employed and the literature review. The results and a comparison with alternative approaches are shown in the "Results" section. The conclusion and some suggestions for further research are stated in the "Discussion" section. The TRIPOD report checklist is followed by the author when presenting this article.

II LITERATURE REVIEW

The problem of skin cancer has been investigated by numerous scientists. This section will offer a succinct overview of the several strategies that have been used in this field. Ali et al. utilized LightNet for classification purposes in their model and it performed admirably when it came to mobile apps. The tests were run on datasets from the International Skin Image Collaboration (ISIC) 2016 and the results were reported on reliability (81.71%), specificity (98.11%), and sensitivity (14.89%) [8]. Dorj et al. utilized deep SVM machines to identify basal cell cancer (BCC), melanoma, and carcinoma of squamous cells (SCC). When compared to melanoma cancers, non-melanoma cancers such as BCC and SCC are easier to identify and are not spreading to other parts of the body [9]. They tested their system on 3,753 photos, including images of four different types of skin cancer. The best findings they obtained were 94.17% (SCC), 98.78% (active keratosis), and 94.9% (SCC) for accuracy, specificity, and sensitivity, respectively. The minimum values for BCC, SCC, and melanoma were 91.8%, 96.9%, and 90.74%, respectively [9].

Esteva divided the photos into three groups using CNN. Benign seborrheic keratosis (SK), benign Keratinocyte Carcinomas, and Melanoma [10]. Reference [11] describes the usage of a hybrid system for skin cancer diagnostics using sparse coding, SVM, and deep learning. They made use of ISIC, which included 2,624 clinical instances of benign lesions [2,146], unusual nevi [144], and melanoma [334). Twenty iterations of two-fold cross-validation were employed for assessment. Two tasks were offered for classification: (I) melanoma against all other types of cancer while (II) melanoma versus atypical lesions exclusively. There were 93.1% accuracy, 94.9% recollection, and 92.8% specific for the first test, and 73.9% precision, 73.8% recollection, and 74.3% specific for the second task.

Harangi et al. combined GoogleNet, VGGNet, and AlexNet using CNN for structures [12]. Kalouche detected skin cancer by focusing on boundaries and using CNN [13] Rezvantalab et al.'s research signify that all deep learning models perform at least 11.5% better than other models [14]. The research employed DenseNet201, Inception v3, ResNet 152, and InceptionResNet v2. 3,6500 lesion photos were chosen by Sagar et al. from the ISIC dataset and applied deep transfer learning to ResNet-50. The suggested model outperformed DenseNet169 [15] and InceptionV3. ResNet152 was utilized for classification after the region of interest in Jojoa Acosta et al.'s work was removed using a mask and region-based CNN. There were 2742 photos utilized from ISIC [16].

Other works based on progressive GAN (PGAN) and generative adversarial networks, or have also been published. GAN is employed in the ISIC 2018 dataset by Rashid et al. They attempted to diagnose vascular lesions, dermatofibromas, melanoma, melanocytic nevus, benign keratosis, etc. Using a deconvolutional network, they achieved an accuracy of 86.1% [17]. Deep convolutional GAN was employed by Bisla et al. for data augmentation, and it was decoupled. Amongst the data sets that were utilized were the ISIC 2017 and the ISIC 2018. They achieved an accuracy of 86.1% [18]. Self-attention based PGAN was utilized by Ali et al. to identify pigmented Bowen's, nevus, dermatofibroma, vascular, and pigmented benign keratosis. We used the ISIC 2018 dataset. A stabilizing method was applied to improve a generative model. It was 70.1% accurate [19]. A framework which consists of two stages for the automatic classification of skin lesion photos was created by Zunair et al. Transfer learning and adversarial training were applied [20].

III METHODOLOGY

The suggested method is covered in the following two subsections. The sections on image processing and augmentation are covered in the first section. The second subsection contains a description of the model.

Preprocessing

An essential component of medical pictures is preprocessing. When preprocessing, two distinct tasks should be taken into account. The first is a picture preparation section that corrects color errors or improves the image; the second is a data augmentation section that increases the total amount of pictures in the data set in various ways. The photos had to be normalized first. The average of the red, green, and blue value was subtracted from each image in the data to achieve this. Krizhevsky et al. proposed this approach [21]. Detecting skin cancer is challenging due to the high rate of positive results images required for testing and training in order to construct a robust CNN. It takes time to create vast datasets with diversity in many cancer kinds. ISIC 2020 [7, 22] is the data set used in this research. 33,130 thermoscopic training pictures from 2,100 patients are included in it. A total of 1,400 (about 4% of these photos) were selected and kept aside for testing. Since real photos are required to verify performance, only the set of training data needs to be extended in order to provide the model more examples and boost its robustness. We rotated each image by 10 degrees to create 35 × 31,726 new photos, increasing the total amount of training images.

Transfer learning

Because CNN requires a huge number of image data to test and train, training it is a difficult task. Along with image augmentation, transfer learning is utilized to solve the lack of images problem. When a network uses transfer learning, it makes use of the parameters of an artificial neural network which was developed on an image dataset that is comparable to the one we wish to exploit and has an adequate sample size. Transfer learning was used in this work in the following ways: Initially, the network is weighted using the CNN parameters for ImageNet [23]. It indicates that the ImageNet dataset was used to fully train a CNN. This network substitutes ImageNet's last layer, which was designed to categorize 1,000 classes, with a softmax layer that divides data into two categories: Melanoma versus non-melanoma. This is a binary classification method; in addition, sigmoid functions can be used

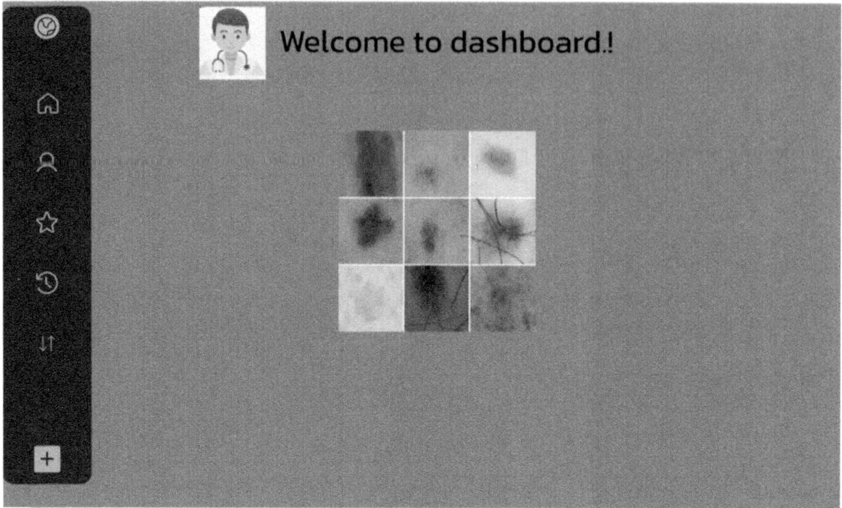

Figure 13.2 Nine images from ISIC 2020 dataset, each image was cropped (75%) and then resized [7]. The licensed material is licensed under this public license: https://creativecommons.org/licenses/by-nc/4.0/legalcode.txt

in place of the softmax layer. The total amount of nodes in the softmax layer is equivalent to the total amount of outcomes, and the quantity indicates the likelihood of each output. The network undergoes back-propagation, which adjusts weights based on output errors to better adapt to new classes. It's crucial to avoid drastically altering the weights, hence a modest learning rate must to be applied. A neural network's learning rate hyperparameter regulates how the model changes in response to errors. Utilizing a high rate of learning will completely destroy the weights. One way to think of applying back-propagation as an a hyperparameter in training is up to a certain point.

This work also employs a similar technique to that found in [24] but with a more sophisticated structure and the addition of transfer learning. Using disparate pictures and transfer learning, three distinct networks were trained. One is to apply the entire procedure outlined above to the skin cancer database for the photos that are true to size. The majority of the surrounding area in a skin cancer diagnosis problem is made up of healthy skin, which could potentially deceive the network. The malignant area is located in the middle of the image. There are two approaches to solving this issue. The suggested approach is the other, and segmentation is the first [25, 26]. Segmentation separates an image into two sections: the healthy portion and the unhealthy portion, but it has disadvantages of its own. The segmentation problem's primary flaw is that it contains flaws that will hinder learning. The proposed strategy involved training two additional networks. One for photos that have been cropped to 50% of their original size and then resized, and a third for photos that have been cropped to 75% of their original size

Figure 13.3 Nine images from ISIC 2020 dataset, each image was cropped (50%) and then resized [7]. The licensed material is licensed under this public license:https://creativecommons.org/licenses/by-nc/4.0/legalcode.txt

and then resized. The Figures below which are represented in the dashboard layout show the original size, 75% and 50% of the cropped pictures.

All three networks were trained and tuned. The global-average-pooling-two-dimensional (2D) pooling layer creates a vector based on the input from three structures. This decreases dimension and concentrates the strongest reaction in small parts of the feature map. With two outputs, an entirely connected layer was produced. The entire construction is shown in the dashboard.

The fine-tuning learning rate was 0.00001, whereas the values for the number of groups, development rate, and periods were 32, 10, and 0.0001. The 53-layer CNN that was utilized was called MobileNetV2.

The following metrics are used to gauge how well the suggested structure performs.

$$\text{Accuracy} = (\text{TP} + \text{TN}) / (\text{TP} + \text{TN} + \text{FP} + \text{FN}) \quad [1]$$

$$\text{Recall} = \text{TP} / (\text{TP} + \text{FN}) \quad [2]$$

$$\text{Precision} = \text{TP} / (\text{TP} + \text{FP}) \quad [3]$$

All the acronyms used above represent true positive, true negative, false positive, and false negative, respectively.

IV CALCULATIONS USED

Initialization:

1. Weight Initialization:

$$W_{ij}^{(l)} \sim N\left(0, \frac{2}{n_{in}^{(l)} + n_{out}^{(l)}}\right)$$

Where, $W_{ij}^{(l)}$ is the weight connecting neuron i in layer $l-1$ to neuron j in layer l, and $n_{in}^{(l)}$ and $n_{out}^{(l)}$ are the input and output neurons in layer $\times$ l.

2. **Softmax Function:**

$$p_i = \frac{e^{z_i}}{\sum_{j=1}^{n} e^{z_j}},$$

p_i represents the probability of class "i", while z_i represents the output of the final layer of class i.

3. **Learning Rate update:**

$$W_{new} = W_{old} - \eta \frac{dl}{dw}$$

Training Propagation:

1. Back Propagation:

$$\frac{\partial L}{\partial W_{ij}^{(l)}} = \delta_j^{(l)} \cdot a_i^{(l-1)}$$

The error term for neuron j in layer l is $\delta_j^{(l)}$ and the activation of neuron I in layer l-1 are represented $a_i^{(l-1)}$

2. **Gradient descent update:**

$$W_{new} = W_{old} - \eta \frac{\partial L}{\partial W}$$

V ALGORITHM

Initialization:

Initialize the network with CNN parameters from ImageNet.
Replace the final layer with a softmax layer for binary classification (melanoma and non-melanoma).
Set the learning rate to a low value to avoid drastic weight changes.

Training:

Train three networks using transfer learning with disparate images.
Apply back-propagation to adjust weights based on output errors.
Utilize a modest learning rate to prevent drastic weight alterations.

Evaluation:

Evaluate the network performance based on accuracy, precision, and recall metrics.
Use a multi-scale structure to distinguish between melanoma and non-melanoma.
Implement image processing techniques to enhance training photo quality.

The algorithm involves initializing the network with pre-trained CNN parameters, training using transfer learning, and evaluating performance based on various metrics. It aims to improve skin cancer detection accuracy through a multi-scale framework and image processing techniques.

CODE SNIPPETS

```
model = Sequential()
model.add(MobileNetV2(weights='imagenet',
  include_top=False))
model.add(GlobalAveragePooling2D())
model.add(Dense(2, activation='softmax'))

model.compile(optimizer=tf.keras.optimizers.
              Adam(learning_rate=0.00001),
              loss='sparse_categorical_crossentropy',
              metrics= ['accuracy'])

import tensorflow as tf
from tensorflow.keras.models import Sequential
from tensorflow.keras.layers import Dense, GlobalAverage
  Pooling2D
from tensorflow.keras.applications import MobileNetV2

loss, accuracy = model.evaluate(test_images, test_labels)
print(f'Loss: {loss}, Accuracy: {accuracy}')
```

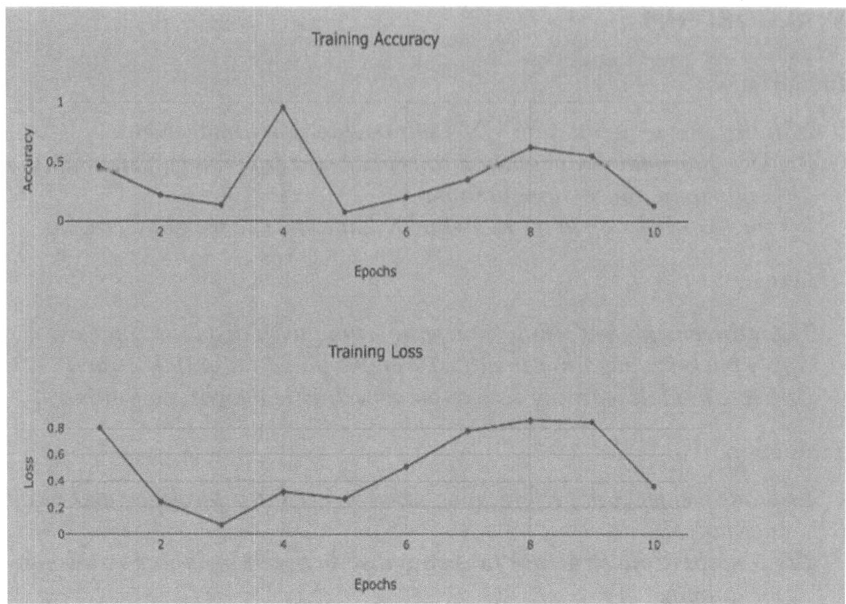

Figure 13.4 Training accuracy and loss graph which is displayed on the dashboard.

VI RESULTS

The GPU used in the studies was an AMD Radeon RX 6801M, the Central processing unit used was an AMD Ryzen 9 5901HX, and the RAM was 32 GB of 3600 MHz DDR4. The laptop used for the experiments was a Lambda model. The accuracy and loss of the suggested method are displayed in dashboard, and the ROC (receiver operating characteristic) curve is displayed in dashboard

The graph for precision as well as loss is displayed in dashboard. Table 13.1 presents the outcome of the suggested approach in comparison to a few different approaches, and Table 13.2 displays the error matrix. It is important to note that some techniques may have different test sets.

Table 13.1 Comparing the proposed approaches to other methodologies

Method	*Accuracy (%)*	*Precision (%)*	*Recall (%)*	*Dataset*
[24]	90.5	–	–	ISIC
[27]	81	75	81	170 skin lesion images
Proposed method	94.39	91.72	88.51	ISIC 2020
[8]	14.8	–	81.7	ISIC 2016
[15]	77	94	93.6	3,600 lesion Images from the ISIC dataset

Table 13.2 Error matrix for test results

Predicted value/actual value	*Cancer*	*Non-cancer*
Cancer	352	48
Non-cancer	30	970

VII FUZZY LOGIC

Prior to processing

The first stage in the CAD system is pre-processing the lesion image by removing noise and irregularities. The initial phase involves data augmentation, color gray scaling, morphological operations, binarization, and intensity correction. This phase enhances raw photos by contrast adjusting and histogram curves, as well as removing noise and other artifacts. Three noteworthy stages This implementation report provides suggestions for pre-processing dermoscopic pictures. Dull Razor algorithm is used in the initial step to remove any digital hair from the lesion image [26]. The hair present in the dermoscopic image using a gray-morphological operation, which is then verified by eliminating neighboring pixels according to the thickness and length of the hair follicle. A interpolation algorithm which is bilinear is employed to restore the impacted pixels and median filter is used to smooth them out. In the next phase, histogram is used to improve the overall quality of the dermoscopic image by adjusting its contrast and brightness.

Segmentation

Segmentation is a technique that uses a previously processed dermoscopic image to identify the edge of a lesion. This phase involves digitally differentiating the affected region from the surrounding normal skin., focusing on the region of interest. Segmentation techniques fall into seven categories: merging threshold, quantization, active contours, clustering, thresholding, merging, and fuzzy methods. However, research indicates that because of their computational and temporal complexity, traditional segmentation techniques like region boosting, thresholding, and clustering have difficulty analyzing complex lesion images, which can result in algorithmic failures. Numerous researchers have resorted to segmentation techniques based on saliency, k-mean algorithms, fuzzy methods, and deconvolutional networks in order to overcome this constraint [27, 28]. To classify COVID-19, a GAN-based technique was utilized for CT scan image segmentation by Juanjuan et al. Chen and colleagues employed the U-Net++ model and feature pyramid to automate the process of segmenting of heart arteries during invasive angiography. Segmentation strategy used in the suggested method that is presented here takes two iterations. Processing is done using a 5 × 5 sub-matrix windows to lessen the computational load.

Phase-1

Step 1:	First, make an input image matrix (I × J) by identifying the position of each pixel as follows: (i,j)∈I≡{(1,2,3,..,I) × (1,2,3,..,J)}
Step 2:	Carry out each step independently on the blue, Red, and green planes. Let b(i,j) represent the blue plane's intensity value where the pixel location is (i,j). Subsequently generate a flag image with dimensions (J × I), setting the starting values of each pixel = 0.
Step 3:	Create a window of the required length, where i = (3, 4,......) and j = (3, 4,....) are the coordinates of the center pixel.
Step 4:	Assemble the pixel values for a certain window with center as (k,l).
Step 5:	Find out if the absolute difference greater than equal to T, where T = σ (the chosen region's standard deviation). In case it is false, assign f_r(k−2,l−2) = 0; if otherwise, assign f_r(k−2,l−2) = 1.
Step 6:	Similarly, set f_r(k+2,l+2) = 1 if the absolute difference is greater than equal to T, and assign f_r(k+2,l+2) = 0 otherwise.
Step 7:	To create updated binary flag images, f_g and f_b, repeat steps 4 through 6 for the Green and Blue planes. The following operation will be used to merge these flag images into the flag image, f_f.
Step 8:	If it is found that there is a segmented region in the pre processed image, set f_f(a,b) to 1 if the condition (f_b(a,b) && f_g(a,b) && f_r(a,b)) == 1 is met. Use the following operation to determine the segmented region's border: In case f_f(a,b) = 0, indicate the area; if not, proceed

Phase Two

We do exact picture segmentation after examining the affected region. Firstly, the phase-1 assessment level is selected relying on the presumed ground truth do note, nevertheless, that the picture segmentation findings still include several non-affected areas. To mitigate this, we include a second threshold assessment that defines the entire impacted zone and has a value less than T. Still unanswered is the following: What percentage of the affected area falls between phase 1,2 assessment rates? Therefore it is very challenging to identify the precise assessment rate threshold that will allow us to handle the impacted area and eliminate the largest non-affected area L-Function fuzzy numbers are used in the second iteration of the picture segmentation algorithm to account for human perception's inherent imprecision and uncertainty. A approach which is the opposite of this is developed to convert L-Function fuzzy numbers to crisp values. The de-fuzzified result sets the minimum evaluation rate for the subsequent phase of segmentation.

A fuzzy number $\tilde{S}$ is said to be an L-R type fuzzy number if

$$\mu_{\tilde{P}}(x) = \begin{cases} L\dfrac{(a-i)}{j}, \text{ for } i \le a,\ j > 0 \\ R\dfrac{(i-a)}{k}, \text{ for } i \ge a,\ k > 0 \end{cases}$$

Here, L is for left and R for right reference functions, ***j***, ***k*** are called left and right spreads, respectively. A fuzzy number $\tilde{S}$ is said to be an L-type fuzzy number if:

$$L(i;j,k) = \begin{cases} 1, & i < j \\ \dfrac{(i-j)}{k-j}, & j \le i < k \\ 0, & i \ge k \end{cases}$$

De-fuzzification of L-Fuzzy Number: A linear L-fuzzy number $\tilde{S}_{FN}$ can be transformed into a crisp number using the area approximation system The mathematical formulation is,

$$\varnothing = A_L(j) + A_R(j)$$

where, $A_L(j)$ = Area of left Zone (Rectangular shape (Figure 6a))

$$= 1.j = j$$

$A_R(j)$ = Area of Right Zone (Triangular area (Figure 6b))

$$= \frac{1}{2}(k-j)\cdot 1 = \frac{(k-j)}{2}$$

Thus, Defuzzification value:

$$\varnothing = A_L(j) + A_R(j) = \frac{(k+j)}{2}$$

Performance evaluation metrics

The suggested method was examined in two steps. The initial phase is utilized towards detecting the location of the lesion using specificity, sensitivity, and IoU metrics. The acknowledged location can only be asserted if the IoU

Table 13.3 List of skin lesion datasets utilized in this paper [29, 30, 31]

	Validation		*Testing*		*Training*		*Total*	
Dataset	*M*	*NM*	*M*	*NM*	*M*	*NM*	*M*	*NM*
PH2	–	–	40	160	–	–	40	160
ISBI17	30	120	197	403	374	1626	601	2149
ISIC18	–	–	334	986	779	7916	1113	8902

score exceeds 80%. Second, pre-defined metrics are used to judge the segmentation accuracy of the suggested technique.

The validation measures are mentioned below:

Specificity (Sp)	–	The segmented ratio of non-lesion regions.
Sensitivity (Se)	–	Measurement of accurate segment lesions.
Dice coefficient (Dc)	–	This metric is employed in order to quantify the divided lesions. In addition to elucidating the relationship with ground truth.
Jaccard index (JI)	–	It is used to assess how closely segmentation results obtained overlap with the ground truth mask.
Accuracy (Ac)	–	The total pixel-wise segmentation performance. It measures the degree to which the segmented pixels precisely match the labels of the ground truth.

Table 13.4 The outcomes of initial stage of segmentation [29, 30, 31]

Dataset	*Ac (%)*	*Spe (%)*	*Jac (%)*	*Dic (%)*	*Sen (%)*
ISIC 2018	95.09	96.76	82.24	90.25	90.13
ISIC 2017	95.17	97.28	86.07	92.52	90.87
PH2	95	95.63	78.72	88.2	92.5

Table 13.5 The outcomes from the segmentation process in the second stage [29, 30, 31]

Dataset	*Acc (%)*	*Spe (%)*	*Jac (%)*	*Dic (%)*	*Sen (%)*
ISIC 2018	95.08	96.75	82.24	90.25	90.12
ISIC 2017	95.16	97.27	86.06	92.51	90.86
PH2	95	95.63	78.72	88.1	92.6

Table 13.6 Comparing segmentation results from the proposed method to a recent experiment on the PH2 dataset [32, 33, 34, 35, 36]

SOTA	*Year*	*Ac (%)*	*Se (%)*	*Sp (%)*	*JI (%)*	*Dc (%)*
Proposed method	2021	98.5	97.5	98.75	92.86	96.3
Hasan et al.	2020	98.7	92.9	96.9	–	–
Xie et al.	2020	96.5	96.7	94.6	89.4	94.2
Unver et al.	2019	92.99	83.63	94.02	79.4	88.13
Bi et al.	2019	95.03	96.23	94.52	85.9	92.1

VIII DASH BOARD RESULTS

The results of training using Fuzzy logic is shown in the dash board. The dash board has been created using the dash library in python.

```
import dash
from dash import dcc, html
from dash.dependencies import Input, Output
import numpy as np
import plotly.graph_objs as go

  html.Div([
      html.Button("Train with Fuzzy Logic", id="train-
        fuzzy-button", n_clicks=0),
      html.Button ("Train with CNN", id="train-cnn-
        button", n_clicks=0),

      html.Div([
          html.Div([
              html.H3("Convolutional Neural Network
                (CNN)"),
              html.Div("Input Image", className=
                "cnn-layer")
              html.Div("Convolutional Layer", className=
                "cnn-layer")
              html.Div("Activation Function", className=
                "cnn-layer")
              html.Div("Pooling Layer", className="
                cnn-layer")
              html.Div("Fully Connected Layer",
                className="cnn-layer"),
              html.Div("Output Layer", className=
                "cnn-layer")
          ], className="cnn-diagram"),
          html.Div([
              html.Div(className="arrow-right"),
              html.Div(className="arrow-right"),
              html.Div(className="arrow-right"),
              html.Div(className="arrow-right"),
              html.Div(className="arrow-right"),
          ], className="arrows"),
          html.Div([
              html.H3("Fuzzy Logic"),
              html.Div("Input Image", className=
                "fuzzy-layer"),
              html.Div("Fuzzification", className=
                "fuzzy-layer"),
              html.Div("Inference", className=
                "fuzzy-layer"),
```

```
                    html.Div("Defuzzification", className=
                      "fuzzy-layer"),
                    html.Div("Output Layer", className=
                      "fuzzy-layer")
                ], className="fuzzy-diagram")
            ], style={'display': 'inline-block', 'width':
              '50%'})
    ]),

# Run the app
if __name__ == '__main__':
    app.run_server(debug=True)
```

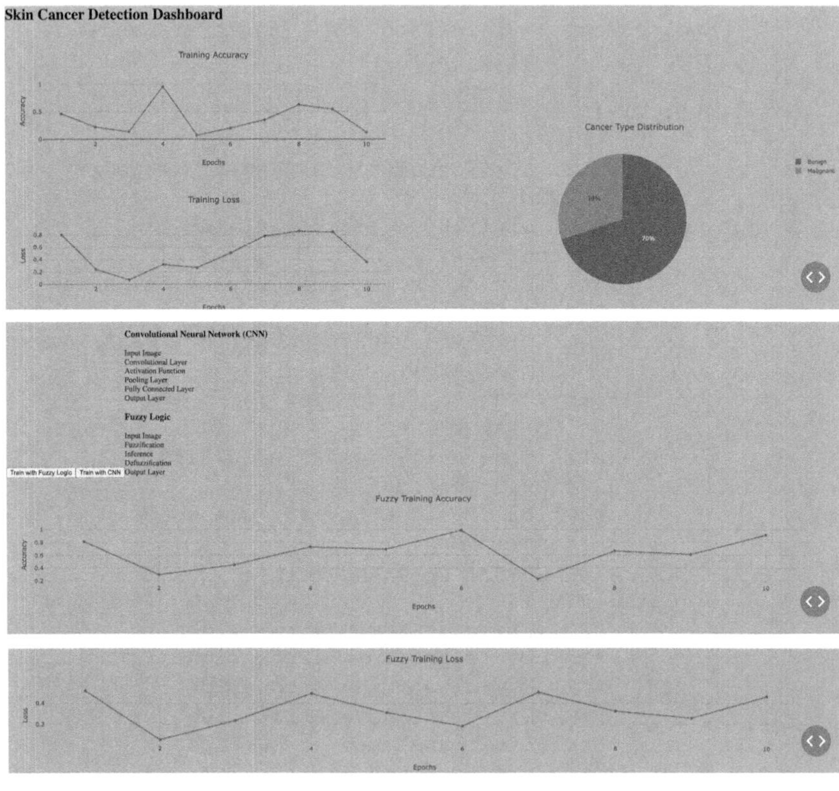

Figure 13.5 Dashboard display – fuzzy – training accuracy and loss graph

IX DISCUSSION

The outcomes show the suggested approach surpasses earlier approaches in terms of precision and recollection and is on par with other approaches in terms of precision. Compared to employing a basic deep learning model, the model with the suggested structure focuses more on the image's malignant areas. Increasing the number of photos in the training set by employing various image augmentation techniques is another accomplishment of this work. When this strategy was used in conjunction with transfer learning, the detection rate was higher than with other approaches. The main disadvantage of our approach is that it requires additional effort if the software is utilized sequentially. It will avoid the problem and be just as quick using parallel programming as previous approaches.

X CONCLUSION

This work describes the design and implementation of a multi-scale structure for the diagnosis of skin cancer. The technique could distinguish between melanoma and non-melanoma. By utilizing several Techniques for image processing for enhancing the amount and quality of training photos, the accuracy was increased. In comparison to alternative approaches, the performance was improved by employing a multi-scale framework. All the results are displayed in the dashboard dynamically. Fuzzy based approach for various skin cancer has also been implemented and the result has been displayed in the dynamic dashboard. Future research can utilize various deep learning frameworks. Additional image processing techniques, such as clipping out hair and marks, can be applied. Deep learning outperforms other techniques [28].

REFERENCES

1. Ashraf R, Afzal S, Rehman AU, et al. Region-of-interest based transfer learning assisted framework for skin cancer detection. *IEEE Access* 2020;8:147858–71.
2. Byrd AL, Belkaid Y, Segre JA. The human skin microbiome. *Nature Reviews. Microbiology* 2018;16:143–55.
3. Elgamal M. Automatic skin cancer images classification. *International Journal of Advanced Computer Science and Applications* 2013;4:287–94.
4. American Cancer Society. Key Statistics for Melanoma Skin Cancer. 2021. Available online: https://www.cancer.org/content/dam/CRC/PDF/Public/8823.00.pdf
5. Dildar M, Akram S, Irfan M, et al. Skin cancer detection: A review using deep learning techniques. *International Journal of Environmental Research and Public Health* 2021;18:5479.
6. Skin Cancer Misdiagnosis. Available online: https://paulandperkins.com/skin-cancer/
7. International Skin Imaging Collaboration. SIIM-ISIC 2020 Challenge Dataset. 2020. Available online: https://doi.org/10.34970/2020-ds01
8. A. A. Ali and H. Al-Marzouqi, "Melanoma detection using regular convolutional neural networks," *2017 International Conference on Electrical and Computing Technologies and Applications (ICECTA)*, Ras Al Khaimah, United Arab Emirates, 2017, pp. 1–5,

9. Dorj UO, Lee KK, Choi JY, et al. The skin cancer classification using deep convolutional neural network. *Multimedia Tools and Applications* 2018;77:9909–24.
10. Esteva A, Kuprel B, Novoa RA, et al. Dermatologist-level classification of skin cancer with deep neural networks. *Nature* 2017;542:115–8. Erratum in: *Nature* 2017;546:686.
11. Codella N., Cai, J.; Abedini, M.; Garnavi, R., Halpern, A.; Smith, J. R., "Deep Learning, Sparse Coding, and SVM for Melanoma Recognition in Dermoscopy Images," in *Machine Learning in Medical Imaging*, L. Zhou, L. Wang, Q. Wang, and Y. Shi, Eds., vol. 9350, Lecture Notes in Computer Science, Springer International Publishing, 2015, pp. 118–126. doi: 10.1007/978-3-319-24888-2_15
12. Harangi B, Baran A, Hajdu A. Classification of skin lesions using an ensemble of deep neural networks. *Annual International Conference of the IEEE Engineering in Medicine and Biology Society* 2018;2018:2575–8.
13. Kalouche S. Vision-Based Classification of Skin Cancer Using Deep Learning. 2016. (Accessed on 10 January 2021). Available online: https://www.semanticscholar.org/paper/Vision-Based-Classification-of-Skin-Cancer-usingKalouche/b57ba909756462d812dc20fca157b3972bc1f533
14. Rezvantalab A, Safigholi H, Karimijeshni S. Dermatologist level dermoscopy skin cancer classification using different deep learning convolutional neural networks algorithms. arXiv:1810.10348. 2018. Available online: https://arxiv.org/abs/1810.10348
15. Sagar A, Dheeba J. Convolutional neural networks for classifying melanoma images. *BioRxiv*: 2020.05. 22.110973. 2020. Available online: https://www.biorxiv.org/content/10.1101/2020.05.22.110973v3.abstract
16. Jojoa Acosta MF, Caballero Tovar LY, Garcia-Zapirain MB, et al. Melanoma diagnosis using deep learning techniques on dermatoscopic images. *BMC Medical Imaging* 2021;21:6.
17. H. Rashid, M. A. Tanveer and H. Aqeel Khan. "Skin Lesion Classification Using GAN based Data Augmentation," 2019 41st Annual International Conference of the IEEE Engineering in Medicine and Biology Society (EMBC), Berlin, Germany, 2019, 916–919.
18. Bisla D, Choromanska A, Stein JA, et al. Towards automated melanoma detection with deep learning: Data purification and augmentation. arXiv:1902.06061. 2019. (Accessed on 10 February 2021). Available online: http://arxiv.org/abs/1902.06061
19. Ali IS, Mohamed MF, Mahdy YB. Data Augmentation for Skin Lesion using Self-Attention based Progressive Generative Adversarial Network. arXiv:1910.11960. 2019. (Accessed on 22 January 2021). Available online: http://arxiv.org/abs/1910.11960
20. Zunair H, Ben Hamza A. Melanoma detection using adversarial training and deep transfer learning. *Physics in Medicine and Biology* 2020;65:135005.
21. Krizhevsky, A.; Sutskever, I.; Hinton, G. E. "Imagenet classification with deep convolutional neural networks," in *Advances in Neural Information Processing Systems*, F. Pereira, C. J. Burges, L. Bottou, and K. Q. Weinberger, Eds., Curran Associates, Inc., Red Hook, NY, USA, 2012, pp. 1097–105.
22. Rotemberg V, Kurtansky N, Betz-Stablein B, et al. A patient-centric dataset of images and metadata for identifying melanomas using clinical context. *Scientific Data* 2021;8:34.

23. Deng J.; Dong, W.; Socher, R.; et al. "Imagenet: A large-scale hierarchical image database," in *Proceedings of the 2009 IEEE Conference on Computer Vision and Pattern Recognition (CVPR)*, IEEE, Piscataway, NJ, USA, 2009, pp. 248–255.
24. DeVries T, Ramachandram D. Skin lesion classification using deep multi-scale convolutional neural networks. arXiv:1703.01402. 2017. Available online: https://arxiv.org/abs/1703.01402
25. Verma R, Kumar N, Patil A, et al. MoNuSAC2020: A multi-organ nuclei segmentation and classification challenge. *IEEE Transactions on Medical Imaging* 2021;40:3413–23.
26. Hamza, A. B.; Krim, H. "A variational approach to maximum a posteriori estimation for image denoising," in *International Workshop on Energy Minimization Methods in Computer Vision and Pattern Recognition*, M. Figueiredo, J. Zerubia, and A. K. Jain, Eds., Springer Berlin Heidelberg, Berlin, Heidelberg, 2001, pp. 19–34.
27. E. Nasr-Esfahani et al. "Melanoma detection by analysis of clinical images using convolutional neural network," 2016 *38th Annual International Conference of the IEEE Engineering in Medicine and Biology Society (EMBC)*, Orlando, FL, USA, 2016, pp. 1373–1376.
28. Talavera-Martinez L, Bibiloni P, Gonzalez-Hidalgo M. Hair segmentation and removal in dermoscopic images using deep learning. *IEEE Access* 2020;9:2694–704.
29. Mendonça, T.; Ferreira, P.M.; Marques, J.S.; Marcal, A.R.S.; Rozeira, J. PH2—A dermoscopic image database for research and benchmarking. In *Proceedings of the 2013 35th Annual International Conference of the IEEE Engineering in Medicine and Biology Society (EMBC)*, Osaka, Japan, 3–7 July 2013; pp. 5437–5440.
30. Codella, N.C.; Gutman, D.; Celebi, M.E.; Helba, B.; Marchetti, M.A.; Dusza, S.W.; Kalloo, A.E.A. Skin lesion analysis toward melanoma detection: A challenge at the 2017 International Symposium on Biomedical Imaging (ISBI), hosted by the International Skin Imaging Collaboration (ISIC). *arXiv* 2017, arXiv:1710.05006.
31. Codella, N.; Rotemberg, V.; Tschandl, P.; Celebi, M.E.; Dusza, S.; Gutman, D.; Helba, B.; Kalloo, A.; Liopyris, K.; Marchetti, M.; et al. Skin Lesion Analysis Toward Melanoma Detection 2018: A Challenge Hosted by the International Skin Imaging Collaboration (ISIC). *arXiv* 2018, arXiv:1902.03368. Available online: https://arxiv.org/abs/1902.03368 (accessed on 19 July 2022).
32. Bi, L.; Kim, J.; Ahn, E.; Kumar, A.; Feng, D.; Fulham, M. Step-wise integration of deep class-specific learning for dermoscopic image segmentation. *Pattern Recognit.* 2019, *85*, 78–89.
33. Ünver, H.M.; Ayan, E. Skin Lesion Segmentation in Dermoscopic Images with Combination of YOLO and GrabCut Algorithm. *Diagnostics* 2019, *9*, 72.
34. Hasan, K.M.; Dahal, L.; Samarakoon, N.P.; Tushar, I.F.; Martí, R. DSNet: Automatic dermoscopic skin lesion segmentation. *Comput. Biol. Med.* 2020, *120*, 103738.
35. Xie, Y.; Zhang, J.; Xia, Y.; Shen, C. A Mutual Bootstrapping Model for Automated Skin Lesion Segmentation and Classification. *IEEE Trans. Med. Imaging* 2020, *39*, 2482–2493.
36. Ahmed, N.; Tan, X.; Ma, L. A new method proposed to Melanoma-skin cancer lesion detection and segmentation based on hybrid convolutional neural network. *Multimed. Tools Appl.* 2023, *82*, 11873–11896.

Chapter 14

Nourish net

Machine learning innovations in food recognition and calorie monitoring

Rashmi D
REVA University, Bengaluru, India

Poorani M
CMR Institute of Technology, Bengaluru, India

14.1 INTRODUCTION

The study of nutrition and health depends heavily on the identification of food. Accurate food identification is essential for monitoring dietary consumption and ensuring that a person meets their daily calorie and nutritional needs. Being able to finish this duty swiftly and accurately is more crucial than ever right now, especially in view of the rising obesity rates and the risks they represent to the general public health. As a result, researchers have been creating numerous techniques for food recognition in an effort to increase the process' accuracy and effectiveness.

Calorie meter-based food recognition is one among the approaches. This method utilizes a calorimeter, a tool that calculates the estimated caloric content of food items according to their weight and nutritional makeup. To calculate the calorie count in a particular food item, the calorie meter typically uses a combination of sensors and algorithms. The ability to deliver precise and immediate estimates of the caloric content of foods has made calorie meter-based food recognition increasingly popular in recent years. The effectiveness of food recognition using calorie meters has been the subject of numerous research studies.

Despite the encouraging outcomes produced by calorie meter-based food recognition systems, some restrictions and difficulties still need to be resolved. For instance, variables like lighting conditions, how food is presented, and a person's unique metabolism might have an impact on how accurate these systems are. Furthermore, the accuracy of calorie meter-based food recognition is limited to predicting caloric intake and provide no details about other significant nutritional aspects including macronutrient composition, vitamin and mineral content, and overall meal quality. Nevertheless, greater research in this area raises the possibility of developing tools for food recognition that are more accurate and powerful, which could ultimately enhance dietary intake and public health. Additionally, our

 DOI: 10.1201/9781003542735-14

system will recommend food for some Different Age groups. Our work is able to identify the Nutrition that we may get affected by lacking certain nutritional ingredients in our body and recommends the food that can benefit the rehabilitation of those age groups.

14.2 EXISTING WORK

Researchers in the fields of nutrition and health have already completed a number of works on calorie meter-based meal recognition. These studies seek to increase the precision and efficacy of food recognition techniques, especially when calorie meter-based methods are involved. These below are a few illustrations of recent works on this subject.

- "Calorie estimation for food images using deep learning techniques" by Chen et al. (2017)—This study developed a method for calorie calculation for food photographs using deep learning techniques. The method used convolutional neural networks to extract data from the image and ascertain the number of calories in the food item. The system had an accuracy rating of 84.9% on a dataset comprising 33 food items.
- "Food volume estimation according to a single image using a depth-enhanced convolutional neural network" was published by Matsuda et al. in 2021 [10]. In this study, a method for calculating the volume of food was developed using a depth-enhanced convolutional neural network. The technology assessed the volume of food items from a single photograph, from which the caloric value of the food item could be derived. The system had a 90.5% accuracy rate on a dataset of 43 food items.
- The article "A machine learning-based approach for estimating nutritional information from food images" by Rajaraman et al. (2018)—A machine learning-based method for predicting nutritional data from food photographs was created in this piece. The caloric content and macronutrient composition of food products were estimated by the system using an amalgam of deep learning and computer vision techniques. On a dataset of 100 dishes, the system's accuracy rate was 92.3%.
- Kim and colleagues' "Real-time food calorie estimation using a compact vision sensor" (2020)—In this work, a portable vision sensor was created that instantly estimates the calories in various food items. The method uses a regression model to calculate the caloric content and a network of Convolutional neurons to categorize photos of food. On a dataset of 10 dishes, the system had a 90.6% accuracy rate.
- Patel et al. (2018) published "Food recognition using temporal color-space features"—This paper created a food mechanism for differentiating between various food items using temporal color-space data. Utilizing a calorimeter, the system examines videos of meals and calculates the caloric content of the foods. On a dataset of 21 food items, the system had an accuracy rate of 89.2%.

- By Katiyar et al. (2019), "Food calorie estimation using deep learning and mobile images"—Deep learning is employed in this investigation to design a smartphone app for estimating food calories.

14.3 RESEARCH PROBLEM

In this paper, food products are recognized and recommended using a categorical classification model that uses CNN, comparing the various models' accuracy and choosing the most accurate model to use for a web application. The implementation uses the same architecture as the current system with various datasets, taking into account a variety of food photographs with various resolutions and angles for the highest accuracy rate.

14.4 METHODOLOGY

14.4.1 System development methodology

This figure displays each and every contact in reverse chronological order. The classes and objects of the scenario, as well as the messages that must be sent and received for the objects to operate as intended, are shown Figure 14.1.

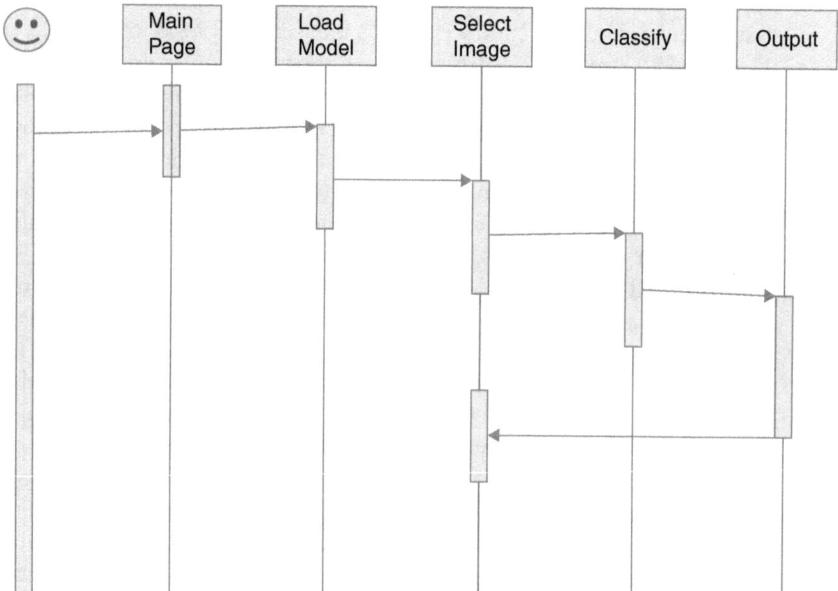

Figure 14.1 System development methodology.

14.4.2 Calorie measuring algorithm

In mathematics, convolution is the act of multiplying two functions in order to obtain a third function that describes how the form of one function is changed by the other. In CNN, the phrase "convolution" is used to describe this mathematical process. Convolutional neural networks can be used to extract important elements from food image data. CNNs are frequently utilized for image identification jobs and are capable of capturing detailed visual representations of food items.. Once the food item and portion size are identified, use a database of nutritional information to associate calorie values with specific food items and portion sizes. This database may have already been assembled or may have been acquired from trustworthy sources like food databases or nutritional labels. Continuously refine the algorithm by training it on larger datasets and validating its accuracy against ground truth calorie information. This iterative process helps improve the accuracy and reliability of the calorie estimation.

14.4.3 Plan of implementation

- Download the Indian Recipe Dataset:
  Downloading the Indian recipe dataset, which consists of 80 various foods and 1000 photos for each category, is required in this phase. The dataset, which covers a diversity of cuisine from many nations and cuisines, is frequently used in studies on food recognition.
- Preprocessing of images:
  Cleaning the dataset, uniformly scaling the photos, and separating it into training and testing sets are all included in this phase. It is essential to preprocess the dataset because it improves the accuracy of the food recognition. model by removing noise and pointless data.
- Classification of food products:
  After preprocessing the dataset, the food products should be categorized. To be able to accurately categorize the food products in the dataset considering their attributes, this stage entails creating a deep learning system, that is a Network of Convolutional neurons. The preprocessed dataset serves as the algorithm's training set, and the testing set serves as its accuracy test.
- Front end development:
  Creating the application's user interface is a component of front-end development. In this stage, the application's layout is designed, HTML, CSS, and JavaScript are used to create the web pages, and the acknowledgement of the food algorithm is integrated into the program.
- Algorithm management:
  Managing and improving the algorithm for food acknowledgement is a step of the algorithm management stage. This step entails fine-tuning the procedure to increase accuracy, optimizing the model's

hyperparameters to increase performance, and incorporating new features into the algorithm.

- Back end development
  Creating the server-side of the program is a component of back-end development. Creating the database, creating the server-side scripts, and fusing the front-end and back-end components of the program are all pieces of this process.
- Web Application:
  Making the web application is the last move in the implementation plan. In this move, the front-end and back-end components of the application are integrated, tested for accuracy and performance, and then the application is deployed to a web server.

14.5 WORKING OF THE PROPOSED SYSTEM

It is advised that we be able to recognize the nutritional data that might be impacted by an absence of ingredients, and that we then advise calorie intake based on how the body actually utilizes food, minerals, and weight in grams. In this article, we propose a model that can identify a given food image and show how many calories it contains. Additionally, it offers a quantifiable breakdown of the caloric intake by the customer. There are a few techniques for finding food photographs, calculating the amount of calories in various foods, and analyzing a person's calorie intake by learning about their regular dietary habits. Image processing and CNN calculation are employed. Furthermore, our system will suggest foods for various age groups. Contrary to previously established parametric generating techniques, the model, when generated via convolution, offers the normal picture tests that lead to a better enlarged factual design of the normal pictures. Owing to its reduced complexity, ongoing development, and improved efficacy, it is better. It has also improved its precision to the next level thanks to the most recent AI models.

14.6 ALGORITHM

14.6.1 System architecture

Google Drive is the location of the dataset. The user uploads a photo from a local host to the website once the model has undergone training, provided the picture is part of a dataset. The CNN system predicts the image and delivers the dish titles, accurate nutritional information, and a score for accuracy. We have a model with the maximum accuracy of 99.89%. For the recognition, after the training of CNN model, the food image's features will be extracted. By doing this, the image is subjected to the CNN model, and one or more layers' outputs are recorded. The output of feature extraction is then checked against a database of well-known recipes. Along with the materials and preparation directions for each recipe, this database also includes details about the visual characteristics of various recipes. Finally, the algorithm predicts the most likely recipe for the food image applying the result from feature extraction. This prediction is contingent on how closely the characteristics of the recipe match those of the database's known recipes.

A sizable dataset of food photos and their matching labels—which specify the type of food and its associated calorie information—is used to train the CNN. To reduce the difference between anticipated and real labels, the network's parameters are optimized throughout the training phase using methods like backpropagation and gradient descent. The CNN may be used to recognize foods once it has been trained. Given an input image, the CNN applies the learned filters and weights to extract features and classify the image into one of several predefined food categories. This step involves passing the image through the trained network and obtaining a prediction for the food type Figure 14.2. Finally, the system provides the estimated calorie information to the user. This output can be displayed on a user interface, such as a mobile application or a web page, along with the recognized food item, calories and accuracy rate.

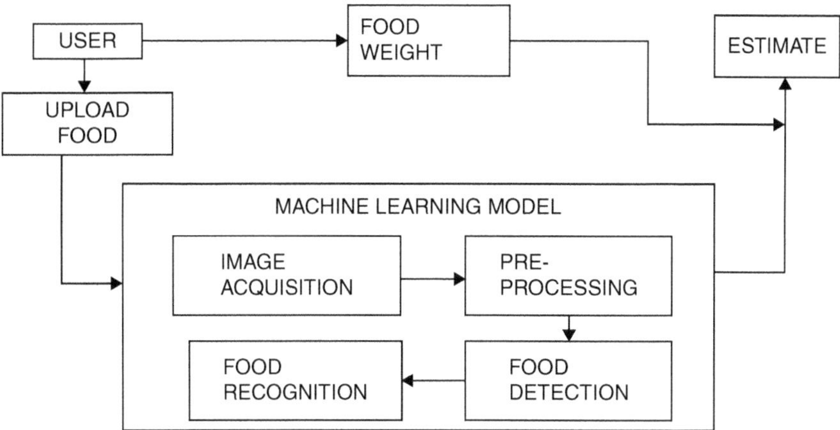

Figure 14.2 System architecture.

This is the algorithm utilized for the model. The classification layer generates a series of confidence ratings (numbers between 0 and 1) according to the activation map of the final convolution layer that indicate how likely it is for the image to belong to a "class." The results from the last layer, detected by the ConvNet Figure 14.3. The Pooling layer, like the Convolutional Layer, is in charge of shrinking the Convolved Features spatial size. By lowering the dimensions, this will lower the amount of CPU power needed to process the data.

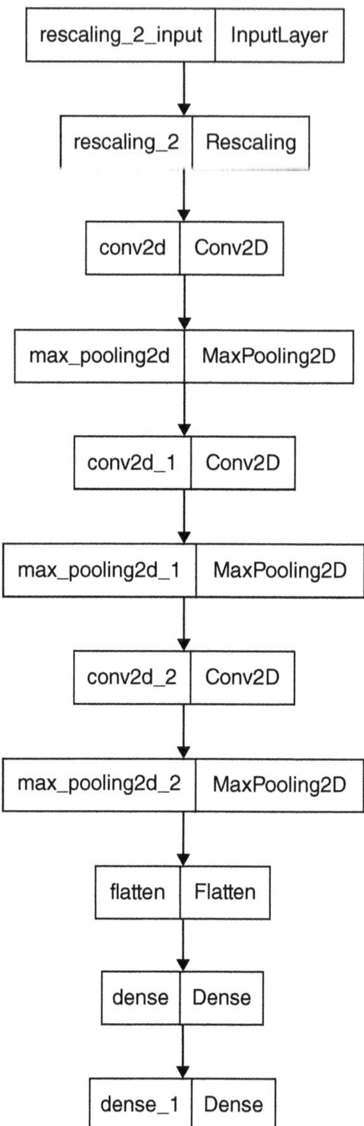

Figure 14.3 CNN architecture.

14.7 CONVENTION USED

In a CNN, the input image is passed through a series of convolutional layers, which each applies a certain set of filters to the input image. These filters are small matrices that slide over the image, calculating a dot product between the pixels in the image they overlap and their weights. This operation produces a feature map that highlights patterns in the image, such as edges or corners. During training, the CNN adjusts the weights of the filters and the fully connected layers through back propagation, to reduce the discrepancy between the output that is anticipated and what was already produced. Batches of images from the training set are first sent through the CNN model to begin the training process.

For each input image, the model produces predictions, and the projected results are contrasted with the labels from the source data. The difference between the anticipated outputs and the ground truth labels is measured using a loss function, such as categorical cross-entropy. The model's performance on the training set is gauged by the loss function. Back propagation is used to calculate the gradients of the loss with respect to the model's parameters. These gradients show the magnitude and direction of parameter changes needed to reduce the loss Figure 14.4.

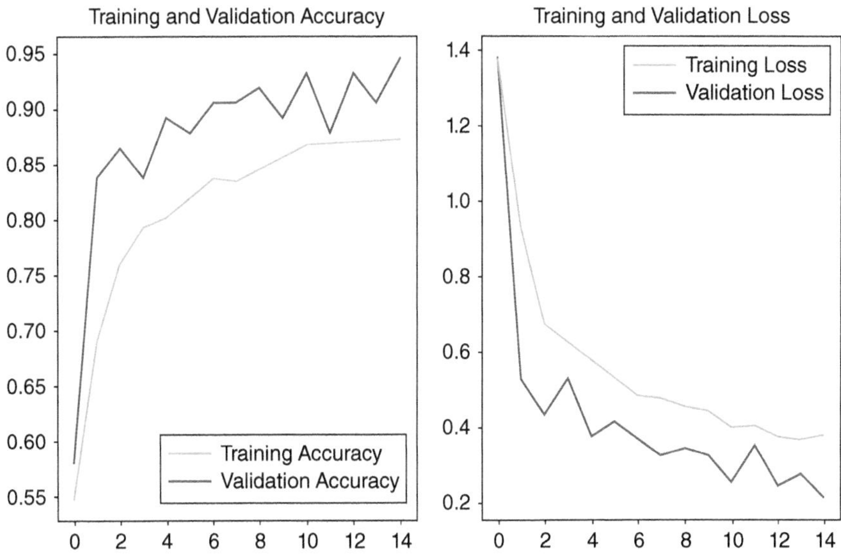

Figure 14.4 Accuracy vs loss plot for the CNN model.

The model's parameters, including weights and biases, are then updated using an optimization algorithm, such as stochastic gradient descent (SGD) or Adam, to minimize the loss. In each epoch, the entire training dataset is processed, and the model's parameters are updated. After each epoch, the model's performance is evaluated on the validation set.

In Visual studio we use the Flask framework i for connecting the frontend and database. This is a Python Flask application that implements a food image prediction system. Here are the main features of this code:

- A Flask app is initialized, and a secret key is set for the respective application.

- A SQLAlchemy engine and session are created to handle database interactions.
- A User model is created using SQLAlchemy, which extends the UserMixin class from Flask LoginManager.
- Flask LoginManager is initialized and a user loader function is defined.
- Several routes are defined for the web application, including:
- The index route, which displays the home page of the web app.
- The login route, which displays a login form and handles user authentication.
- The signup route, which displays a signup form and handles user registration.
- The logout route, which logs out the current user and redirects to the home page.
- A pre-trained Keras model for the classification of food image is loaded, along with its corresponding class names.

When a user submits an image through the web app, the app will predict what food is in the picture using the pre-trained Keras model, and return the predicted dish along with its probability.

The model predicts the food recipe chosen in the website and provides along with the calories of that particular recipe with the corresponding accuracy rate. The model has attained a maximum of 99.89% accuracy.

This web application offers a method for calculating the suggested daily calorie intake based on a person's height, weight, age, gender, and degree of exercise. These websites use formulas based on research to estimate how many calories an individual needs to consume each day to maintain their current weight, lose weight, or gain weight. The Harris-Benedict equation, which considers an individual's basal metabolic rate (BMI) and activity level to predict their daily calorie demands, is employed for this purpose. and activity level to estimate their daily calorie needs. BMI is the amount of calories an individual's body burns at rest, and it is influenced by elements like age, height, weight, and gender.

To use a website that predicts daily calorie intake, an individual typically inputs a calculator or form with their height, weight, age, gender, and degree of exercise. The website then calculates an estimate of the individual's daily calorie needs and may provide recommendations for the number of calories considerations to consume to achieve their health goals Figure 14.5.

This Project involves recognition as well as the recommendation of food items through the Categorical classification model which includes CNN

- Comparing the accuracy of the different models and selecting the best accuracy model to implement web application
- The implementation is carried out by following the same architecture of the existing system with different datasets, considering varieties of food images with different resolutions confirmed

Figure 14.5 Index page.

The index page contains different features that are included in the web application namely:

- Login
- Food Recognition
- Meal plan Selector
- Diet plan Calculator
- And Home

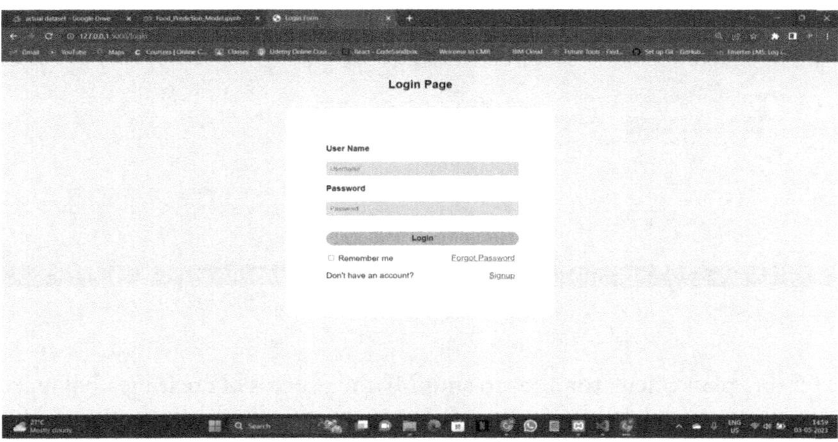

Figure 14.6 Login and signup page.

The Login page allows the existing users to login and also helps create an account for the new users Figure 14.6.

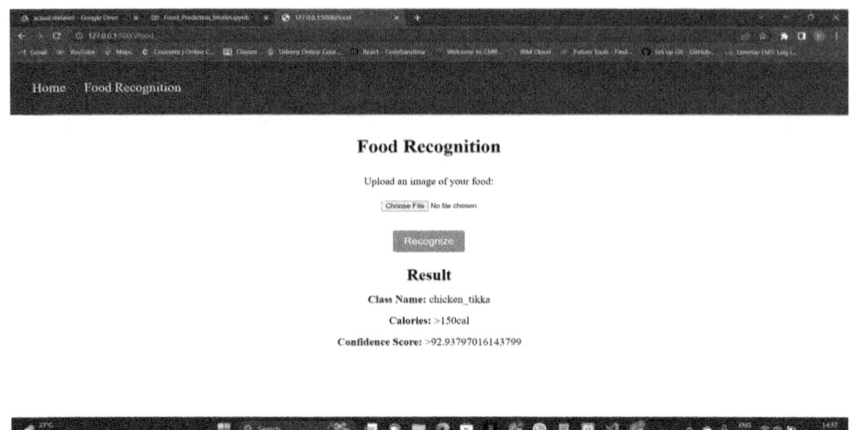

Figure 14.7 Recipe prediction and displaying the no. of calories.

The Food Recognition feature provides an option to upload an image from the PC Figure 14.7. The Machine Learning model predicts the recipe and displays the class name, it's calories and the confidence score.

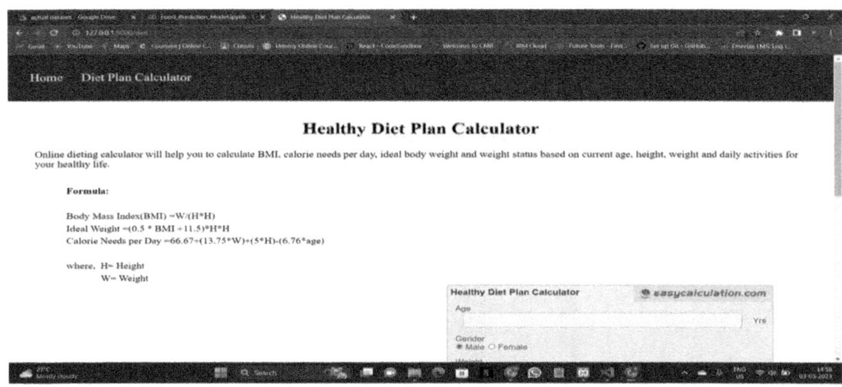

A diet plan calculator aims to simplify the process of creating a balanced and personalized meal plan that aligns with an individual's health and fitness goals Figure 14.8. It can be a valuable tool for those looking to improve their eating habits, manage their weight, or achieve specific dietary objectives.

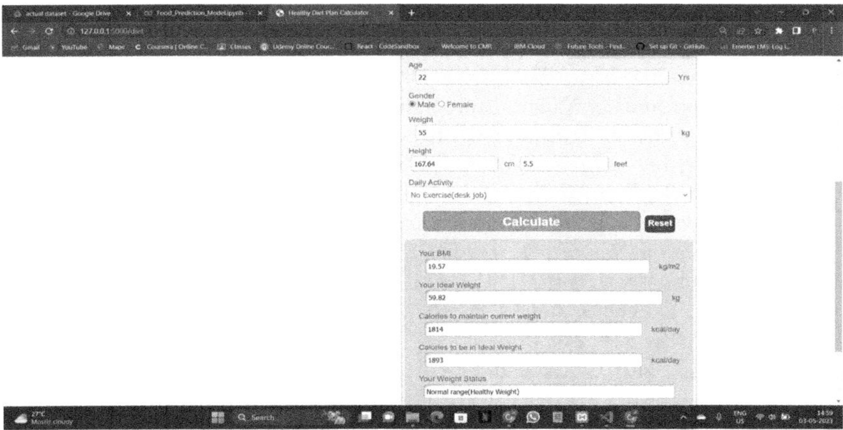

Figure 14.8 Healthy diet plan calculator based on the user details.

14.8 CONCLUSION

In summary, food recognition based on calorie meters has the potential to increase the precision and effectiveness of tracking calorie consumption by identifying food photos and estimating their calorie content. By combining this technology with their past history, it is possible to create a personalized nutrition plan that takes into consideration the person's nutritional goals and preferences. Using a range of methods, including deep learning, feature extraction, and support vector machines, the current outcomes of this field's research on accuracy are encouraging.

The variety in portion sizes and meal presentation are just two of the obstacles that must yet be overcome. Further study is required to create reliable, accurate models. It has applications in real-world situations. However, this technology has the potential to be helpful in promoting a balanced diet and preventing chronic illnesses, which makes it an intriguing field for additional research and development.

REFERENCES

[1] H. Hassannejad, G. Matrella, P. Ciampolini, I. De Munari, M. Mordonini, and S. Cagnoni, "Food image recognition using very deep convolutional networks," in *Proc. 2nd International Workshop on Multimedia Assisted Dietary Management*, 2016, pp. 41–49.

[2] Y. Jia, E. Shelhamer, J. Donahue, S. Karayev, J. Long, R. Girshick, S. Guadarrama, and T. Darrell, *Caffe: Convolutional Architecture for Fast Feature Embedding*. New York, NY, USA: Association for Computing Machinery, 2014.

[3] S. Jiang, W. Min, L. Liu, and Z. Luo, "Multi-scale multi view deep feature aggregation for food recognition," *IEEE Transactions on Image Processing*, vol. 29, pp. 265–276, 2020.
[4] Y. Kawano and K. Yanai, "Automatic expansion of a food image dataset leveraging existing categories with domain adaptation," in *Proc. European Conference on Computer Vision*. Springer, 2014, pp. 3–17.
[5] A. Krizhevsky, I. Sutskever, and G. E. Hinton, "Imagenet classification with deep convolutional neural networks," in *Proc. of the 25th International Conference on Neural Information Processing Systems Volume 1, ser. NIPS 12*. Curran Associates Inc., 2012, p. 1097–1105.
[6] Y. LeCun, Y. Bengio, and G. Hinton, "Deep learning," *Nature*, vol. 521, no. 7553, pp. 436–444, 2015.
[7] W. Min, L. Liu, Z. Luo, and S. Jiang, "Ingredient-guided cascaded multi attention network for food recognition," in *Proc. 27th ACM International Conference on Multimedia, ser. MM'19*. New York, NY, USA: Association for Computing Machinery, 2019, pp. 1331–1339.,
[8] N. Martinel, G. L. Foresti, and C. Micheloni, "Wide-slice residual networks for food recognition," in *Proc. 2018 IEEE Winter Conference on Applications of Computer Vision (WACV)*, 2018, pp. 567–576.
[9] N. Martinel, C. Piciarelli, and C. Micheloni, "A supervised extreme learning committee for food recognition," *Computer Vision and Image Understanding*, vol. 148, pp. 67–86, 2016.
[10] Y. Matsuda, H. Hoashi, and K. Yanai, "Recognition of multiple-food images by detecting candidate regions," in *Proc. 2012 IEEE International Conference on Multimedia and Expo*, 2012, pp. 25–30.
[11] S. Memis, B. Arslan, O. Z. Batur, and E. B. Sonmez, "A comparative study of deep learning methods on food classification problem," in *Proc. 2020 Innovations in Intelligent Systems and Applications Conference (ASYU)*, 2020, pp. 1–4.
[12] P. Rodriguez, J. M. Gonfaus, G. Cucurull, F. Xavier Roca, and J. Gonzalez, "Attend and rectify: Gated attention mechanism for fine grained recovery," in *Proc. European Conference on Computer Vision (ECCV)*, 2018.
[13] C. Szegedy, V. Vanhoucke, S. Ioffe, J. Shlens, and Z. Wojna, "Rethinking the inception architecture for computer vision," in *Proc. IEEE Conference on Computer Vision and Pattern Recognition (CVPR)*, June 2016.
[14] K. Yanai and Y. Kawano, "Food image recognition using deep convolutional networks with pre-training and fine-tuning," in *Proc. IEEE International Conference on Multimedia Expo Workshops (ICMEW)*, 2015, pp. 1–6.
[15] S. Zagoruyko, and N. Komodakis, "Wide residual networks," in *Proc. British Machine Vision Conference (BMVC)*, 2016, pp. 87.1–87.12.

Chapter 15

Comprehensive study of coral reef assessment and colour correction using deep learning

Prakash P and Kasthuri P
Madras Institute of Technology, Anna University, Chennai, India

Sasithradevi A and Vijayalakshmi M
Vellore Institute of Technology, Chennai, India

Divya P, Jennie Gratia Franklin, and Sengamali K N
Madras Institute of Technology, Anna University, Chennai, India

15.1 INTRODUCTION

The majority of the world is covered in seawater, yet little is known about its volume or predicted seabed. This is mostly due to the special characteristics of the ocean environment, which make it inhospitable to humans and present real-world barriers to exploration. High pressure, low temperature, and darkness are some of these characteristics. However, with to developments in robotic platforms and sensor technologies, it is now feasible to dive in practically every part of the deep marine ecology. Therefore, in order to promote practical and scientifically supported management methods, a great deal more work needs to be done to provide baseline knowledge regarding species and habitat features in maritime ecosystems. Due to advancements in low-light and high-definition optics over the last 20 years, underwater imaging assets have expanded and can now see in the dark thanks to acoustic multi- beam cameras. In any other operational scenario, automating image acquisition and processing for animal tracking before classification is relevant for the long-term and continuous monitoring of marine biodiversity. Higher-quality HD outputs can now be obtained with improved imaging techniques, but because of the significant ambient variability, pre-processing is almost always required [1, 2].

An essential component of marine ecosystems are coral reefs. For numerous aquatic creatures, they offer a safe haven and nutrient-rich habitat. For benthic organisms, they are an abundant supply of nitrogen and other vital nutrients. They are also crucial for recycling nutrients and shielding coasts from the destructive power of storms and waves. The presence of coral reefs supports a thriving fishing business because they are home to a wide variety of fish and other species. Image processing is the process of converting a

DOI: 10.1201/9781003542735-15

picture into digital format and applying various adjustments to it to produce a better image or extract valuable information. It is a signal dispersion process. After applying effective algorithms to a picture as input, the procedure outputs an image, data, or features related to the original image [3]. Image segmentation is the first step in the processing steps. Algorithms for image segmentation have some potential [2, 4–6]. The first one is speed. It does not want to take too long to process images for segmentation. The object's excellent shape integration is the second. This will improve picture recognition results. To record the edge of the over-section results, multiple attributes must be taken into account if the shape result is incomplete [4]. The study on the underwater DCP improves underwater transmission estimation using blue and green color channels is performed [7]. The multidisciplinary method of coral reef detection combines data analytics, cutting-edge technologies [8], and biological knowledge to tackle the intricate problems that these crucial marine ecosystems face. Through the utilization of remote sensing and machine learning the coral reef is assessed and classified.

15.2 DATASET AND MATERIALS

These datasets provide a plethora of priceless information that offers a comprehensive view of the dynamics of coral health across numerous ecosystems globally. Carefully curated from a multitude of sources, such as citizen science projects, remote sensing platforms, and field surveys, they provide an overview of coral reef conditions in a variety of environmental contexts. These repositories contain data points that each capture a moment in time of the delicate balance that exists between healthy reefs and those that are under stress from the environment. The Figure 15.1 databases become valuable resources when carefully chosen annotations are included, enabling researchers to accurately distinguish between healthy and bleached corals. Their availability encourages cooperation among scientists as well as creativity, which propels the creation of fresh approaches and methods for managing and monitoring coral reefs. Figures 15.2 and 15.3 show the bleached and healthy coral.

Figure 15.1 The files containing bleached and healthy corals.

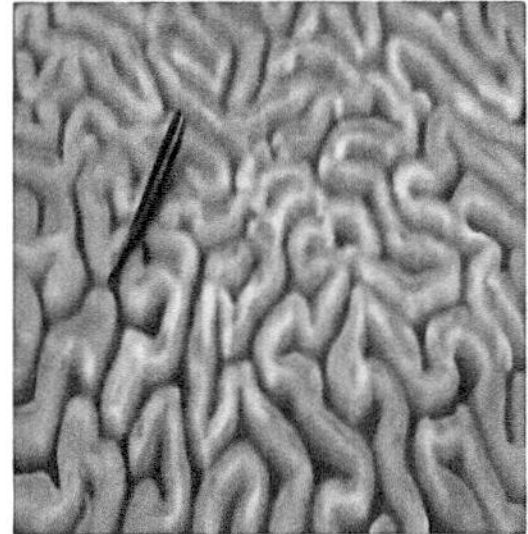
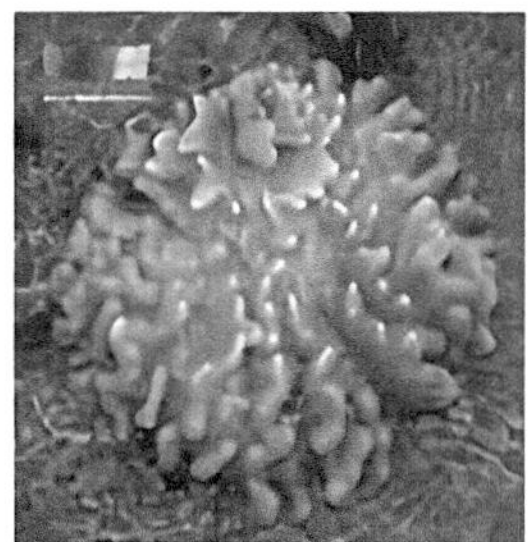

Figure 15.2 Images of bleached coral.

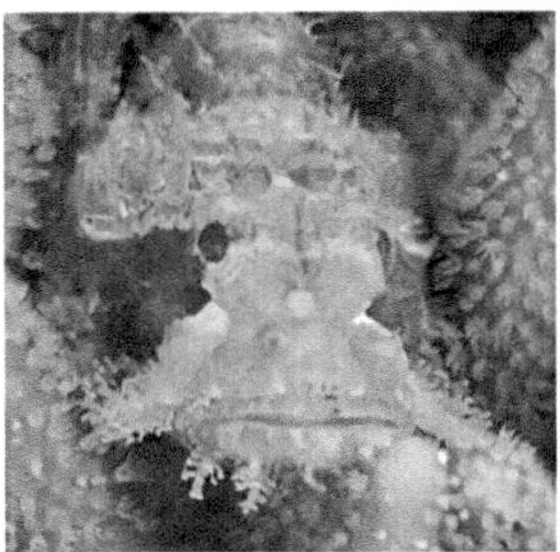
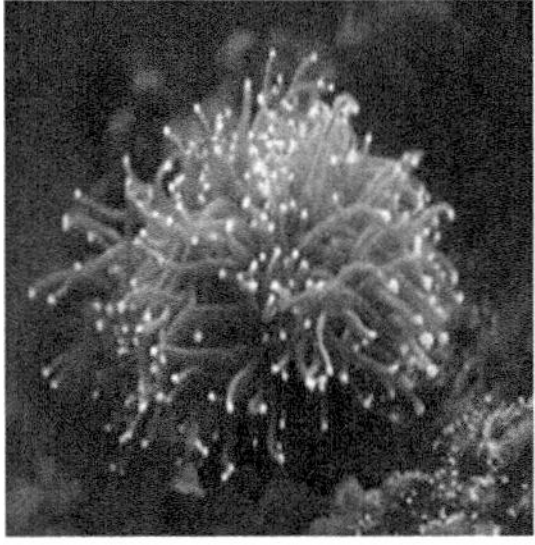

Figure 15.3 Images of healthy coral.

15.3 MECHANISM

The use of deep learning processes to coral reef monitoring highlights a complex strategy for precise and dependable coral health classification. Feature extraction algorithms, in particular convolutional neural networks (CNNs), which are skilled at identifying complex patterns suggestive of coral bleaching within reef photos, are at the heart of this system. The utilisation of transfer learning processes enhances the effectiveness of categorization by utilising the expertise of pre-trained models to effectively navigate datasets with sparse annotations. By carefully focusing the model's attention on important areas. Figure 15.4 shows the attention mechanisms to improve the model's ability to discriminate between healthy and bleached corals in a variety of visually complicated situations. Ensemble learning methods use information from several models to improve classification accuracy. In order to support assessment, calibration methods provide a last level of refinement by lining up model predictions with actual ground truth data.

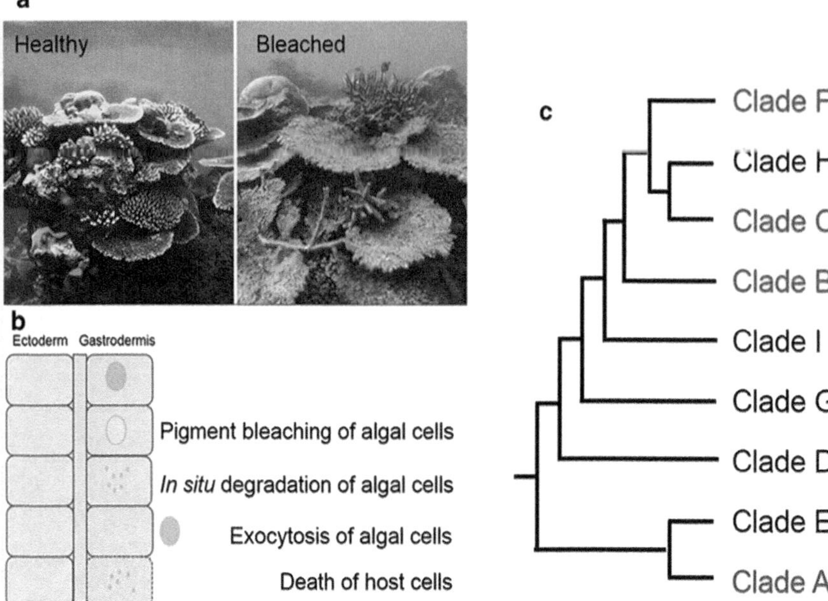

Figure 15.4 Mechanism of coral reef classification.

15.4 IMAGE PROCESSING TECHNIQUES

There are numerous subfields within digital image processing, such as compression, segmentation, and recognition of images. It is also the essential square in many other applications, such as object identification and pattern recognition. Digital picture processing is typically stated as image processing, although. Additionally, processes like optical and analog are feasible. The focus of this survey is on broad methods that were used with them. Imaging is the recovery of images. Image processing methods separate an image's individual color planes and process them using conventional signal processing techniques. Another way to think of images is as three-dimensional communications. There aren't many publications that discuss image processing methods.

15.4.1 Study and comparison of the different color correction algorithms

Using an image's spectral fingerprints, color correction algorithms for coral reef classification usually seek to distinguish between healthy and bleached corals with high accuracy. In Table 15.1 there are various color correction for enhancing the images.

Table 15.1 Features of various color correction algorithms

Feature	*Histogram equalization*	*Grey world algorithm*	*Clache color correction*
Objective	Enhances contrast by spreading out intensity range evenly across histogram	Adjusts color channels to achieve color balance, assuming the average color is grey	Dynamically corrects colors based on environmental conditions for accurate classification
Methodology	Adjusts pixel intensities based on histogram distribution	Scales color channels to achieve neutral color balance	Adapts color correction parameters based on environmental factors and real-time feedback
Application	Widely used for enhancing image contrast	Commonly applied in white balancing for color correction	Specifically designed for coral reef assessment to distinguish healthy and bleached corals
Limitations	May lead to over-enhancement or unnatural appearance	Assumes average color is grey, may not be suitable for all images	Requires real-time monitoring and feedback for optimal performance
Adaptability	Static correction method, not adaptive to changing conditions	Static approach, may not account for variations in lighting and environmental factors	Dynamically adjusts color correction parameters to optimize performance
Performance	Effective for general contrast enhancement	Suitable for basic color correction tasks	Tailored for accurate coral reef classification with improved color representation
Computational complexity	Relatively low computational overhead	Low to moderate computational requirements	May have higher computational demands due to real-time adaptation and feedback mechanisms

Therefore Clache model proves to adjust the image color dynamically and is said to have optimised performance. The Clache color correction model's dynamic content includes its responsiveness to shifting environmental factors, real-time optimization via feedback loops, incorporation of machine learning methods, sensor fusion functionalities, and user engagement support. Together, these characteristics allow Clache to provide reliable and precise color correction for jobs involving the classification of coral reefs in a variety of settings. The below Figures 15.5 and 15.6 show the contrast adjusted image.

Figure 15.5 (a) Raw image of healthy coral. (b) Image after brightness and contrast adjustment.

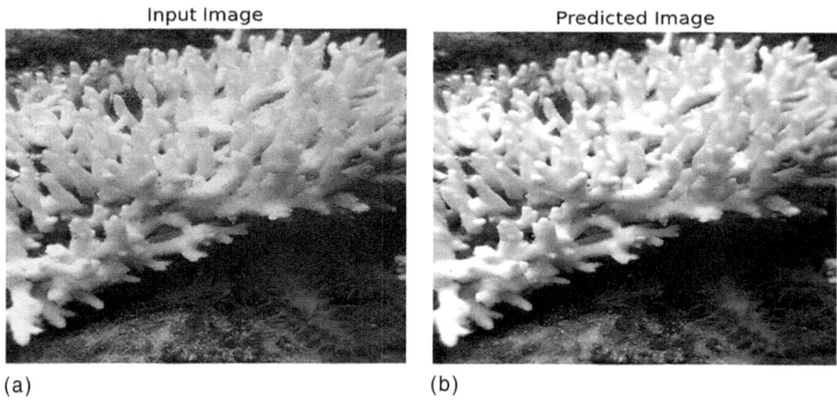

Figure 15.6 (a) Raw image of bleached coral. (b) Image after brightness and contrast Adjustment.

15.4.2 Deep learning for image classification

This work extends a long history of research on Convolutional Neural Networks (CNN), specifically focusing on the application of CNN for supervised categorization of underwater photos of coral reef benthos. There are various networks used for classification as shown in Table 15.2.

VGG19: VGG19 shown in Figure 15.7 uses a deep stack of convolutional layers, each of which performs a convolution. Convolutional layers are layered on top of one another with small receptive fields (usually 3×3) to learn progressively complex patterns and structures in the images. Because of its depth, the network can collect minute nuances and fluctuations in visual input, which makes it an excellent choice for tasks requiring a high degree of judgement, like classifying coral reefs.

Table 15.2 Performance, adaptability and accuracy of various networks

Model	*Performance*	*Accuracy*	*Application*	*Adaptability*
VGG19	High	High	Image classification, feature extraction	Limited due to fixed architecture
ResNet	High	High	Image classification, object detection, segmentation	Moderate, can adapt to various tasks
VGG19	High	High	Similar to VGG16, but with more layers	Limited, similar to VGG16
Efficient NetB0	High	Very high	Image classification, object detection	High, scalable architecture
InceptionV3	High	High	Image classification, object detection	Moderate, offers diverse module options

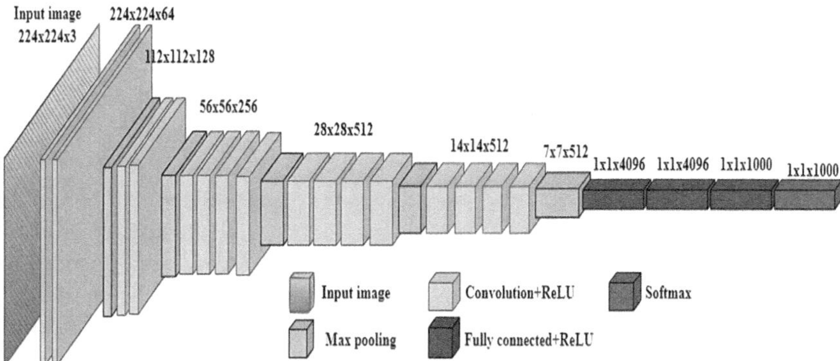

Figure 15.7 Architecture of VGG19.

Tables show architecture's convolutional layers, pooling layers, fully connected layers, and activation functions are sequentially operated upon to produce a final softmax activation for classification.

Step number	*Operation*
1	Input image (224×224×3)
2	Convolutional layer with 64 filters (3×3×3)
...	
42	Fully connected layer with 4096 units
43	ReLU activation
44	Fully connected layer with 1000 units
45	Softmax activation

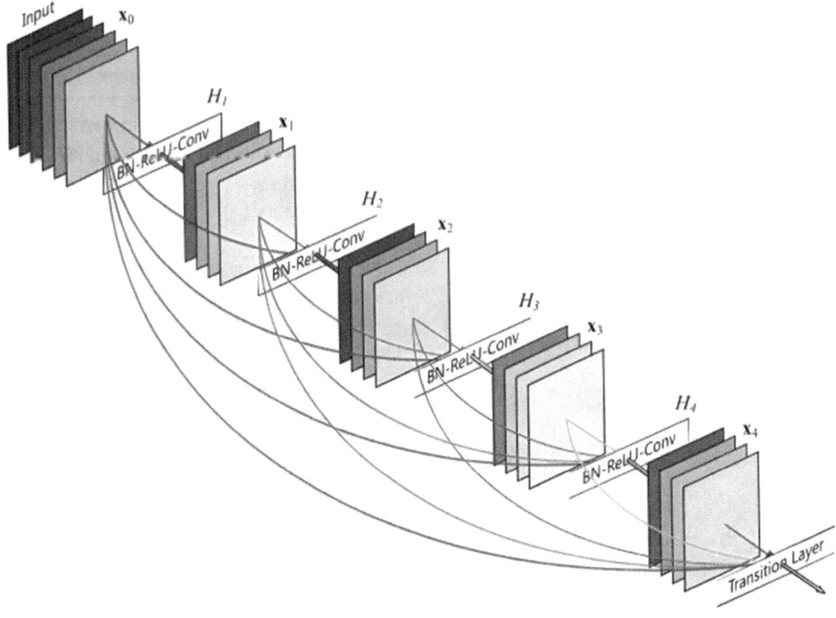

Figure 15.8 Architecture of Resnet.

Resnet: ResNet's fundamental principle is the usage of residual blocks as shown in Figure 15.8, which are made up of rectified linear unit (ReLU) activation functions and convolutional layers followed by batch normalization. By learning residual functions, these blocks facilitate the network's ability to learn the identity mapping when needed. ResNet is an effective tool for coral reef categorization due to its creative architecture, which uses residual connections to enable training of deep networks.

The Resnet algorithm table are shown below.

Step number	*Operation*
1	Input image (224×224×3)
2	Convolutional layer with 64 filters (7×7×3)
...	
33	Global average pooling
34	Fully connected layer with 1000 units
35	Softmax activation

VGG16: After a series of convolutional layers, VGG16 as shown in Figure 15.9 employs a series of max-pooling layers to progressively extract and abstract information from input images. The convolutional layers filter the input image by capturing different aspects of the spatial hierarchy of the image. The architecture of VGG16 is defined by a dense stack of

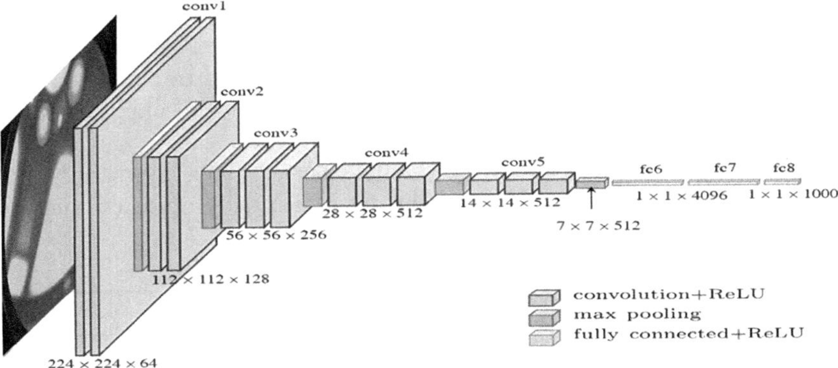

Figure 15.9 Architecture of VGG16.

convolutional layers with small receptive fields (3×3) and max- pooling layers to minimize spatial dimensions.

Step number	*Operation*
1	Input image (224×224×3)
2	Convolutional layer with 64 filters (3×3×3)
...	
34	Fully connected layer with 4096 units
35	ReLU activation
36	Fully connected layer with 1000 units
37	Softmax activation

EfficinetNetB0: EfficientNet as shown in Figure 15.10 design has numerous levels of processes, including activation functions, batch normalization,

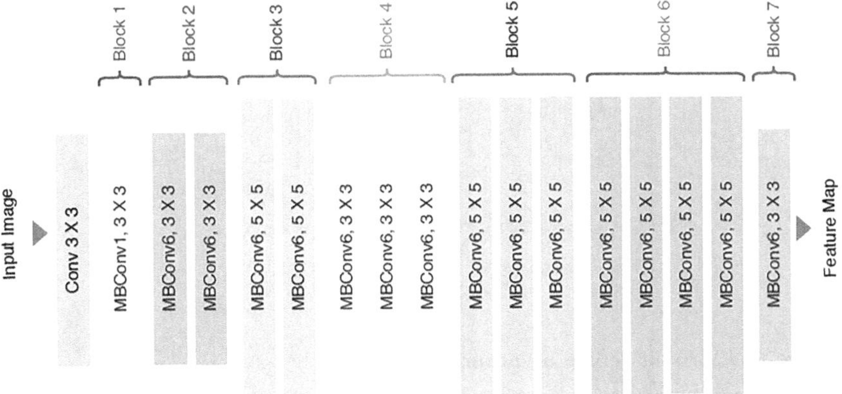

Figure 15.10 Architecture of EfficientNet.

and convolutions. The scaling factors that are applied to these blocks—which regulate the network's depth, width, and resolution—are the primary source of innovation. EfficientNet accomplishes a good trade-off between model complexity and performance over a broad range of applications by methodically scaling these elements.

Tables show architecture's convolutional layers, pooling layers, fully connected layers, and activation functions are sequentially operated upon to produce a final softmax activation for classification.

Step number	*Operation*
1	Input image (224×224×3)
2	Convolutional layer with 32 filters (3×3×3)
...	
36	Global average pooling
37	Fully connected layer with 1000 units
38	Softmax activation

InceptionV3: One of the key benefits of InceptionV3 as shown in Figure 15.11 is its ability to balance model complexity and computing performance. 1×1 convolutions help maintain representational capacity while reducing the number of parameters and computing cost by decreasing the dimensionality of feature maps. To increase the efficiency of the model even further, factorized convolutions and batch normalization are incorporated.

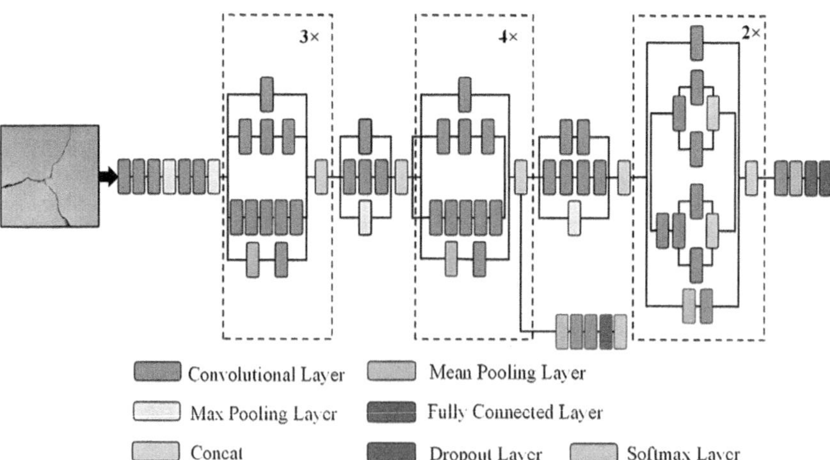

Figure 15.11 Architecture of InceptionV3.

The InceptionV3 algorithm table are shown below.

Step number	*Operation*
1	Input image (299 × 299 × 3)
2	Convolutional layer with 32 filters (3 × 3 × 3)
...	
93	Global average pooling
94	Fully connected layer with 1000 units
95	Softmax activation

15.4.3 Interactive visualization technique

Visualization is essential for clarifying complicated data correlations and patterns in deep learning algorithms for coral reef monitoring, such as color correction and assessment. The effectiveness of deep learning models in differentiating between healthy and bleached corals can be efficiently communicated by researchers using visualization techniques including heatmaps, scatter plots, and superimposing classification results onto photos of coral reefs. Accurate classification is made possible even in the presence of changing environmental conditions by color correction algorithms, which dynamically modify the color representation of images. To show how the model can adjust to different conditions, such as illumination and water clarity, these corrections can be shown graphically. Additionally, visualization helps to highlight how coral health statuses are distributed spatially throughout reef ecosystems, providing information about specific stressors and enabling focused conservation efforts. All things considered, visualization is a potent instrument for communicating the potential and results of deep learning-based coral reef monitoring, encouraging comprehension and participation from stakeholders in marine conservation programs.

15.4.4 Tool analysis

Visualization is a critical component in understanding and streamlining the software development process when examining Visual Studio as a development tool. Developers can obtain more profound understanding of the composition and interdependencies of their codebase by employing visualization tools such as dependency graphs, code maps, and debugging visualizers. Code relationships are visualized to help find possible bottlenecks, improve code maintainability, and make architectural refactoring easier. Furthermore, real-time insights into application performance and debugging procedures are provided by Visual Studio's visualization tools, such as the Performance Profiler and IntelliTrace. Developers can identify performance problems and track the application's execution flow via interactive charts, timelines, and call trees, which facilitates more effective debugging and optimization.

15.4.5 Colab library

The Colab (Google Colaboratory) platform provides a flexible and easy-to-use environment for creating and implementing deep learning models for coral categorization. State-of-the-art neural network architectures like ResNet, VGG19, VGG16, InceptionV3, and EfficientNetB0 are accessible through utilizing Colab libraries like TensorFlow and Keras. Within a cloud-based collaborative workspace, these libraries provide extensive functionality for data preprocessing, model training, evaluation, and inference. Moreover, Colab's smooth interface with Google Drive makes it simple to save and share data, which supports group research projects. Colab is a great platform for coral classification jobs because it speeds up model training and inference with its free GPU and TPU support. Additionally, researchers can experiment and document using its interactive notebook interface, which speeds up iteration and allows them to share their discoveries with the scientific community.

15.5 EVALUATION

Deep learning techniques in coral reef monitoring require thorough testing to determine the effectiveness and reliability of created models. Basic measures like recall, accuracy, precision, and F1 score are essential benchmarks for measuring how well classification algorithms work in differentiating between corals that are bleached and those that are healthy. By verifying performance across a variety of datasets, methods like holdout testing and cross-validation strengthen the robustness of the model even more. Evaluation also takes into account factors like scalability, computing efficiency, and generalization to unknown data in addition to accuracy. A supplementary function is provided by qualitative evaluation via visual examination of classification results, which makes it easier to spot incorrect classifications and clarify how the model behaves in practical situations. To ensure that coral reef monitoring efforts are accurate and relevant, they must be continuously evaluated and refined based on input from stakeholders and domain experts. This will support the efficacy of conservation measures.

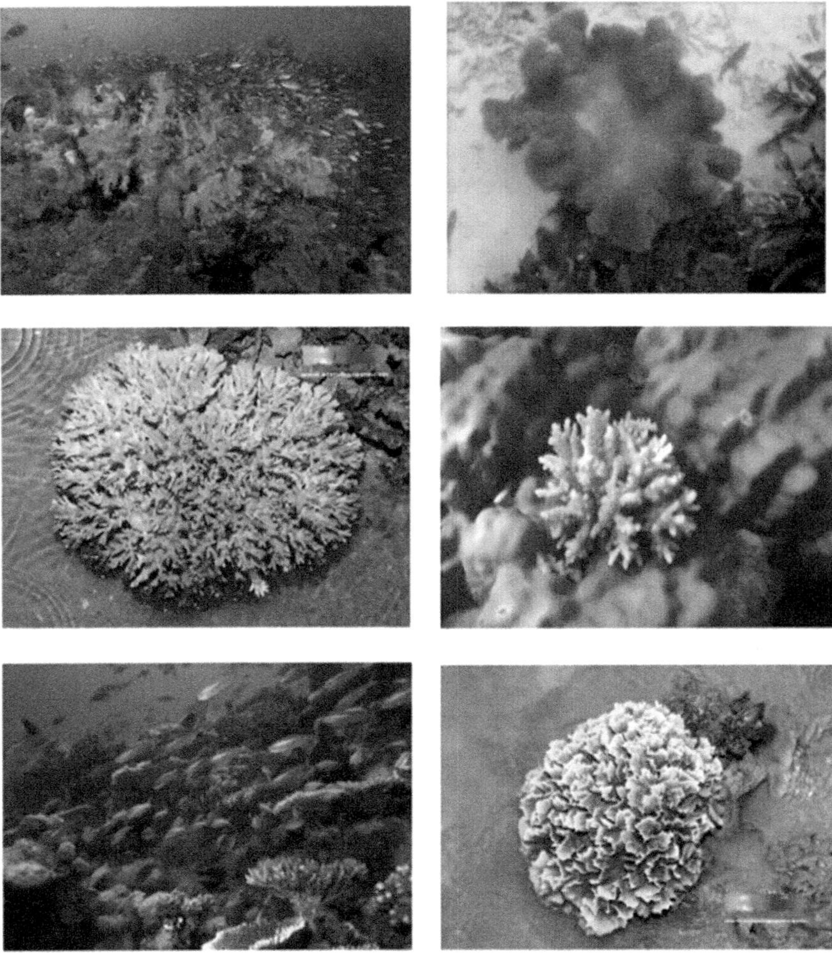

Figure 15.12 The input images given to the different nets.

Figure 15.12 shows the input image and Figure 15.13(a), (b) shows the classified image as bleached and healthy coral.

When it comes to coral reef classification tasks, a number of deep learning architectures—such as ResNet, VGG19, VGG16, InceptionV3, and Efficient NetB0—show excellent accuracy and performance. Scalability is provided by EfficientNetB0, while gradient vanishing problems are handled by ResNet. Intricate feature extraction is a strength of VGG19 and VGG16, while InceptionV3 demonstrates effectiveness in challenging visual scenarios.

Figure 15.13 (a) The output predicted result-classification of images into bleached and healthy coral. (b) The output predicted result-classification of images into bleached and healthy coral.

15.6 ANALYSIS TABLE

Together, the evaluated publications shown in Table 15.3 present state-of-the-art methods for underwater imaging, such as object detection, color correction, and image enhancement. These developments make it possible to observe and analyze data precisely, which is essential for comprehending marine ecosystems and assisting with conservation efforts. Underwater imaging technology could be significantly improved by integrating deep learning approaches.

Table 15.3 Publications analysis

Serial number	*Paper title*	*Techniques*	*Addressed issues*
1	"WaterGAN: Unsupervised Generative Network to Enable Real-time Color Correction of Monocular Underwater Images" [8]	Unsupervised Generative Network, Color Correction	Underwater image color adjustment in real time
2	"Underwater image enhancement with latent consistency learning-based color transfer" [4]	Latent Consistency Learning, Color Transfer	Improvement of underwater picture quality and uniformity of color
3	"An In-Depth Survey of Underwater Image Enhancement and Restoration" [2]	Image Enhancement, Image Restoration	thorough analysis of methods for underwater image enhancement and restoration
4	"Computer vision and deep learning for fish classification in underwater habitats: A survey" [9]	Deep Learning, Fish Classification	An overview of deep learning and computer vision methods for classifying fish in underwater settings
5	"High-Resolution Underwater Robotic Vision-Based Mapping and Three Dimensional Reconstruction for Archaeology" [5]	Robotic Vision, Mapping, 3D Reconstruction	3D reconstruction and high-resolution mapping for underwater archaeology
6	"Underwater Image Restoration Based on A New Underwater Image Formation Model" [2]	Image Restoration, Formation Model	underwater picture restoration using a novel formation model
7	"Initial results in underwater single image dehazing" [4]	Single Image Dehazing	preliminary findings for dehazing photos taken underwater
8	"A Real-Time Fish Target Detection Algorithm Based on Improved YOLOv5" [6]	Real-Time Detection, YOLOv5	YOLOv5-based real-time fish target detection algorithm
9	"High-resolution underwater robotic vision based mapping and 3D reconstruction for archaeology" [10]	Robotic Vision, Mapping, 3D Reconstruction	3D reconstruction and high-resolution mapping for underwater archaeology
10	"A Deep-Learning Based Pipeline for Estimating the Abundance and Size of Aquatic Organisms in an Unconstrained Underwater Environment from Continuously Captured Stereo Video" [11]	Deep Learning, Stereo Video, Abundance Estimation, Size Estimation	Deep learning pipeline for using stereo footage in unrestricted environments to estimate the size and abundance of aquatic organisms

(*Continued*)

Table 15.3 (Continued)

Serial number	*Paper title*	*Techniques*	*Addressed issues*
11	"High-resolution underwater robotic vision based mapping and 3D reconstruction for archaeology" [12]	Robotic Vision, Mapping, 3D Reconstruction	3D reconstruction and high-resolution mapping for underwater archaeology
12	"Underwater image restoration based on image blurriness and light absorption" [13]	Image Restoration, Blurriness, Light Absorption	Restoration of underwater photos with light absorption and blurriness in mind
13	"Bounded activation functions for enhanced training stability of deep neural networks on visual pattern recognition problems" [14]	Activation Functions, Deep Neural Networks	Improved deep neural network training stability for visual pattern recognition tasks
14	"Underwater depth estimation and image restoration based on single images" [3]	Depth Estimation, Image Restoration	Depth estimation underwater and image restoration using individual images
15	"Underwater Imaging and the effect of inherent optical properties on image quality" [15]	Imaging, Optical Properties	Impact of innate optical characteristics on the quality of underwater images
16	"Underwater image processing: State of the art of restoration and image enhancement methods" [16]	Image Processing, Restoration, Enhancement	evaluation of the most recent techniques for underwater image repair and enhancement
17	"Deriving inherent optical properties from background color and underwater image enhancement" [17]	Optical Properties, Image Enhancement	Deriving intrinsic optical characteristics from the color of the background and improving underwater photos
18	"Transmission estimation in underwater single images" [18]	Transmission Estimation	Calculating transmission in single underwater photos
19	"Underwater Image Restoration Based on A New Underwater Image Formation Model" [19]	Image Restoration, Formation Model	underwater picture restoration using a novel formation model

15.7 CONCLUSION

The coral reef detection project is a major advancement in our knowledge of and attempts to conserve these priceless marine ecosystems. By utilising sophisticated image processing methods, such as convolutional neural networks and deep learning algorithms, we have made significant strides towards automating the monitoring and identification of coral reef species and their environments. It is clear from much testing and assessment that all of the chosen deep learning architectures have demonstrated good performance in differentiating between healthy and bleached corals. ResNet, which is renowned for its residual connections, has proven to be a dependable option for precise classification tasks due to its robust performance and generalization over a variety of coral reef datasets. With their deep convolutional layers, VGG19 and VGG16 performed consistently, though at marginally higher computing costs. Prominent for its inception modules, InceptionV3 demonstrated effectiveness in capturing fine-grained characteristics, enabling precise coral categorization.

REFERENCES

1. M Vijayalakshmi, A Sasithradevi and P Prakash, *Variants of generative adversarial networks for underwater image enhancement*. Data Analytics for Intelligent Systems, IOP Publishing, pp. 9-1–9-19, February 2024.
2. Miao Yang, Jintong Hu, Chongyi Li, Gustavo Rohde, Yixiang Du, Ke Hu, "An in-depth survey of underwater image enhancement and restoration," *IEEE Access*, vol. 7, pp. 123638–123657, 2019.
3. Y. T. Peng, P. C. Cosman, "Underwater image restoration based on image blurriness and light absorption," *IEEE Transactions on Image Processing*, vol. 26, no. 4, pp. 1579–1594, 2017.
4. Hua Yang, Fei Tian, Qi Qi, Q. M. Jonathan Wu, Kunqian Li, "Underwater image enhancement with latent consistency learning-based color transfer," *IET Image Processing*, vol. 16, no. 6, pp. 1594–1612, Feb. 2022.
5. Miao Yang, Jintong Hu, Chongyi Li, Gustavo Rohde, Yixiang Du, Ke Hu, "High-resolution underwater robotic vision-based mapping and three dimensional reconstruction for archaeology," *Journal of Field Robotics*, vol. 27, pp. 702–717, Nov. 2010.
6. M. Zhang, J. Peng, "Underwater image restoration based on a new underwater image formation model," *IEEE Access*, vol. 6, pp. 58634–58644, 2018.
7. P. Drews, Jr., E. do Nascimento, F. Moraes, S. Botelho, M. Campos, "Transmission estimation in underwater single images," in *Proc. IEEE Int. Conf. Comput. Vis. Workshops*, Sydney, NSW, Australia, June 2013, pp. 825–883.
8. Jie Li, Katherine A. Skinner, Ryan M. Eustice, Matthew Johnson-Roberson, "WaterGAN: Unsupervised generative network to enable real-time color correction of monocular underwater images," *IEEE Robotics and Automation Letters*, vol. 3, pp. 387–394, 2018.

9. Miao Yang, Jintong Hu, Chongyi Li, Gustavo Rohde, Yixiang Du, Ke Hu, "Computer vision and deep learning for fish classification in underwater habitats: A survey," *IEEE Access*, vol. 23, pp. 1–1, August 2019.
10. N. Carlevaris-Bianco, A. Mohan, R. M. Eustice, "Initial results in underwater single image dehazing," *OCEANS 2010 MTS/IEEE SEATTLE*, pp. 1–8, 2010.
11 Wanghua Li, Zhenkai Zhang, Biao Jin, Wangyang Yu, "A real-time fish target detection algorithm based on improved YOLOv5," *Journal of Marine Science and Engineering*, vol. 11, no. 3, 2023. https://doi.org/10.3390/jmse11030572
12. Matthew Johnson-Roberson, Mitch Bryson, Ariell Friedman, Oscar Pizarro, Giancarlo Troni, Paul Ozog, Jon Henderson, "High-resolution underwater robotic vision-based mapping and three-dimensional reconstruction for archaeology," *Journal of Field Robotics*, vol. 34, pp. 625–643, 2016.
13. Gordon Böer, Joachim Paul Gröger, Sabah Badri-Höher, Boris Cisewski, Helge Renkewitz, Felix Mittermayer, Tobias Strickmann, Hauke Schramm, "A deep-learning based pipeline for estimating the abundance and size of 42 aquatic organisms in an unconstrained underwater environment from continuously captured stereo video," *Sensors*, vol. 23, no. 6, 2023. https://doi.org/10.3390/s23063311
14. M. Johnson-Roberson et al., "High-resolution underwater robotic vision based mapping and 3D reconstruction for archaeology," *Journal of Field Robotics*, vol. 34, 2016, 625–643
15. S. S. Liew, M. Khalil-Hani, R. Bakhteri, "Bounded activation functions for enhanced training stability of deep neural networks on visual pattern recognition problems," *Neurocomputing*, vol. 216, pp. 718–734, 2016.
16. P. L. J. Drews, E. R. Nascimento, S. S. C. Botelho, M. F. M. Campos, "Underwater depth estimation and image restoration based on single images," *IEEE Computer Graphics and Applications*, vol. 36, no. 2, pp. 24–35.
17. K. Ingrid, "Underwater Imaging and the effect of inherent optical properties on image quality," M.S. Thesis, Dept. Bio., Norwegian Univ. Sci. Technol., Trondheim, Norway, 2014.
18. R. Schettini, S. Corchs, "Underwater image processing: State of the art of restoration and image enhancement methods," *EURASIP Journal on Advances in Signal Processing*, vol. 2010, no. 1, Dec. 2010, Art. no. 746052, pp. 1–14.
19. X. Zhao, T. Jin, S. Qu, "Deriving inherent optical properties from background color and underwater image enhancement," *Ocean Engineering*, vol. 94, pp. 163–172, January 2015.

Index

Pages in *italics* refer to figures and pages in **bold** refer to tables.

abundance estimation, **281**
actionable insights, 2–3, 24, 28, 49, 58, 62, 112, 142–143, 175, 211
activation functions, 273–276, **282**
Adaptive Neuro Fuzy (ANFIS), 69, 74–75
adaptability, 23, 26, 66, 111, 146, 271, 273
advanced analytics, **4**, 42, 47, 136, 144, 206–208
agglomerative clustering, 223
ARIMA, 70, 75, 80
artificial intelligence, *11*, **15**, 42, 71, 88, 169–*170*, 172, 178, 189, 212

bibliometric, 1, 5, *6*, 7–8, 18
breaking boundaries, 236
blurriness, **282**
Bokeh, 115, 187–188

calorie monitoring, 168, 254
Clique, 224–225
cluster algorithms, 218, 220, 223–224, 226, 234
cluster structure, 219–220
CNN models, *11*, 158, *261*
Colab library, 278
comparative analysis, 166, 169, 171, 173
complex datasets, 1–2, 64, 120–121, 133
computer vision techniques, 84, 86
computationally intensive, 132
Convolutional Neural Network, 74, 89, 110–111, 113, 122, 194, 237, 249, 255, 257, 272, 283
crowding problem, 124, 132

data analysis, 1–5, *6*, 7–8, 10, 14, **15**, 18, 21, 23, 30, 37, 39, 43, 45–46, 55, *56*, 58, 62, 66, 69, 100, 115–116, 119, 130, 147, 185, 188, 199, 203, 212, 218–219, 226
data collection, 1, 77, 90, 98, 137, 189, 207–208
data quality, 6, 42, 72, 136
data mining, **15**, 77–78
deep learning technologies, 255, 278
deep neural network, 71, **282**
depth estimation, **282**
density based clustering, 219, 224
dimensions, 1, 24, 63, 119–120, 122, *129*, 200, 203, 207, 220, 225–226, 234, 246, 259, 275
dimensionality reduction, 1, 7, 21, 63–64, 66–67, 110, 119, 121, 124, 130, 132–133, 224–225
dimension reduction, 121–122, 224
divisive clustering, 223–224
dynamic dashboard, *6*, 7–8, 14, 18, 22, 34, 100–101, 104, 107, 116, 135–137, *138*, 141–144, *200*, 218, 220, 251

ECOC, 194
ecosystems, 69, 267–268, 277, 280, 283
EfficientNetB0, 278–279
empowering, 2–4, 143, 146, 168–169, 189, 207, 218
exploratory analysis, 2, **16**
extraction, 76–77, 95, 98, 103, 110–111, 113, 185, 187, 193, 198–199, 202, 213, 236–237, 259, 265, 269, **273**, 279

facial expression, 86, 98
Feed Forward Neural Network (FFNN), 69, 74
fish classification, **281**
Flyr tools, 185
food quality analysis, 116
food recognition, 254–255, 257, *259*, 263–265
fuzzy logic, 74, 245, 249

GLCM, 195, 197, 203
Graphical User Interface (GUI), **16**, 115

human evaluators, 86–87

image classification, 272, **273**
image enhancement, 180, 280, **281–282**
image restoration, **281–282**
InceptionV3, 238, **273**, *276*, 277–279, 283
interactive computer systems, **16**
interactive dashboards, 1–3, **4**, 10, 14, 23–25, 28, 30, *34*, 37, 39, 42–43, 45, 47, *48*, 49–50, 53, 55, 58–59, 62, 64, 86, 105, 112, 116, 119, 146, 150, 159, *163*, 166, 199–200

Key Performance Indicators (KPIs), 3, 22, 141, 207
key insights, 143
k-Medoids, 219
K-Nearest Neighbour (KNN), 69, 73, 224

latent consistency learning, **281**
light absorption, **282**
linguistic features, 86
linear discriminative analysis, 71
loading, 140, 213, 257
Long Short-Teram Memory (LSTM), 70, 74–76

machine learning, 63, 71, 110
machine learning algorithms, 100, 109, 112, **113**, 116, 206, 211, 219
machine learning innovations, 254
Magnetic Resonance Imaging (MRI), 193, *194*, 195, 198–199, 202–203
mapping, 4, 18, 111, 274, **281–282**
MDS, 227
melanoma, 236, 238–239, 243, 251
memory requirements, 132
meteorological, 16, 70–76, 78
model optimization, 92
morphological process, 198
multidimensional scaling, *227*

NIR imaging, 100–101, **103**, 104–107, *108*, 109, 111–112, **113**, 115–116, *128*
non-destructive, 101–102, 111, 116
Nourish net, 254–255

ophthalmology, 168–169, *170*, 171–173, 175, 178–180, 190
optimal cluster representation, 132
optimized, 92, 259

pathological, 102, 193
PCA, 63, 66, **103**, 110–111, 119–121, 130, **131**, 132, 224, *230–233*
Power BI, **4**, 7, 30, 35–37, *38*, 102, 112, 126, 135–137, *141*, 143–144, 146–147, 206–208, *211*, 212, 214, *215*
principal component analysis, *63*, 110–111, 119–120, 130, 132, 224
PROCLUS, 225–226, 229–234, 275
precipitation, 69–71, 74–75, 77, 80, 84
prediction, 69–71, 74–*81*, 84, 87, 89, 92, 110–111, 113–*114*, 208, 259, 261, 264, 269

random forest (RF), 71–72, 89, 110–112, **113**
real time detection, 102, **104**, **281**
ReLu activation, 273–275
robotic vision, **281–282**

scleritis, 175, 178
segmentation, 110, 121, **181**, 193–196, 198–199, 202, 208, 232, 240, 245–246, **248**, 270, **273**

size estimation, **281**
skin care, 236–240, 243, 251
softmax activation, 273–277
stochastic nature, 132
stock market, 206–208, 212–215
stereo video, **281**

Tableau, **4**, 7, 25–26, 28, *29*, 30, 126, 147, 218–220, 226, *230–233*, 234
temperature variations, 70, 171–173, 176–177, 179–180
texture analysis, 193–194, 197
thermal imaging, 106, 168–169, 171–173, **174**, 175, 178–180, **181**, 183, 189–190
time lines, 159–160, *163*, 278
tools analysis, 277
transfer learning, 236–237, 239–240, 243, 251
transformation, *32*, 207, 212–213, 215

vector regression, 69, 71, *72*, 74–75
VGG16, **273**, 274–275, 278–279, 283
VGG19, 272, **273**, 278–279, 283
visual data analysis, 1, 3–4, *6*, *11–12*, *14*, 21, 67, 69

YOLO, 102, **104**, **281**